DICTIONARY OF AERONAUTICAL ENGLISH

Titles in the series

English Dictionary for Students   1-901659-06-2

Dictionary of Accounting   0-948549-27-0
Dictionary of Aeronautical English   1-901659-10-0
Dictionary of Agriculture, 2nd ed   0-948549-78-5
Dictionary of American Business   0-948549-11-4
Dictionary of Automobile Engineering   0-948549-66-1
Dictionary of Banking & Finance, 2nd ed   1-901659-30-5
Dictionary of Business, 2nd ed   0-948549-51-3
Dictionary of Computing, 3rd ed   1-901659-04-6
Dictionary of Ecology & Environment, 3rd ed   0-948549-74-2
Dictionary of Government & Politics, 2nd ed   0-948549-89-0
Dictionary of Hotels, Tourism, Catering Management   0-948549-40-8
Dictionary of Human Resources & Personnel, 2nd ed   0-948549-79-3
Dictionary of Information Technology, 2nd ed   0-948549-88-2
Dictionary of Law, 2nd ed   0-948549-33-5
Dictionary of Library & Information Management   0-948549-68-8
Dictionary of Marketing   0-948549-73-4
Dictionary of Medicine, 2nd ed   0-948549-36-X
Dictionary of Military Terms   1-901659-24-0
Dictionary of Printing & Publishing, 2nd ed   0-948549-99-8
Dictionary of Science & Technology   0-948549-67-X

Professional Series
Dictionary of Astronomy   0-948549-43-2
Dictionary of Multimedia, 2nd ed   1-901659-01-1
Dictionary of PC & the Internet, 2nd ed   1-901659-12-7

Workbooks

Check your:
Vocabulary for Banking & Finance   0-948549-96-3
Vocabulary for Business, 2nd ed   1-901659-27-5
Vocabulary for Colloquial English   0-948549-97-1
Vocabulary for Computing, 2nd ed   1-901659-28-3
Vocabulary for Hotels, Tourism   0-948549-75-0
Vocabulary for Law, 2nd ed   1-901659-21-6
Vocabulary for Medicine   0-948549-59-9

Visit our web site for full details of all our books
http://www.pcp.co.uk

SUPPLEMENT

Aircraft registration codes

These codes are painted on all aircraft, showing their country of registration

3A	Monaco	A7	Qatar
3B	Mauritius	A9C	Bahrain
3C	Equatorial Guinea	AP	Pakistan
3D	Swaziland	B	China (People's Republic)
3X	Guinea	B	China/Taiwan (R o C)
4K	Azerbaijan	C,CF	Canada
4R	Sri Lanka	C2	Nauru
4U	United Nations Organization	C3	Andorra
4X	Israel	C5	The Gambia
5A	Libya	C6	The Bahamas
5B	Cyprus	C9	Mozambique
5H	Tanzania	CC	Chile
5N	Nigeria	CCCP	Former USSR
5R	Madagascar	CN	Morocco
5T	Mauritania	CP	Bolivia
5U	Niger	CS	Portugal
5V	Togo	CU	Cuba
5W	Western Samoa	CX	Uruguay
5X	Uganda	D	Germany
5Y	Kenya	D2	Angola
6O	Somalia	D4	Cape Verde Islands
6V	Senegal	D6	Comoros Islands
6Y	Jamaica	DQ	Fiji
7O	Yemen	EC	Spain
7P	Lesotho	EI	Eire
7Q	Malawi	EL	Liberia
7T	Algeria	EP	Iran
8P	Barbados	ER	Moldova
8Q ·	Maldives	ES	Estonia
8R	Guyana	ET	Ethiopia
9A	Croatia	EW	Belarus
9G	Ghana	EY	Tajikistan
9H	Malta	EZ	Turkmenistan
9J	Zambia	F	France
9K	Kuwait	G	Great Britain
9L	Sierra Leone	H4	Solomon Islands
9M	Malaysia	HA	Hungary
9N	Nepal	HB	Switzerland & Lichtenstein
9Q,9T	Zaire	HC	Ecuador
9U	Burundi	HH	Haiti
9V	Singapore	HI	Dominican Republic
9XR	Rwanda	HK	Colombia
9Y	Trinidad and Tobago	HL	Republic of Korea
A2	Botswana	HP	Panama
A3	Tonga	HR	Honduras
A40	Oman	HS	Thailand
A5	Bhutan	HV	The Vatican
A6	United Arab Emirates	HZ	Saudi Arabia

Aircraft registration codes *(continued)*

I	Italy	TI	Costa Rica
J2	Djibouti	TJ	Cameroon
J3	Grenada	TL	Central African Republic
J5	Guinea Bissau	TN	Congo Brazzaville
J6	St Lucia	TR	Gabon
J7	Dominica	TS	Tunisia
J8	St Vincent and the grenadines	TT	Chad
JA	Japan	TU	Cote d'Ivoire
JY	Jordan	TY	Benin
LN	Norway	TZ	Mali
LV	Argentina	UK	Uzbekistan
LX	Luxembourg	UR	Ukraine
LY	Lithuania	V2	Antigua and Barbuda
LZ	Bulgaria	V3	Belize
MT	Mongolia	V4	St Kitts and Nevis
N	USA	V5	Namibia
OB	Peru	V7	Marshall Islands
OD	Lebanon	V8	Brunei
OE	Austria	VH	Australia
OH	Finland	VN	Vietnam
OK	Czech Republic	VP-F	Falkland Islands
OO	Belgium	VP-LA	Anguilla
OY	Denmark	VP-LM	Montserrat
P	Korea (PDR)	VP-LV	British Virgin Islands
P2	Papua New Guinea	VQ-T	Turks and Caicos Islands
P4	Aruba	VR-B	Bermuda
PH	Netherlands	VR-C	Cayman Island
PJ	Netherland Antilles	VR-G	Gibraltar
PK	indonesia	VR-H	Hong Kong
PP,PT	Brazil	VT	India
PZ	Suriname	XA,XB,XC	Mexico
RA	Russia	XT	Burkina Faso
RDPL	Laos,People's Democratic Republic	XU	Cambodia
		XV	Vietnam
RP	Philippines	XY	Myanmar (Burma)
S2	Bangladesh	YA	Afghanistan
S5	Slovenia	YJ	Vanuatu
S7	Seychelles	YK	Syria
S9	Sao Tome	YL	Latvia
SE	Sweden	YN	Nicaragua
SP	Poland	YR	Romania
ST	Sudan	YS	El Salvador
SU	Egypt	YU	Yugoslavia
SX	Greece	YV	Venezuela
T2	Tuvalu	Z	Zimbabwe
T3	Kiribati	ZA	Albania
T7	San Marino	ZK	New Zealand
TC	Turkey	ZP	Paraguay
TF	Iceland	ZS	South Africa
TG	Guatemala		

International Airline Designators

Designator	Airline	Designator	Airline
2J	Azerbaijan Airlines	FI	Icelandair
9U	Air Moldova	FJ	Air Pacific
AA	American Airlines	FO	Expedition Airways
AC	Air Canada	FQ	Air Aruba
AF	Air France	FR	Ryanair
AH	Air Algerie	GA	Garuda Indonesia
AI	Air India	GC	Centrafrican Airlines
AM	Aeromexico	GF	Gulf Air
AN	Ansett Australia	GH	Ghana Airways
AQ	Aloha Airlines	GL	Greenlandair
AR	Aerolineas Argentinas	GN	Air Gabon
AS	Alaska Airlines	GR	Aurigny Air Services
AT	Royal Air Maroc	GY	Guyana Airways
AY	Finnair	HA	Hawaiian Airlines
AZ	Alitalia	HM	Air Seychelles
BA	British Airways	HP	America West Airlines
BB	Seaborne Aviation	HV	Transavia Airlines
BD	British Midland	HY	Uzbekistan Airways
3G	Biman Bangladesh Airlines	IB	Iberia
BH	Transtate Airlines	IC	Indian Airlines
BI	Royal Brunei Airlines	IE	Solomon Airlines
BL	Pacific Airways	IL	Istanbul Airways
BM	Air Sicilia	IR	Iran Air
BO	Bouraq Indonesia Airlines	IV	Fujian Airlines
BP	Air Botswana	IY	Yemenia Yemen Airways
BU	Braathens ASA	JE	Manx Airlines
BW	BWIA International	JG	Air Greece
CA	Air China	JL	Japan Airlines
CB	Suckling Airways	JM	Air Jamaica
CI	China Airlines	JP	Adria Airways
CJ	China Northern Airlines	JU	JAT Jugoslovenski Aerotransport
CM	COPA (Compania Panamena de Aviación)	JY	Jersey European Airways
		KE	Korean Air
CO	Continental Airlines	KL	KLM Royal Dutch Airlines
CP	Canadian Airlines International	KM	Air Malta
CU	Cubana	KP	Kiwi international Airlines
CX	Cathay Pacific Airways	KQ	Kenya Airways
CY	Cyprus Airways	KU	Kuwait Airways
CZ	China Southern Airlines	KV	Kavminvodyavia
DL	Delta Air Lines	KX	Cayman Airways
DS	Air Senegal	KY	Air Sao Tome
DT	TAAG Angola Airlines	LA	Lan-Chile
DU	Hemus Air	LG	Luxair
DY	Air Djibouti	LH	Lufthansa
EI	Aer Lingus	LN	Jamahiriya Libyan Arab Airlines
EK	Emirates	LO	LOT Polish Airlines
ET	Ethiopian Airlines	LU	Air Atlantic Dominicana
FC	Tempelhof Express Airlines	LY	El Al Israel Airlines
FG	Ariana Afghan Airlines	LZ	Balkan Bulgarian Airlines

Designator	Airline
MA	Malev Hungarian Airlines
MD	Air Madagascar
MH	Malaysia Airlines
MK	Air Mauritius
MN	Commercial Airways
MR	Air Mauritanie
MS	Egyptair
NF	Air Vanuatu
NG	Lauda Air
NH	All Nippon Airways
NO	Aus Air
NQ	Georgian Airways
NV	Northwest Territorial Airways
NW	Northwest Airlines
NZ	Air New Zealand
OA	Olympic Airways
OB	Shepparton Airlines
OG	Go
OK	Czech Airlines
OM	Mongolian Airlines
ON	Air Nauru
OO	Skywest Airlines
OS	Austrian Airlines
OU	Croatia Airlines
OV	Estonian Air
PB	Provincial Airlines
PC	Air Fiji
PH	Polynesian Airlines
PK	Pakistan International Airlines
PR	Philippine Airlines
PS	Ukraine International Airlines
PU	PLUNA (Premieras Lineas Uruguayas de Navegación Aerea)
PX	Air Niugini
PY	Surinam Airways
PZ	Transportes Aereos del Mercosur
QF	Qantas Airways
QM	Air Malawi
QR	Qatar Airways
QU	Uganda Airlines
QV	Lao Aviation
QX	Horizon Air
RA	Royal Nepal Airlines
RB	Syrian Arab Airlines
RG	Varig Brazilian Airlines
RJ	Royal Jordanian
RK	Air Afrique
RO	TAROM
SA	South African Airways
SD	Sudan Airways

Designator	Airline
SH	Air Toulouse
SK	SAS
SN	SABENA
SQ	Singapore Airlines
SR	Swissair
SU	Aeroflot
SV	Saudia
SW	Air Namibia
TC	Air Tanzania
TE	Lithuanian Airlines
TG	Thai Airways International
TK	Turkish Airlines
TM	LAM Lineas Aereas de Moçambique
TN	Air Tahiti Nui
TP	TAP Air Portugal
TU	Tunis Air
TW	TWA
U2	Easyjet
UA	United Airlines
UB	Myanmar Airways
UI	Alaska seaplane services
UL	Air Lanka
UM	Air Zimbabwe
UP	Bahamasair
US	US Airways
UY	Cameroon Airlines
VE	AVENSA
VH	Aeropostal
VJ	Royal Air Cambodge
VN	Vietnam Airlines
VO	Tyrolean Airlines
VR	Transportes Aereos de Cabo Verde
VS	Virgin Atlantic
VU	Air Ivoire
VX	ACES (Aerlineas Centrales de Colombia)
WG	Wasaya Airlines
WJ	Labrador Airways
WN	Southwest Airlines
WR	Royal Tongan Airlines
WY	Oman Air
YJ	National Airlines
YK	Horizon Airlines
YN	Air Creebec
YU	Dominair
YZ	Transportes Aereos da Guiné Bissau
ZB	Monarch Airlines
ZQ	Ansett New Zealand

The Phonetic Alphabet

Certain letters of the alphabet sound very similar, especially when a person is talking on a telephone or radio. The phonetic alphabet is designed to prevent confusion, by using a distinctive word to represent each letter.

Aa	Alpha*	['ælfə]
Bb	Bravo	['brɑːvəʊ]
Cc	Charlie	['tʃɑːlɪ]
Dd	Delta	['deltə]
Ee	Echo	['ekəʊ]
Ff	Foxtrot	['fokstrot]
Gg	Golf	[gɒlf]
Hh	Hotel	[həʊ'tel]
Ii	India	['ɪndɪə]
Jj	Juliet	['dʒuːlɪət]
Kk	Kilo	['kiːləʊ]
Ll	Lima	['liːmə]
Mm	Mike	[maɪk]
Nn	November	[nə'vembə(r)]
Oo	Oscar	['ɒskə(r)]
Pp	Papa	['pɑpə]
Qq	Quebec	[kwɪ'bek]
Rr	Romeo	['rəʊmɪəʊ]
Ss	Sierra	[sɪ'eərə]
Tt	Tango	['tæŋgəʊ]
Uu	Uniform	['juːnɪfɔːm]
Vv	Victor	['vɪktə(r)]
Ww	Whisky**	['wɪskɪ]
Xx	X-ray	['eksreɪ]
Yy	Yankee	['jæŋkɪ]
Zz	Zulu	['zuːluː]

Alfa in US English
**Whiskey in US English*

Some Standard Words and Phrases

Word/Phrase	Meaning (Text in italics is in simplified form)
Acknowledge	Let me know that you have received and understood this message.
Affirm	Yes.
Approved	*I give you permission for what you asked.*
Cancel	*Cancel the last clearance I gave to you.*
Check	Examine a system or procedure.
Cleared	*I give permission for you to continue, bearing in mind the conditions already given.*
Confirm	Have I correctly received the following … ? or Did you correctly receive this message ?
Contact	*Contact by radio …*
Correct	That is correct.
Correction	*An error was made in the last transmission.. What follows is correct.*
Disregard	*Assume that the last transmission was not sent.*
How do you read?	*Tell me how good this transmission is on a 1 to 5 scale where 1 = unreadable (cannot understand) to 5 = excellent reception (no difficulty at all in understanding)*
I say again	*I am repeating in order to make my meaning very clear.*
Over	*My transmission is finished and I want a response from you.*
Out	*This exchange of transmissions is finished. I do not want a response from you.*
Pass your message	Proceed with your message
Read back	Repeat all, or the specified part of this message back to me exactly as received.
Request	*I want to know. or I want to have.*
Roger	I have received all of your last transmission.
Say again	Repeat all, or the following part of your last transmission.
Speak slower	*Speak more slowly*
Standby	Wait and I will call you
Verify	*Check and confirm with me.*
Wilco	I understand your message and will comply with it
Words Twice	(a) as a request = Communication is difficult. Please send every word or group of words twice.
	(b) as information = *Because communication is difficult, every word or group of words in this message will be sent twice.*

Standard Symbols & Abbreviations

+	....	positive	lb.	....	pound
-	....	negative	m	....	metre
Ω	....	ohm	pmb	....	milibar
°	....	degree	mf or μf	....	microfarad
"	....	inch(es)	mHz	....	millihertz
'	....	foot(feet)	MHz	....	megahertz
amp	....	ampere	mi.	....	mile(s)
Btu	....	British thermal unit	mm	....	millimetre
C	....	Celsius	MPH	....	miles per hour
cal	....	calorie(s)	mv	....	millivolt
cal.	....	large calorie(s)	neg	....	negative
cm	....	centimetre	oz.	....	ounce(s)
cos	....	cosine	pf or μμf	....	picofarad
cu. cm	....	cubis centimetre(s)	pos	....	positive
cu. in.	....	cubis inch(es)	PPH	....	pounds per hour
cu. ft.	....	cubic foot(feet)	PPM	....	parts per million
cu. m	....	cubic metre(s)	PSI	....	pounds per square inch
dB	....	decibel	PSIA	....	pounds per square inch absolute pressure
deg	....	degree			
ESHP	....	equivalent shaft horsepower	PSID	....	pounds per square inch differential pressure
f	....	farad			
F	....	Fahrenheit	PSIG	....	pounds per square inch gage
ft.	....	foot (feet)	pt.	....	pint
ft.-lb.	....	foot-pound	qt	....	quart
ft.-lbs.	....	foot-pounds	R	....	Rankine
g	....	gram	rev.	....	revolution(s)
gal	....	gallon	RPM	....	revolutions per minute
HP	....	horsepower	sec.	....	second
hr	....	hour	SHP	....	shaft horsepower
Hz	....	hertz	sin	....	sine
in	....	inch(es)	sq. cm	....	square centimetre(s)
in. hg.	....	inch(es) of mercury	sq. in	....	square inch(es)
IPS	....	inches per second	sq. ft	....	square foot (feet)
k	....	kilo	sq. m	....	square metre
K	....	Kelvin	sq. mi.	....	square mile(s)
kg	....	kilogram	sq. mil	....	square mil
kHz	....	kilohertz	tan	....	tangent
km	....	kilometer	TEHP	....	thrust equivalent horsepower
kM	....	kilomega	THP	....	thrust horsepower
kw	....	kilowatt	U.S.	....	United States
kw-hr	....	kilowatt hour	V	....	volt
l	....	litre	yd.	....	yard

Common measures

Metric units		Imperial equivalent		Imperial units		Metric equivalent
Length						
	1 millimetre (mm)	0.03937 in		1 inch		2.54 cm
10 mm	1 centimetre (cm)	0.39 in	12 in	1 foot		30.48 cm
10 cm	1 decimetre (dm)	3.94 in	3 ft	1 yard		0.9144 m
100 cm	1 metre (m)	39.37 in	1760 yd	1 mile		1.6093 km
1000 m	1 kilometre (km)	0.62 mi				
Area						
				1 square inch		6.45 cm^2
	1 square millimetre	0.0016 sq in	144 sq in	1 square foot		0.0929 m^2
	1 square centimetre	0.155 sq in	9 sq ft	1 square yard		0.836 m^2
100 cm^2	1 square decimetre	15.5 sq in	4840 sq yd	1 acre		0.405 ha
10 000 cm^2	1 square metre	10.76 sq ft	640 acres	1 square mile		259 ha
10 000 m^2	1 hectare	2.47 acres				
Volume						
				1 cubic inch		16.3871 cm^2
	1 cubic centimetre	0.016 cu in	1728 cu in	1 cubic foot		0.028 m^2
1000 cm^3	1 cubic decimetre	61.024 cu in	27 cu ft	1 cubic yard		0.765 m^2
1000 dm^3	1 cubic metre	35.31 cu ft / 1.308 cu yd				
Liquid volume						
				1 pint		0.571
			2 pt	1 quart		1.141
	1 litre	1.76 pt	4 qt	1 gallon		4.551
100l	1 hectolitre	22 gal				
Weight						
				1 ounce		28.3495 g
	1 gram	0.035 oz	16 oz	1 pound		0.4536 kg
1 000 g	1 kilogram	2.2046 lb	14 lb	1 stone		6.35 kg
1 000 kg	1 tonne	0.0842 ton	8 st	1 hundredweight		50.8 kg
			20 cwt	1 ton		1.016 t

Conversion factors

Imperial to metric

Length			Multiply by
inches	→	millimetres	25.4
inches	→	centimetres	2.54
feet	→	metres	0.3048
yards	→	metres	0.9144
statute miles	→	kilometres	1.6093
nautical miles	→	kilometres	1.852

Area			Multiply by
square inches	→	square centimetres	6.4516
square feet	→	square metres	0.0929
square yards	→	square metres	0.8361
acres	→	hectares	0.4047
square miles	→	square kilometres	2.5899

Volume			Multiply by
cubic inches	→	cubic centimetres	16.3871
cubic feet	→	cubic metres	0.0283
cubic yards	→	cubic metres	0.7646

Capacity			Multiply by
UK fluid ounces	→	litres	0.0284
US fluid ounces	→	litres	0.0296
UK pints	→	litres	0.5682
US pints	→	litres	0.4732
UK gallons	→	litres	4.546
US gallons	→	litres	3.7854

Weight			Multiply by
ounces (avoirdupois)	→	grams	28.3495
ounces (troy)	→	grams	31.1035
pounds	→	kilograms	0.4536
tons (long)	→	tonnes	1.016

Conversion factors

Metric to imperial

Length		Multiply by
millimetres	→ inches	0.0394
centimetres	→ inches	0.3937
metres	→ feet	3.2806
metres	→ yards	1.9036
kilometres	→ statute miles	0.6214
kilometres	→ nautical miles	0.54

Area		Multiply by
square centimetres	→ square inches	0.155
square metres	→ square feet	10.764
square metres	→ square yards	1.196
hectares	→ acres	2.471
square kilometres	→ square miles	0.386

Volume		Multiply by
cubic centimetres	→ cubic inches	0.061
cubic metres	→ cubic feet	35.315
cubic metres	→ cubic yards	1.308

Capacity		Multiply by
litres	→ UK fluid ounces	35.1961
litres	→ US fluid ounces	33.8150
litres	→ UK pints	1.7598
litres	→ US pints	2.1134
litres	→ UK gallons	0.2199
litres	→ US gallons	0.2642

Weight		Multiply by
grams	→ ounces (avoirdupois)	0.0353
grams	→ ounces (troy)	0.0322
kilograms	→ pounds	2.2046
tonnes	→ tons (long)	0.9842

Temperature		Operation (in sequence)
Celsius	→ Fahrenheit	x 9, ÷ 5, + 32
Fahrenheit	→ Celsius	- 32, x 5, ÷ 9

DICTIONARY OF
AERONAUTICAL
ENGLISH

David Crocker

PP

PETER COLLIN PUBLISHING

First published in Great Britain 1999

Published by Peter Collin Publishing Ltd
1 Cambridge Road, Teddington, Middlesex, TW11 8DT

British Library Cataloguing-in-Publication Data

A catalogue record for this book is available from the British Library

ISBN 1-901659-10-0

Text computer typeset by PCP
Printed and bound in Finland by WSOY
Cover artwork by Gary Weston

PREFACE

English is the universal language of communication used in civil aviation. This dictionary aims to provide the basic vocabulary of terms used by pilots, cabin staff, maintenance crews and ground staff world-wide. The terms are those used in everyday work on aircraft, and cover parts of the aircraft, manipulating the aircraft on the ground and in the air, instructions to passengers, conversations with air traffic control, weather, emergencies, etc.

Unlike conventional aeronautical dictionaries, the Dictionary of Aeronautical English defines vocabulary often found in conjunction with the purely technical terms as well as the technical terms themselves. Simple explanations are presented in simple language, making the dictionary ideal for those working towards a private or commercial pilot's licence, as well as trainee maintenance engineers and more experienced professionals. We also give examples to show how the words are used in context.

We have selected quotations from various specialized magazines to show the words and phrases as they are used in real life situations. The supplement at the back gives further information in the form of tables.

We are particularly grateful to the staff at Qatar Aeronautical College for their assistance. To Gwen, Lionel, Carol, Lara and Cathy; thanks for everything.

The information contained in this dictionary is not to be regarded as a substitute for formal training in a given discipline.

Aa

AAIB = AIR ACCIDENT INVESTIGATION BRANCH

abbreviate [ə'bri:vɪeɪt] *verb* to shorten a word or a text; *Air Traffic Control is usually abbreviated to ATC;* **abbreviated weather report** = shortened weather report

abbreviation [əbri:vɪ'eɪʃn] *noun* short form of a word or text; *aeronautical charts use abbreviations and symbols; km is the abbreviation for kilometre*

COMMENT: abbreviations can cause confusion. They may range from those which have a very specific meaning as defined by an authoritative body, to others which may come about because of personal usage in note-making, etc. ICAO approved abbreviations may differ from those used in JARs. AC can mean 'alternating current' or 'altocumulus'. CPL is generally taken to mean Commercial Pilot's Licence but the ICAO definition is Current Flight Plan. Advances in technology have significantly increased the number of abbreviations with which pilots and engineers must be familiar. Abbreviations in the dictionary include those with generally accepted definitions and others with specific ICAO definitions

ability [ə'bɪlɪti] *noun* the power, knowledge or skill needed to do something; *strength is the ability of a material to support a load;* **he has great ability** = he has good skills; he's very clever

able ['eɪbl] *adjective* capable, or competent at something; **although he's** young, he's a very able pilot = he's a very good pilot, in spite of his age; **to be able to** = to have the power, knowledge, skill or strength to do something; *a pilot must be able to check the condition of the electrical circuit in flight; is she able to carry this heavy suitcase?*

able-bodied ['eɪbl'bɒdɪd] *adjective* without physical handicap or disability; *physically disadvantaged, as well as able-bodied people can gain a PPL*

abnormal [æb'nɔ:ml] *adjective (often in a negative or unwanted sense)* not normal; **abnormal load** = load which is heavier than normal; **abnormal weather conditions** = unusual or unfavourable weather conditions

abnormality [æbnɔ:'mælɪti] *noun* something not normal (and therefore possibly worrying); *any abnormality in engine performance should be checked;* **physical abnormalities** = aspects of somebody's health or physique which may prevent them from doing something or even from passing a medical check-up

abort [ə'bɔ:t] *verb* to stop something taking place; to end something before time; *to abort landing because of a violent storm*

absolute ['æbsəlu:t] *adjective* complete; total; **absolute necessity** = something which is 100% necessary; **absolute silence** = complete silence; **absolute trust** = complete trust

absolute pressure ['æbsəlu:t 'preʃə] *noun (of a gas)* unit of force per unit of area without comparison to other pressure;

aircraft show absolute pressure in inches of mercury on the inlet manifold pressure gauge

absolute value ['æbsəluːt 'væljuː] *noun* size or value of a number regardless of its sign; *the absolute value of -64.32 is 64.32*

absolute zero ['æbsəluːt 'ziːrəʊ] *noun* the lowest temperature possible, 0 °K (Kelvin), or -273.15 °C (Celsius)

absorb [əb'zɔːb] *verb* (a) to take in; *the atmosphere absorbs radio waves*; *warm air absorbs moisture more easily than cold air*; *our bodies absorb oxygen* (b) to **absorb information** = to take in and understand something; *only a few passengers absorb the pre-departure safety information*

absorption [əb'zɔːpʃən] *noun* taking in; *there is absorption of energy by the tyre when the aircraft lands* (NOTE: compare the spelling of the verb **absorb** and that of the noun **absorption**)

AC = ALTERNATING CURRENT; *(ICAO)* ALTOCUMULUS

accelerate [æk'seləreɪt] *verb* to increase or gather speed; *after start-up, the engine accelerates up to idling speed*; *the aircraft accelerated down the runway and took off* (NOTE: the opposite is **decelerate**)

acceleration [ækselə'reɪʃn] *noun* (a) the act of accelerating; increase in speed; *acceleration is the rate of change of velocity of a body*; *acceleration can be felt as the aircraft begins its take-off run* (b) outward force caused by change in direction without changing speed; *acceleration forces can be felt during aerobatic manoeuvres* (NOTE: the opposite is **deceleration**)

accept [ək'sept] *verb* (a) to be able to take or receive; *some units accept electrical inputs from the autopilot* (b) to take or receive; **to accept a gift or a prize** = to take a prize which is handed to you (c) **to accept the blame** = to be willing to take the blame for something; **to accept**

responsibility = to be willing to be answerable for something; *the copilot accepted responsibility for the incident*

acceptable [ək'septəbl] *adjective* which is allowed or approved of, although it may not be perfect; **acceptable level of safety** = good enough standard of safety; **acceptable limits** = limits generally regarded as correct; **there must be a continuous flow of clean oil at an acceptable temperature** = the temperature of the oil must be within certain maximum and minimum figures

acceptance [ək'septəns] *noun* (a) willingness to believe something; agreeing to something; *there is a growing acceptance that safety is the main priority* (b) willingness to do or use something; **acceptance of new technology** = willingness to use new technology

accepted [ək'septɪd] *adjective* believed or recognized; *it is accepted that incorrect use of English played a part in the accident*; *it is generally accepted that flying is one of the safest forms of transport*

access ['ækses] **1** *noun* way to find or get at something; **to gain access to** = to manage to enter; **access panel** = part of the aircraft skin which can be easily removed for inspection of internal components; **access to information** = the means to get at, retrieve and use information **2** *verb* to find and use; **to access data** *or* **information** = to find, retrieve and use data *or* information

accessibility [əksesɪ'bɪlɪti] *noun* ease with which something can be reached or found; *accessibility of components and equipment during servicing enables work to be done more quickly*

accessible [ək'sesɪbl] *adjective* easy to get at; *it is a good idea to have a set of emergency charts in an accessible place in the cockpit*; *instruments which need resetting in flight must be accessible to the crew*

accessory [ək'sesəri] *noun* **1** system or piece of equipment of secondary importance; *a camera **with** several accessories* **2** *adjective* of secondary importance; *there are many accessory systems which need engine power to operate them - pumps, generators, magnetos, etc.* (NOTE: the noun **accessory** is not connected with the noun **access** or the verb **to access**)

accident ['æksɪdənt] *noun* **(a)** something which happens which seems to have no cause; *it was an accident* = it was not deliberate or intentional; *by accident* = by chance; *we met by accident* = we met by chance **(b)** unfortunate or harmful event; something causing damage; *an accident must be reported*; *the flight attendant was injured in the accident*

QUOTE Mr Skidmore lost both arms in an accident while serving in the army as a young man, and is believed to be the first pilot in the UK - and possibly the world - to go solo with two artificial arms
Pilot

accidental [æksɪ'dentl] *adjective* **(a)** not deliberate or intentional; which happens by accident; *there is a safety device to prevent accidental retraction of the undercarriage* **(b)** referring to an accident; *we were told of his accidental death*

accompanied [ə'kʌmpnid] *adjective* which goes with; which is found together with; **accompanied luggage** = luggage which belongs to one of the passengers and is carried on the same aircraft; *see also* UNACCOMPANIED

accompany [ə'kʌmpni] *verb* to go with; to find together with something else; *engine failure is sometimes accompanied by fire*; *Mr Smith was accompanied by his wife and children on the flight to New York* = Mr Smith's wife and children were with him on the flight (NOTE: **accompanied by** someone *or* something)

accomplish [ə'kʌmplɪʃ] *verb* to do something; *feathering is accomplished by moving the pilot's control lever*; *retraction of the undercarriage is accomplished by electrical power*; **to accomplish a task** = to successfully finish doing something demanding (NOTE: formal technical texts often use the verb **to accomplish** simply to mean **to do**)

accomplishment [ə'kʌmplɪʃmənt] *noun* **(a)** achievement; *Charles Lindbergh's flight across the Atlantic in May 1927 was a great accomplishment* **(b)** *(in physics)* work done; *power is measured by units of accomplishment correlated with time*

accordance [ə'kɔːdəns] *noun* **in accordance with** = in agreement with; following (rules, instructions or laws); *fuels must be used in accordance with instructions*; **in accordance with Buys Ballot's Law** = as described by Buys Ballot's Law

QUOTE use full heat whenever carburettor heat is applied, partial hot air should only be used if an intake temperature gauge is fitted and only then in accordance with the Flight Manual or Pilot's Operating Handbook
Civil Aviation Authority, General Aviation Safety Sense Leaflet

according to [ə'kɔːdɪŋ 'tʊ] *preposition* **(a)** as determined by; in relation to; *the force exerted by the pilot on the control column will vary according to a number of factors* **(b)** as written or said by somebody else; *according to the copilot, engine vibration was detected in engine number one* **(c)** in agreement with (instructions, etc.); **according to instructions** = exactly as said in the instructions; **according to requirements** = as required

accordingly [ə'kɔːdɪŋli] *adverb* as needed; *check for increasing manifold pressure and reduce power accordingly*

account [ə'kaʊnt] *noun* **to take something into account** = to consider something; to remember something and think about carefully; *when planning a flight, wind speed and direction must be taken into account*; *in the event of an in-flight emergency, the aircraft should be landed at the nearest suitably equipped airport, taking into account fuel available*; **on no account** = under no circumstances, never; *on no account should anybody fly an aircraft without carrying out pre-flight checks*

account for [əˈkaʊnt ˈfɔː] *verb* **(a)** to make up or constitute; *Kevlar and carbon fibre account for a large percentage of the materials used in modern aircraft* **(b)** to provide the main reason for something; *high humidity accounted for the longer take-off run*

accrete [əˈkriːt] *verb* to increase in amount by slow external addition; to accumulate; **ice accretes on the rotor** = ice builds up on the rotor

accretion [əˈkriːʃn] *noun* increase by slow external addition; accumulation; *ice accretion can cause loss of lift and significantly increase the weight of the aircraft*

accumulate [əˈkjuːmjəleɪt] *verb* to collect and increase; *due to katabatic effects, cold air flows downwards and accumulates over low ground*

accumulation [əkjuːmjəˈleɪʃn] *noun* collection and increase of something; *fire in a toilet could present difficulties due to the confined space and possible smoke accumulation*

accumulator [əˈkjuːmjəleɪtə] *noun* **(a)** device for storing energy in hydraulic systems; *an accumulator is fitted to store hydraulic fluid* **(b)** electric circuit in a calculator or computer, in which the results of arithmetical and logical operations are formed

accuracy [ˈækjərəsi] *noun* **(a)** state of being correct; **to check for accuracy** = to make certain that the result is correct **(b)** precision; *the accuracy of modern*

navigational equipment is much greater than older systems

accurate [ˈækjərət] *adjective* **(a)** correct; *skill in accurate flying can only be achieved by practice*; **accurate results** = results which are exactly correct **(b)** precise; *this watch is very accurate*

ACFT = AIRCRAFT

achieve [əˈtʃiːv] *verb* **(a)** to manage to do something demanding; *in order to achieve a safe landing in a crosswind, the correct techniques must be used* **(b)** to obtain; *in wind shear conditions, a fly-by-wire system allows the pilot to achieve maximum lift by pulling hard back on the stick without risk of a stall*

achievement [əˈtʃiːvmənt] *noun* successful completion of something demanding; *for most trainee pilots, your first solo flight is a great achievement*; *see also* ACCOMPLISHMENT

acid [ˈæsɪd] *noun* chemical substance which reacts with a base to form a salt - an acid turns a litmus indicator red and has a sour taste; *sulphuric acid* (H_2SO_4); **acid-proof** = able to resist the harmful effects of an acid

acidity [əˈsɪdɪti] *noun* having an acid content; **the acidity of a substance** = the amount of acid in a substance

acid test [ˈæsɪd ˈtest] *noun* difficult or exacting test of worth or quality; *the pilot's ability to react appropriately in an emergency situation is the acid test of his professionalism*

ACN = AIRCRAFT CLASSIFICATION NUMBER

acoustic [əˈkuːstɪk] *adjective* referring to sound; **acoustic ear muffs** = ear protectors or defenders

acquire [əˈkwaɪə] *verb* to get or to obtain; to buy; *speed control is used to acquire and maintain a selected airspeed*; *to acquire a new aircraft*

acquisition [ækwɪˈzɪʃn] *noun* getting or obtaining; buying; *each computer checks* *data acquisition; the image of the airline improved after the acquisition of the new aircraft*

acronym ['ækrənɪm] *noun* word, which is made up of the initial letters of a name, and is pronounced as a word; *NASA is the acronym for National Aeronautics and Space Administration*; *VASI is the acronym for visual approach slope indicator*

act [ækt] *verb* **(a)** to behave in a particular way; *the crew must act with authority* **(b)** to take the role of; *mountain ranges act as a barrier; the governor spill valve also acts as a safety relief valve* **(c)** *(of force)* **to act on** = to produce an effect; *bending and twisting forces act on a propeller; gravity acts vertically downwards*

acting ['æktɪŋ] *adjective* temporarily taking on the responsibilities of somebody; *Captain Smith will be acting Chief Flying Instructor while Captain White is absent from work*

action ['ækʃn] *noun* **(a)** something (to be) done; *if there is a risk of collision, the crew should take the appropriate action*; **corrective action** = something done to improve a bad situation **(b)** effect; **braking action good** = ATIS report about the effectiveness of brakes

activate ['æktɪveɪt] *verb* to make a system or a piece of equipment or a procedure start to work or to operate; *the system is activated by the pilot or copilot; the alarm will activate the emergency services*

activation [æktɪ'veɪʃn] *noun* making something start to work or to operate; *the normal activation method is automatic but the system can be activated manually; activation may be mechanical or electrical*

active ['æktɪv] *adjective* **(a)** live; in action; being used, in use; **the system is active** = the system is on and working **(b)** not passive; **in a secondary radar system, the target is active** = in a secondary radar system the target transmits a signal while in a primary radar system it does not **(c)**

active Cb clouds = developing cumulonimbus clouds **(d) active runway** = runway being used

COMMENT: large airports often have more than one runway, arranged to cope with varying wind directions. Some busy airports have parallel runways which can be used simultaneously

QUOTE never cross an active runway without permission from the tower: there may be more than one active runway
Civil Aviation Authority, General Aviation Safety Sense Leaflet

activity [æk'tɪvɪti] *noun* movement or action of some kind; being active; *sunspot activity can affect the amount of solar radiation*; **lightning activity** = period of time when there are a lot of lightning (flashes); **thermal activity** = period of time when there is a lot of vertical movement of air caused by heating; *cumulus clouds may develop because of thermal activity resulting from the warming of the surface*; **thunderstorm activity** = occurrence of weather conditions (rain, thunder, wind, lightning) associated with thunderstorms

actual ['æktʃʊəl] *adjective* real; *the actual path of the aircraft over the ground is called its track, which may not be the same as the desired course*

actually ['æktʃʊəli] *adverb* in fact, in reality; *the design is such that, although the aircraft loses altitude rapidly, it does not actually stall*

actuate ['æktʃueɪt] *verb* **(a)** to move a device or a part; *the fore and aft movement of the control column actuates the elevators* **(b)** to switch on or to put into operation a system or a piece of equipment; *a lever actuates the fire deluge system* **(c)** to put a procedure into action; *receipt of the distress signal will actuate the support facilities at the airport*

actuation [æktʃʊ'eɪʃn] *noun* **(a)** movement of a device or a part; **electrical actuation** = movement caused by an electric motor; **mechanical actuation** = movement caused by a rod, arm or lever, etc. **(b)** causing a device or a part to move

actuator ['æktʃueɪtə] *noun* device which changes electrical or hydraulic energy into mechanical motion; *the actuator control is sensitive to engine rpm*; *actuators are classified as either linear or rotary*; **linear actuator** = actuator which operates in a straight back and forth manner (eg to open undercarriage doors); **rotary actuator** = actuator which rotates and operates a screw jack (eg to extend flaps)

adapt [ə'dæpt] *verb* **(a)** to change or modify for special use; *the turboprop engine is often used in transport aircraft and can be adapted for use in single-engine aircraft* **(b)** to change to suit new conditions; *crew flying long-haul routes have to adapt to time changes*

adaptation [ædæp'teɪʃn] *noun* **(a)** changing or modifying something for special use; *Doppler VOR is an adaptation of VOR to reduce errors caused by location* **(b)** adjustment to new conditions; *adaptation to time changes when travelling west to east takes time*

adapter [ə'dæptə] *noun* piece of equipment or device which allows a change or modification; device which allows two incompatible devices to be connected; *a 'T' piece adapter* = a device for connecting two inputs to one output or vice versa

add [æd] *verb* **(a)** to put (figures) together to form a sum, to make a total; *add the two numbers together to find the sum* **(b)** to put together to make a larger group or a group with different properties; *there are only nine chairs, add another one*; *a substance is added to the fuel to clean fuel injectors*

addition [ə'dɪʃn] *noun* **(a)** mathematical operation consisting in putting numbers together; *addition is normally taught before subtraction, multiplication and division; the addition sign is +* **(b)** the act of adding something; *with the addition of methanol, the turbine inlet temperature is restored* **(c)** *in addition* = also; *in addition to* = as well as

additional [ə'dɪʃənl] *adjective* added or extra; *additional fuel is carried for holding en route* = extra fuel is carried in case it is needed

additive ['ædɪtɪv] *noun* chemical substance (often liquid) added to another substance (often liquid) to give extra qualities; *additives are used in engine oils to prolong the life of the engine*; *anti-icing additives are used in radiator coolants*

adequate ['ædɪkwət] *adjective* enough, sufficient; *the compressor must provide an adequate airflow through the engine*; **adequate fuel** = enough fuel; **adequate protection** = enough protection

ADF = AUTOMATIC DIRECTION FINDER

adhere [əd'hɪə] *verb* to stick as if glued; *clear ice adheres strongly to airframes*

adhesive [əd'hiːzɪv] **1** *noun* glue; *'superglue' is an all-purpose adhesive* **2** *adjective* having the sticking quality of glue; *adhesive bonding of aluminium parts is widely employed*; *adhesive tape*

adiabatic [ædɪə'bætɪk] *adjective* (i) (process) through which heat cannot be lost or gained; (ii) referring to a change in temperature (in a mass of air) due to compression or expansion caused by an increase or decrease in atmospheric pressure without loss or gain of heat to or from its surroundings; **adiabatic compression** = compression caused by atmospheric factors (causing warming of descending air); **adiabatic cooling** = cooling (of ascending air) caused by a decrease in atmospheric pressure (as when pressure is released from a pressurized can of drink) without heat transfer; **adiabatic expansion** = expansion caused by atmospheric factors (causing cooling of ascending air); *cooling by adiabatic expansion may result in cloud formation*; **adiabatic heating** = heating (of descending air) caused by an increase in atmospheric pressure (as when a bicycle pump is used) without heat transfer; **adiabatic lapse rate** = rate at which air temperature decreases as it rises above the earth's surface (as altitude increases, temperature decreases); *see also* LAPSE RATE

adjacent [ə'dʒeɪsənt] *adjective* next to; near by; *fire extinguishers should be positioned adjacent to the aircraft during all ground-running operations*; adjacent **areas** = areas next to each other

adjust [ə'dʒʌst] *verb* to change and improve the position or setting of a piece of equipment; *the pilot adjusts the throttle or propeller controls*; **to adjust the seat** = to move the seat into a position suitable for yourself; **to adjust the volume** = to increase or decrease the volume to improve the sound quality

adjustable [ə'dʒʌstəbl] *adjective* designed to be adjusted; *an adjustable stop on the throttle control ensures a positive idling speed*

adjustment [ə'dʒʌstmənt] *noun* change to improve the setting or position of something; *maximum system pressure is often controlled by adjustment of the main engine-driven pump*

admit [əd'mɪt] *verb* to allow to enter; *cold air can be admitted to the cabin through adjustable louvres or shutters*

adopt [ə'dɒpt] *verb* to choose to use or have something as standard equipment or procedure; *a policy of no smoking on short-haul flights has been adopted by many airlines*; **widely adopted** = now in standard use with many companies, institutions and organizations

adoption ə'dɒptʃn] *noun* the act or instance of using something as standard equipment or procedure; *in spite of the adoption of the axial flow type compressor, some engines retain the centrifugal type*

advance [əd'vɑːns] **1** *noun* **(a)** developments; progress; **enormous advances in aircraft design** = great progress or developments in aircraft design

(b) in advance of = ahead of; *the Gulf region is three hours in advance of GMT* **2** *verb* to move forwards; *(action)* **the throttle lever is advanced** = the throttle lever is moved forwards; *(in time)* **to advance the ignition** = to adjust the timing of the ignition so that the spark occurs earlier

advanced [əd'vɑːnst] *adjective* **(a)** modern and sophisticated; *the A340 is an advanced type of aircraft* **(b)** **advanced level exams** = high level exams, after several years of studies

a Seattle-based modification company specializing in advanced winglet designs is developing a lightweight winglet for the Boeing 747 200F

Flight International 1-7 May 1996

advantage [əd'vɑːntɪdʒ] *noun* good or beneficial factor; *the multi-wheel combination has the advantage of smaller and lighter undercarriage structures*; **mechanical advantage** = the ratio of the output force produced by a machine to the input force; **to take advantage of** = to get the most benefit from a situation; **to take advantage of favourable winds** = to use tailwinds to increase groundspeed and thus save time and money (NOTE: the opposite is **disadvantage**)

advantageous [ædvɑːnt'teɪdʒəs] *adjective* better; **the most advantageous** = the best; *the minimum time path is the most advantageous for economy*

advect [əd'vekt] *verb (of air)* to move in a horizontal direction due to convection; *dispersal of hill fog takes place when surface heating lifts the cloud base or drier air is advected*

advection [əd'vekʃn] *noun* movement of air in a horizontal direction; **advection fog** = fog which forms when warmer moist air moves over a colder surface (land or sea); *compare* CONVECTION

advent ['ædvənt] *noun* coming or arrival, especially of something very important; *with the advent of satellite* *navigation systems, pilots of light aircraft have a more accurate means of knowing their position*

adverse ['ædvɜːs] *adjective* **(a)** *(conditions)* bad or poor; *only in extremely adverse conditions should the crew evacuate the aircraft*; **adverse handling characteristics** = aspects of an aircraft's handling which are poor; **adverse weather (conditions)** = bad weather **(b)** which blows or goes against you; **adverse yaw** = yaw, caused by aileron drag, in the opposite direction to the direction of the intended turn

advice [əd'vaɪs] *noun* useful or helpful information; *the instructor's advice was of great help to the student pilot* (NOTE: there is no plural form)

advisability [ədvaɪzə'bɪlɪti] *noun* statement or opinion on whether a particular action is a good idea or not; *flying manuals often contain guidance on the advisability of flying with a cold*

advisable [əd'vaɪzəbl] *adjective* recommended, suggested; *where possible, it is advisable to closely check the condition of the tyres*

advise [əd'vaɪz] *verb* **(a)** to inform, to notify; *the flight deck will advise cabin crew that descent will start in 20 minutes* **(b)** to recommend, to suggest; *because of the bad weather, the instructor advised the trainee pilot not to fly*; **to advise against** = to recommend or to suggest that something should not be done

advisory [əd'vaɪzri] *adjective* giving advice and information; **advisory routes** = published routes served by advisory service; **advisory service** = facility providing advice and information to assist a pilot in the safe conduct of the flight; *see also* AIRSPACE

aerate ['eəreɪt] *verb* to put a gas - especially carbon dioxide or air - into a liquid so that bubbles are formed; *aerated fuel causes problems*; *see also* DE-AERATE

aeration [eə'reɪʃn] *noun* putting a gas - especially carbon dioxide or air - into a liquid; *the purpose of the booster pump is to prevent fuel aeration*; *see also* DE-AERATION

aerator [eə'reɪtə] *noun* device to put a gas - especially carbon dioxide or air - into a liquid; *see also* DE-AERATOR

aerial ['eəriəl] **1** *adjective* **(a)** referring to the air **(b)** referring to an aircraft in the air; **aerial photography** = photography done from an aircraft in the air; **an aerial display** = a display of flying skills and aircraft performance **2** *noun* device to send or receive radio or TV signals; *ice-covering reduces the effectiveness of aerials* (NOTE: American English in this meaning is **antenna**)

aero- ['eərəʊ] *prefix* **(a)** referring to the air; *see also* AERODYNAMIC **(b)** referring to aircraft; *see also* AERO-ENGINE; AERO-TOW

aerobatic [eərə'bætɪk] *adjective* referring to aerobatics; *loops and rolls are aerobatic manoeuvres*; **aerobatic aircraft** = aircraft which is designed to perform aerobatics; **aerobatic display** = show in which aircraft perform aerobatic manoeuvres; **aerobatic team** = team of pilots and aircraft who perform aerobatics

COMMENT: one of the most famous competition aerobatic aircraft is the Pitts Special which first flew in 1944

aerobatics [eərə'bætɪks] *noun* science or art of performing advanced manoeuvres, spectacular controlled movements in an aircraft, in the air, for the purposes of entertainment or competition; *the Russian pilot gave a great display of aerobatics* (NOTE: the noun **aerobatics** is singular)

aerodrome ['eərədrəʊm] *noun* any area of land or water designed for the taking off and landing of aircraft; *airports and military air bases or stations are types of aerodrome*; *all aerodromes are marked on charts*; **disused aerodrome** = an aerodrome which is no longer in use for the purpose of taking off and landing aeroplanes; **aerodrome boundaries** = the physical or geographical limits of the aerodrome; **aerodrome circuit** = the pattern and direction of aircraft movement in the air around the aerodrome; **aerodrome elevation** = height of the aerodrome above mean sea level; **aerodrome QFE** = barometric pressure setting at which the altimeter reads zero when the aircraft is on the runway; **aerodrome QNH** = barometric pressure setting at which the altimeter reads aerodrome elevation when the aircraft is on the runway; *(in general aviation)* **aerodrome traffic zone (ATZ)** = airspace up to circuit height and within 1.5 nm of the aerodrome boundary

aerodynamic [eərədaɪ'næmɪk] *adjective* **(a)** referring to the way in which objects are affected when they move through the atmosphere; **aerodynamic braking** = braking effect of drag; **aerodynamic forces** = forces of the air which act on an aircraft in flight **(b)** referring to a smooth rounded shape which moves easily through the air; **aerodynamic design** = design of aircraft which moves easily through the air because of its shape

aerodynamics [eərədaɪ'næmɪks] *noun* science of dynamics and interaction of moving objects with the atmosphere; *aerodynamics is one of the major areas of study for a trainee pilot* (NOTE: the noun **aerodynamics** is singular)

aero-engine ['eərəʊ'endʒɪn] *noun* engine used in aircraft; *most piston aero-engines are cooled by air*

aerofoil ['eərəʊfɔɪl] *noun* surface which is shaped to produce more lift than drag when moved through the air; *wings, ailerons, elevators, fins and propellers are all examples of aerofoils* (NOTE: American English is **airfoil**)

aeronautical [eərə'nɔːtɪkl] *adjective* referring to aeronautics; **aeronautical charts** = map used in air navigation which may include topographic features, hazards and obstructions, navigational aids and routes, designated airspace and airports;

see also CHART; ENGINEER; ENGINEERING

aeronautics [eərə'nɔːtɪks] *noun* **(a)** science of aircraft design, construction, operation or navigation **(b)** theory and practice of aircraft navigation (NOTE: the noun **aeronautics** is singular)

aeroplane ['eərəpleɪn] *noun* power-driven, heavier-than-air craft with fixed wings; **aeroplane performance** = description in figures regarding the speed of the aircraft, rate of climb, length of take-off run, etc. (NOTE: many people use the words **aeroplane** and **aircraft** synonymously. However, aeroplanes, hot-air balloons, helicopters, airships and gliders are all **aircraft**; the American English is **airplane**)

aero-tow ['eərə'təʊ] *noun* technique of using a powered aircraft to pull a glider into the air; *an aero-tow to 2,000 feet costs $25*

affect [ə'fekt] *verb* to have an influence on (something); to make a change; *humidity and air density are factors which affect the output of the engine; wind direction and speed only affect the movement of the aircraft over the ground* (NOTE: compare **effect**)

aft [ɑːft] **1** *adjective* towards the rear part of the aircraft; *the rear part of the fuselage is called the aft section*; **aft cabin** = passenger compartment at the back of the aircraft **2** *adverb* rearwards; backwards; **to move the control column aft** = to move the control column backwards (NOTE: the opposite is **fore** or **forward**)

agent ['eɪdʒənt] *noun* **(a)** a chemical substance which causes a change; *if de-icing fluid is used as an anti-icing agent it should be sprayed onto the aircraft before the onset of icing*; **extinguishing agent** = one of several substances used to put out fires **(b)** person who represents a company; person who arranges something for a company; *the agent for British Airways; a travel agent*

aggregate ['ægrəgət] **1** *noun* total obtained by adding; *the aggregate of the* *capacity of all the fuel tanks is 50 gallons* **2** *verb* to come together to form a mass; to make up a whole or total; *ice crystals aggregate to form snowflakes*

agree [ə'griː] *verb* **(a)** to have the same idea or opinion about something; *the crew agreed with the findings of the investigation* **(b)** to come to an understanding; *after hours of discussion, the cabin staff agreed to call off the planned strike*

agreed [ə'griːd] *adjective* generally accepted; *the millibar is an agreed unit of pressure*

agreement [ə'griːmənt] *noun* **(a)** having the same idea or opinion as somebody; *we are in agreement* = we agree **(b)** contract or document between two or more parties which explains how they will act; *Regional Air Navigation Agreements*

ahead [ə'hed] *adverb* in front; **look ahead** = look some distance in front of you; **straight ahead** = directly in front

ahead of [ə'hed 'ɒv] *preposition* **(a)** in front of; *air ahead of a cold front is warmer than air behind a cold front* **(b)** in advance of; at an earlier time than; *the flight from Paris arrived 10 minutes ahead of schedule*

AI = ATTITUDE INDICATOR

aid [eɪd] **1** *noun* something which helps someone do something; **first aid kit** = small pack containing plasters, bandages, antiseptic cream, etc., to be used in case of an emergency; **navigational aid (NAVAID)** = electronic devices to help a pilot navigate **2** *verb* to help; *computers can aid students in their studies*

aileron ['eɪlərɒn] *noun* *(aerofoil)* horizontal control surface hinged to the mainplane (wing) which provides movement around the longitudinal axis of the aeroplane; *by rotating the yoke the ailerons are moved and the aircraft rolls into a turn* (NOTE: the word comes from the French 'aile', meaning 'wing')

aim [eɪm] **1** *noun* goal or objective; *a 100% safe operation is the aim of all airline companies* **2** *verb* to intend or to try to do something; **we aim to succeed** = we intend to succeed

air [eə] *noun* mixture of gases which form the earth's atmosphere; *air enters the cabin through an inlet*; **air density** = density of the atmosphere

airborne ['eəbɔːn] *adjective (of aircraft)* lifted and kept in the air by aerodynamic forces; *shortly after the aircraft becomes airborne, the undercarriage is retracted*; **airborne installation** = radar equipment in an aircraft which is used together with equipment in the ground installation; **airborne weather radar (AWR)** = radar installation in an aircraft to give the flight crew information about the en route weather

air conditioner ['eə kən'dɪʃnə] *noun* device which filters and cools the air in a room, in an aircraft, etc.; *in order to obtain maximum engine power, the air conditioner should be switched off for take-off*

air conditioning ['eə kən'dɪʃnɪŋ] *noun* system of controlling the temperature of the air in a building, in an aircraft, etc.

air-cooled ['eə'kuːld] *adjective* cooled by means of a flow of air; **air-cooled engines** = piston aero-engines cooled by air, not water

aircraft ['eəkrɑːft] *noun* man-made flying transport; *aeroplanes, gliders, balloons, airships, helicopters, etc., are all aircraft*; **aircraft classification number (ACN)** = number expressing the relative effect of an aircraft on a pavement for a specified sub-grade strength; **fighter aircraft** = small, single-seat or two-seat aircraft for use in military conflict; **light aircraft** = (generally speaking) small, single engine aircraft for private (not commercial) use; **passenger aircraft** = aircraft specially designed for carrying people; **transport aircraft** = aircraft designed for carrying ten or more passengers or equivalent cargo and having a maximum take-off weight greater than 5,670 kg; *see also* AEROPLANE (NOTE: the plural of **aircraft** is **aircraft**: the abbreviation for aircraft is **ACFT**)

airfield ['eəfiːld] *noun* area of land given over to runways, taxiways and aprons; *when the pressure setting on the altimeter is set to 1013.25 millibars, the pressure altitude of the airfield is known as QNE*

airflow ['eəfləʊ] *noun* **(a)** movement of air over the aircraft caused by the movement of the aircraft through the air; **relative airflow** = airflow over an aerofoil often related to the chord line of the aerofoil **(b)** current of air flowing through or past an object or body; *the compressor must provide an adequate airflow through the engine*

airfoil ['eəfɔɪl] *noun US* = AEROFOIL

airframe ['eəfreɪm] *noun* body of the aircraft; aircraft without the engine(s), the instruments, etc.; *the airframe has to be built to very specific requirements*; **airframe icing** = ice forming on the aircraft structure as opposed to components such as carburettors

air gap ['eə 'gæp] *noun* space between two things; **air gap type spark plug** = spark plug with a space between the electrodes, across which the spark jumps

airline ['eəlaɪn] *noun* company which manages air transport services for passengers or goods; *which airline is he working for, Air France or Air Canada?*; *some airlines do not allow passengers to smoke during flight*; **airline representative** = someone who acts on behalf of an airline; *passengers should assemble in the departure lounge where an airline representative will meet them*

airliner ['eəlaɪnə] *noun* aeroplane designed to carry large numbers of passengers (like those operated by an airline such as KLM or QANTAS, etc.); *Concorde is the world's fastest airliner*

Airline Transport Pilot's Licence (ATPL) *noun* licence required for pilot-in-command or co-pilot of public transport aircraft (two crew operations)

airman ['eəmən] *noun* someone who serves in the Air Force; an aviator

airmanship ['eəmənʃɪp] *noun* all-round skill in piloting an aircraft which includes academic knowledge, common sense, quick reactions, awareness, experience, consideration for other people and property; *keeping a careful lookout for other aircraft in the circuit is good airmanship*

QUOTE I was always told by my airmanship instructor, in an emergency, to find the largest piece of asphalt with the biggest fire trucks
INTER PILOT

airplane ['eəpleɪn] *US* = AEROPLANE

airport ['eəpɔːt] *noun* civil aerodrome designed for the take-off and landing of passenger-carrying aircraft for the general public and/or cargo aircraft; *London Heathrow is one of the busiest airports in the world*

airship ['eəʃɪp] *noun* powered, gas-filled balloon which can be steered; *an airship is classified as a lighter-than-air craft*

airspace ['eəspeɪs] *noun* part of the atmosphere above the surface, subject to the laws of a particular country or controlling authority; *the Korean 747 flew into Soviet airspace and was shot down*; **advisory airspace** = airspace containing advisory routes in which Air Traffic Control (ATC) provide an advisory service but not full control; **controlled airspace** = airspace which is governed by rules and regulations which pilots must comply with; **uncontrolled airspace** = airspace in which a pilot may fly where he wants to without being controlled but must follow simple rules; *controlled airspace, advisory airspace and uncontrolled airspace are categories of ATC airspace which provide a specific service*

airspeed ['eəspiːd] *noun* speed of the aircraft relative to the air around it; *maintain a constant airspeed on final approach*; **calibrated airspeed (CAS)** = indicated airspeed corrected for instrumentation and installation errors; **indicated airspeed (IAS)** = airspeed shown on the cockpit or flight-deck instrument; **rectified airspeed = CALIBRATED AIRSPEED**

COMMENT: the position of the pitot tube and the attitude of an aircraft can affect the accuracy of the airspeed indicator. Aircraft operating handbooks usually have a table to help pilots calculate calibrated airspeed (CAS)

airspeed indicator (ASI) ['eəspiːd 'ındıkeıtə] *noun* primary cockpit or flight deck instrument which shows the pilot the speed of the aircraft in relation to the air around it; *airspeed is shown in knots on the airspeed indicator*

airstream ['eəstriːm] *noun* flow of air caused by the movement of the aircraft through the air; *pressure is built up inside the pitot tube by the airstream*

COMMENT: unlike airflow, airstream does not refer to the movement of air around the airframe and its aerodynamic effect

air-to-air ['eətə'eə] *adjective* **air-to-air communications** = communications between one airborne aircraft to another

air traffic ['eə 'træfik] *noun* aircraft operating in the air or on the airport surface; *students practising circuit flying need to keep a very careful lookout especially at times when there is a lot of air traffic*

air traffic control (ATC) ['eə 'træfik kən'trəul 'sɜːvɪs] *noun* service provided for the safe and efficient flow of air traffic; *controllers in the tower provide an air traffic control service for aircraft in the air around the airfield*

COMMENT: ATC's main function is to maintain separation between aircraft operating within Instrument Flight Rules (IFR), but it also provides a service to aircraft using Visual Flight Rules (VFR). Ground control is for aircraft taxiing to and from runways. The tower controls aircraft around an airport, clearing them for take-off or landing. Departure and approach controls monitor and control aircraft around the airport, and en route centres control traffic between airports

air traffic controller ['eə 'træfik kən'trəulə] *noun* person whose job it is to ensure correct separation of aircraft in all phases of flight; the tower; *the air traffic controller approved the emergency landing*; *see also* CONTROLLER

airway ['eəweı] *noun* area of the sky, usually rectangular in cross-section, along which civil aircraft fly from place to place - usually 10 nm wide with a centre-line joining navigational beacons; *airways provide a high degree of safety by ensuring adequate separation between aircraft*; *aircraft inside an airway are controlled by ATC*

airworthiness [eə'wɜːðınəs] *noun (of aircraft)* in good flying condition, as determined by the national certifying authority; **certificate of airworthiness (C of A)** = certificate given by an aviation authority stating that an aircraft meets specific safety, etc., requirements which allow it to be used in service

airworthy ['eəwɜːði] *adjective (aircraft)* which meets the standards of a national certifying authority; *it is the pilot's responsibility to ensure that the aircraft is airworthy*

aisle [aıl] *noun* long passageway between seats in the passenger cabins of airliners; **aisle seat** = seat which is by an aisle (as opposed to 'window seat' which is next to a window)

alarm [ə'lɑːm] **1** *noun* **(a)** fear or worry; *if the ammeter shows a high level of* *charge after start-up, it is quite normal and no cause for alarm* **(b)** warning sound or light; *in the event of fire or overheat, the control unit will produce an alarm*; **smoke alarm** = warning system that will go off if there is smoke somewhere; *washrooms are fitted with smoke alarms* **2** *verb* to frighten or worry; *severe turbulence may alarm passengers*

alert [ə'lɜːt] **1** *adjective* fully awake, watchful and ready to deal with any situation; *the crew must be alert at all times to the possibility of hijacking, bombs and stowaways* **2** *noun* signal, warning everyone to be alert; **to be on the alert** = to be watchful and ready for anything that may happen **3** *verb* to warn; *it is the cabin staff's responsibility to alert the flight crew if they see smoke coming from an engine*

alight [ə'laɪt] **1** *adjective* on fire; *although the passenger thought he had extinguished his cigarette, it was still alight when he threw it into the waste disposal bin* **2** *verb (formal)* **(a)** *(of people)* to leave or get off an aeroplane; *at some airports, passengers alight onto the apron when they leave the aircraft* **(b)** *(of plane)* to land; *an aeroplane may not fly over a city below such a height as would allow it to alight in the event of an engine failure*

align [ə'laɪn] *verb* **(a)** to position along an axis or line; *the nose wheel must be aligned in a fore and aft direction during retraction* **(b)** to set in a correct position in relation to something else; *aligned white marks on the wheel and tyre indicate that there is no creep*

alignment [ə'laɪnmənt] *noun* **(a)** positioning along an axis or line; **to check the alignment of something** = to make sure it is in the correct position relative to an axis or line; **to maintain alignment with the runway** = to keep the aircraft on the imaginary extended centre line of the runway **(b)** setting in a correct position in relation to something else; **out of alignment** = not aligned as it should be

alkaline ['ælkəlaɪn] *noun* substance with a pH value of more than 7

alleviate [ə'liːvieɪt] *verb* to reduce or lessen the harmful effect of something; *anti-icing additives are available to alleviate the problem of icing*

alleviation [əliːvi'eɪʃn] *noun* reduction or lessening of the harmful effect of something; *deep, regular breathing may provide some alleviation from stress*

allocate ['æləkeɪt] *verb* to provide something particular for a given purpose; *special seats are allocated to mothers with small children*

allocation [ælə'keɪʃn] *noun* provision of something particular for a given purpose; *at the check-in desk, airline staff are responsible for the allocation of seats to passengers*; **frequency allocation** = frequency or range of radio frequencies set aside for particular use; *the frequency allocation for VOR is 108-117.975 MHz*

allow [ə'laʊ] *verb* to enable, to permit, to authorize; *an engine should be run at low rpm after flight to allow engine components to cool; additional fuel is carried to allow for holding en route; passengers are not allowed to smoke on some aeroplanes*

allowable [ə'laʊəbl] *adjective* permitted, authorized; *maximum allowable weight*; *maximum allowable tyre pressure*

allowance [ə'laʊəns] *noun* **(a)** consideration for possibilities or changing circumstances; **to make allowances for** = take into account; *when estimating flight duration, make allowances for taxiing time* **(b)** something (such as money) given at regular intervals or for a specific purpose; *a travel allowance to cover hotel and restaurant bills* **(c)** something which is allowed; **baggage allowance** = amount of baggage (in weight) each air passenger is allowed to take free; *there is an accompanied baggage allowance of 18 kilos*

QUOTE with many four and six seat aircraft, it is not possible to fill all the seats, use the maximum baggage allowance, fill all the fuel tanks and remain within the approved centre of gravity limits
Civil Aviation Authority, General Aviation Safety Sense Leaflet

alloy ['ælɔɪ] *noun* mixture of metals; *an alloy of aluminium and lithium*

aloft [ə'lɒft] *adjective* up in the air; *a pressure gradient occurs aloft from land to sea*

alter ['ɔːltə] *verb* to change, to modify; to adjust; *if there is a risk of collision, alter course to the right; if the rate of descent is too low, alter the throttle setting accordingly; the rudder linkage was altered to comply with certification requirements*

alteration [ɔːltə'reɪʃn] *noun* (a) change, modification; adjustment; *it was discovered that alterations had been made to the log book; as a result of the accident, alterations were made to the design of the carburettor heat system* (b) making changes, modifications, adjustments; **heading alteration** = the making of heading corrections

alternate 1 [ɔːl'tɜːnət] *adjective* every other (day, month, etc.); *a, c, e, g are alternate letters so are b, d, f, h, etc.*; **alternate days** = every other day; *there are outward flights on alternate days, ie on Mondays, Wednesdays and Fridays* **2** [ɔːl'tɜːnət] *noun* an aerodrome of second choice to be used if the aircraft cannot be landed at the aerodrome of first choice because of bad weather, etc.; *the point of no return is calculated before departure to cover the chance that both the terminal airfield and its alternate become unavailable during flight* **3** ['ɔːltəneɪt] *verb* to happen in turns; **Captain Smith and Captain Jones alternate as CFI on a daily rota** = each CFI has one day on duty followed by a day off, while the other CFI works

alternating current (AC) ['ɔːltəneɪtɪŋ 'kʌrənt] *noun* electric current which reverses its direction at regular intervals; *resistance to alternating current remains almost constant and is independent of frequency*

alternative [ɔːl'tɜːnətɪv] **1** *adjective* referring to another or a second possibility; *a turbine bypass, in the form of an alternative exhaust duct is fitted with a valve*; **an alternative means of doing something** = another or different way of doing something **2** *noun* another possibility; a choice; *in some emergency situations the pilot may have no alternative but to force land the aircraft as soon as possible*

alternator ['ɔːltəneɪtə] *noun* type of generator designed to produce AC power

altimeter ['æltɪmiːtə *US* æl'tɪmɪtər] *noun* pressure or radio instrument for measuring vertical distance or altitude; **pressure altimeter** = altimeter which operates using atmospheric pressure; *the altitude indicated by a pressure altimeter is the height of the altimeter above the level of the pressure set on the sub-scale*; **radio altimeter** = altimeter which operates using a radio signal directed vertically downwards, giving height above ground level; **altimeter check** = routine check to ensure that altimeter pressure setting is correct; **altimeter display** = analogue or digital appearance of altitude information; *see also* POINTER

altitude ['æltɪtjuːd] *noun* the vertical distance between an aircraft - or a point or a level - and mean sea-level; *(of aircraft)* **to lose altitude** = to descend from higher to lower altitude; **cabin altitude** = artificial altitude created in the cabin by pressurization; **cruising altitude** = altitude at which most of the flight is flown en route to a destination, from top of climb to top of descent; **transition altitude** = altitude near an airfield at, or below, which aircraft control is by true altitude; **true altitude** = real or actual height above sea level (calibrated altitude corrected for air temperature)

alto- ['æltəʊ] *prefix meaning* moderate or high altitude

altocumulus [æltəʊ'kjuːmjʊləs] *noun* small white cumulus clouds which form as a layer at moderate altitude (above 3,000 m), usually meaning fair weather; *compare* STRATOCUMULUS (NOTE: can be shortened to **AC** (ICAO))

altostratus [æltəʊ'strɑːtəs] *noun* uniform layer cloud at moderate altitude (above 3,000 m)

aluminium [ælju:'mɪnɪəm] *noun* strong, light metal used in the construction of aircraft (NOTE: American English is **aluminum**)

COMMENT: in recent years, aluminium has been increasingly replaced by the use of composite materials in the construction of different types of aircraft, from small home-built light aircraft to transport aircraft such as the Airbus A320

aluminum [ə'luːmɪnəm] *noun US* = ALUMINIUM

amber ['æmbə] *adjective* orange/yellow colour often used to describe the colour of the yellow light in traffic signals; *an amber light flashes on the instrument panel*

ambient ['æmbɪənt] *adjective* referring to the surrounding atmospheric conditions; *fresh ambient air is routed into the cabin*; **ambient temperature** = the temperature outside the aircraft; **ambient pressure** = the pressure outside the aircraft

ambiguity [æmbɪ'gjuːəti] *noun* something heard or seen which can be understood in more than one way, thus resulting in possible confusion; **to avoid ambiguity** = to avoid misunderstanding or confusion; *correct use of R/T phraseology avoids ambiguity*

ambiguous [æm'bɪgjuəs] *adjective* referring to ambiguity; *it is important that R/T transmissions are not ambiguous* (NOTE: in general English, the term 'current flow' is ambiguous because it can be understood in two ways: electrical flow or flow at the present time)

AMD = AMENDMENT

amend [ə'mend] *verb* to change; to update, to improve or to correct a document or procedure, etc.; **he amended the entry in his log book** = he corrected or changed the entry in his log book

amendment [ə'mendmənt] *noun* change, updating, improvement or correction to a document or procedure, etc.; *when a terminal aerodrome forecast requires amendment, the amended forecast is indicated by inserting AMD after TAF*

ammeter ['æmiːtə] *noun* instrument for measuring amperes in order to give the strength of an electric current; *the centre-zero ammeter tells the pilot the status of the aircraft battery*

amp [æmp] = AMPERE

amperage ['æmprɪdʒ] *noun* strength of an electric current expressed in amperes; *measuring the amperage of a motor can give a rough estimate of the load on the motor*

ampere (amp) ['æmpeə ˘æmp̄] *noun* unit of electric current - a current flowing through an impedance of one ohm which has one volt across; **current flow is measured in amperes**; *a 13-amp fuse*; **ampere hours** = number of amperes per hour; *battery capacity is rated in ampere hours*

ample ['æmpl] *adjective* plenty of; *during the course you will have ample opportunity to demonstrate your skill*; **ample time** = plenty of time

amplification [æmplɪfɪ'keɪʃn] *noun* act or instance of increasing the strength of an electrical signal; *amplification of the signal increases the volume*

amplifier ['æmplɪfaɪə] *noun* electronic device for increasing the strength of an electrical signal; *if the power supply from* *the amplifier to the gauge fails, the needle slowly falls to zero*

amplify ['æmplɪfaɪ] *verb* to increase the strength of an electrical signal; *an electric current is amplified and then transmitted* (NOTE: amplifying - amplified)

amplitude ['æmplɪtjuːd] *noun* the maximum variation of a vibration or oscillation from the position of equilibrium; *a transformer is a static device that changes the amplitude or phase of alternating current*

anabatic [ænə'bætɪk] *adjective* **anabatic wind** = wind currents which are caused by solar heating of the land, rising up south-facing mountainsides; *south-facing slopes are most suitable for the anabatic wind* (NOTE: the opposite is **katabatic**)

analog ['ænəlɒg] = ANALOGUE

analogous [ə'næləgəs] *adjective* similar to; comparable to; *isobars are analogous to contour lines*

analogue ['ænəlɒg] *adjective* **(a)** quantity or signal which varies continuously; *the electronic centralized aircraft monitor (ECAM) does not have analogue presentation of engine information* **(b)** analogue display **(on a clock)** = traditional hands and face display; *compare* DIGITAL DISPLAY

analysis [ə'næləsɪs] *noun* breaking down of a substance into its parts in order to study them closely; *at a crash site, samples of materials are removed for analysis*; **chart analysis** = careful study of charts (NOTE: the plural is **analyses** [ə'næləsiːz]. Note the difference in pronunciation between the singular and plural forms)

analyze ['ænəlaɪz] *verb* to break down into parts and study very closely; **to analyze fuel** = to separate fuel into its different parts to find out what it consists of; **to analyze a chart** = to examine a chart in detail (NOTE: also written: **analyse**)

anchor ['æŋkə] **1** *noun* device connected to and dropped from a boat in order to prevent the boat from moving in the water; **sea-anchor** = a device under a raft to provide stability; *each life raft is equipped with a flame orange coloured canopy and a sea-anchor* **2** *verb* to drop an anchor to prevent the boat from moving

anemograph [ə'neməgrɑːf] *noun* instrument which maintains a continuous recording of wind direction and speed on a graph; *the anemograph gives a continuous recording of wind velocity which is displayed on a chart and reveals gusts, squalls and lulls*

anemometer [ænɪ'mɒmɪtə] *noun* instrument, usually attached to a building, with three or four 'cups' which rotate with the wind thus providing wind-speed information; *the strength of the wind can be seen by the speed with which the anemometer rotates*

aneroid ['ænərɔɪd] *adjective* **aneroid barometer** = barometer which uses an aneroid capsule to sense atmospheric pressure changes; **aneroid capsule** = thin flexible cylindrical box, usually made of metal, which has most of the air removed from it and which expands and contracts with changes in atmospheric pressure; *the aneroid capsule in the barometer is connected to a system of levers which operate a pointer*; **aneroid switch** = switch operated by an aneroid capsule

angle ['æŋgl] *noun* difference in direction between two lines or surfaces measured in degrees; **angle of attack** = angle formed between the relative airflow and the chord line of the aerofoil; **angle of incidence** = angle formed between the chord-line of the mainplane (wing) and the horizontal when the aircraft is in the rigging position

COMMENT: the angle of attack is related to the flight path of the aircraft, not to the angle the wing makes with the horizontal. If angle of attack becomes too great, the smooth airflow over the upper surface of the wing will break down. If no corrective action is taken by the pilot, there will be a sudden loss of lift and the aircraft will stall

angular ['æŋgjʊlə] *adjective* referring to an angle; forming an angle; *the angular difference between the direction of magnetic north and compass north is called variation*

anneal [ə'niːl] *verb* to heat and allow to cool slowly in order to strengthen; *sheet and plate magnesium are annealed at the rolling mill*

annotate ['ænəteɪt] *verb* to add notes to an existing document, book, chart, etc.; *he annotated his report after he was asked to give the exact time of the incident*; *variation is annotated east or west according to the direction of change*

annotation [ænə'teɪʃn] *noun* note added to a document, book, chart, etc.; adding notes to a book, etc.; *wind directions are measured from magnetic north so, always take care to look at the annotations*

announce [ə'naʊns] *verb* to state something publicly or officially; *British Airways announce the departure of flight BA152 to New York*

announcement [ə'naʊnsmənt] *noun* public statement; *the captain made a public address (PA) system announcement asking passengers to remain seated*

annual ['ænjuəl] *adjective* **(a)** which happens or is done once a year; yearly; **annual inspection** = yearly inspection **(b)** over a period of one year; *overload operations should not exceed 5% of annual departures*

annular ['ænjʊlə] *adjective* shaped like a ring; *annular inner and outer air casings form a tunnel around the spine of the engine*

annunciation [ənʌnsi'eɪʃn] *noun* announcement or indication on the annunciator panel; **failure annunciation** =

signals on the annunciator panel indicating the failure of a system

annunciator [ə'nʌnsieɪtə] *noun* device which gives off a sound or light to indicate which of several electrical circuits is active; *an annunciator panel may contain a precise warning*

anode ['ænəʊd] *noun* positive pole or electrode; *the positive connector of a battery is usually called the anode and is indicated by the sign +*

anodize ['ænədaɪz] *verb* to coat or cover by using electrolysis; *anti-corrosion treatment includes bead or shot-blasting of large areas, chromic acid anodizing of aluminium parts followed by use of an epoxy based primer rich in chromate*

anomalous [ə'nɒmələs] *adjective* referring to something departing from what is the normal order or range; unusual; unexpected; **an anomalous instrument reading** = an unusual instrument reading which may require further investigation

anomaly [ə'nɒməli] *noun* something departing from what is the normal order or range; something unusual; something unexpected; *any anomalies in the localizer will be detected during calibration*

anoxia [æ'nɒksiə] *noun* complete lack of oxygen to the body, resulting in death; *the investigation established that the cause of death was anoxia*; *see also* HYPOXIA

COMMENT: anoxia is a complete lack of oxygen and can, of course, be fatal. Hypoxia is a lack of sufficient oxygen, the symptoms of which are sometimes difficult to detect

antenna [æn'tenə] *US noun* device to send or receive radio or TV signals; *long-range radars require a large antenna* (NOTE: plural is **antennas** or **antennae**; **aerial** is preferred in British English)

anti- ['ænti] *prefix meaning* against, opposing; *see also* ANTI-ICING; ANTICLOCKWISE; ANTI-CORROSION

anticipate [æn'tɪsɪpeɪt] *verb* to realise what may happen and do what is necessary in readiness; **during take-offs, pilots should anticipate an engine failure** = pilots should be ready to act immediately in the event of an engine failure

anticipation [æntɪsɪ'peɪʃn] *noun* realising what may happen and do what is necessary in readiness; **anticipation of landmarks** = being ready to see landmarks, which you know from flight planning, should be visible at a particular stage of a flight

anticlockwise [ænti'klɒkwaɪz] *adjective & adverb* referring to a circular movement in the opposite direction to the hands of a clock; *turn the nut anticlockwise to loosen it* (NOTE: the opposite is **clockwise**)

anti-collision [æntikə'lɪʒn] *adjective* which help prevent collisions; **anti-collision light** = flashing white light on an aircraft

anti-corrosion [æntikə'rəʊʒn] *adjective* which protects against corrosion especially rust; *an anti-corrosion treatment*

anticyclone [ænti'saɪkləʊn] *noun* area of high atmospheric pressure, usually associated with fine dry weather in summer and fog in winter; *winds circulate round an anticyclone clockwise in the northern hemisphere and anticlockwise in the southern hemisphere*

anti-icing ['ænti'aɪsɪŋ] *adjective* which prevents icing; *anti-icing additive*; **anti-icing fluid** = fluid which prevents icing; *see also* ICING; SKID

anvil ['ænvɪl] *noun* metal block which ends in a point, has a rounded bottom and a flat top, and on which horseshoes, etc., are made; *a cumulonimbus cloud has a characteristic anvil shape*; **anvil cloud** = cloud formation, in a dark thundercloud, which has the shape of an anvil; **anvil-shaped** = having the shape of an anvil; *an anvil-shaped cumulonimbus*

apart [ə'pɑːt] *adverb* away from one another; separated; *the jets were only 200 feet vertically apart, and only 200 yards horizontally apart*

aperture ['æpətʃə] *noun* opening; *any aperture or cut-out in the fuselage structure must be specially strengthened*

APP = APPROACH *or* APPROACH CONTROL

apparent [ə'pærənt] *adjective* **(a)** clearly understood because of what our senses, or understanding tell us; obvious, clear; *it became apparent that carbon monoxide was entering the cabin*; **from the above, it will be apparent that ...** = from the above, it will be clear that ... **(b)** seeming; which appears to be; *an apparent failure of the system*; *the ILS showed an apparent deflection to the right*

appear [ə'pɪə] *verb* **(a)** to be seen; to come into view; *another aircraft appeared on the radar screen* **(b)** to seem to be; *although air may appear to be still, it is in fact, moving*

appearance [ə'pɪərəns] *noun* **(a)** instance of being seen or coming into view; *the appearance of the passenger on the flight deck surprised the crew* **(b)** the way something looks; the look of something; *it may be difficult to recognize a particular stretch of coast in an area simply by its appearance*

appendix [ə'pendɪks] *noun* section containing additional information, often found at the end of a book, etc.; *charts are reproduced as an appendix to the map section* (NOTE: plural is **appendices** [ə'pendɪsiːz])

applicable [ə'plɪkəbl] *adjective* **(a)** relevant; **rule 24 is not applicable in this case** = rule 24 cannot be used in this case **(b)** suitable, necessary, appropriate; *emergency systems are checked when applicable*

application [æplɪ'keɪʃn] *noun* **(a)** formal request, often on paper, for employment; **application form** = form to be filled out by a person looking for a job, and sent back to the organization offering the job **(b)** putting on of a (liquid, etc.) substance; **the application of a coat of paint** = the painting of something **(c)** the act or instance of using an existing ability; *when an accident occurs, the application of knowledge and skills is important*

apply [ə'plaɪ] *verb* **(a) to apply for a job** = to formally ask for employment; *he applied for the post of chief engineer but was not successful* **(b)** to put on; *apply a plaster to the skin*; *to apply a coat of paint* **(c)** to use existing knowledge or skills; *apply the same method as in the example* **(d)** *(of rules, regulations, orders, instructions, etc.)* to be relevant; to relate to; *the rules which apply to the measurement of wind velocities on isobaric charts apply equally to contour charts* (NOTE: **applying - applied**)

appreciable [ə'priːʃəbl] *adjective* **(a)** possible to measure; *appreciable weakening may be permitted without risk of failure* **(b)** considerable, large in size; a lot; a large amount; **there is an appreciable difference between statute miles and nautical miles** = there is a big difference between statute miles and nautical miles

appreciate [ə'priːʃieɪt] *verb* **(a)** to recognize the importance or significance of something; to understand; *the map reader is in a position to appreciate the relative values of the features seen on the ground* **(b)** to increase in value; *the value of the building has appreciated by 100% in 10 years* (NOTE: in this meaning the opposite is **depreciate**) **(c)** to be thankful or grateful for something; *the student appreciated the extra help given by the instructor*

appreciation [əpriːʃi'eɪʃn] *noun* **(a)** understanding; *it is essential to have an appreciation of the basic gas laws* **(b)** increase in value; *there has been an appreciation of 100% in the value of the building in 10 years* (NOTE: in this meaning the opposite is **depreciation**) **(c)** thankfulness, gratitude; *after gaining his private pilot's licence, the newly-qualified*

pilot showed his appreciation by sending a letter of thanks to his instructor

approach (APP) [ə'prəʊtʃ] **1** *noun* **(a)** path towards something; *the approach to the terminal was blocked by an overturned lorry* **(b)** the act of approaching (for landing); **approach to land** = the final stage of the flight when the aircraft is manoeuvred into position, relative to the landing area, in preparation for landing; *on the approach to land, the aircraft reduces speed and height*; *(VFR)* **final approach** = flight path in direction of landing along extended runway centre line; *the aspect of the runway on final approach helps the pilot to judge height and progress*; **intermediate approach** = part of the approach from arriving at the first navigational fix to the beginning of the final approach; **missed approach** = approach which does not result in a landing, followed by a go-around; **missed approach procedure** = action and flight path to be followed after missed approach at a particular aerodrome; *see also* PRECISION **(c)** way of achieving or doing something; **to take a different approach to a situation** = to take a different way to deal with or to manage a situation **2** *verb* **(a)** to move nearer in place or time to something; *the aircraft is approaching a danger area*; **nightfall is approaching** = it will soon be dark **(b)** to have a particular mental attitude towards something; *he approaches his studies with great enthusiasm* **(c)** to speak to or get in touch with someone; *you must approach the chief flying instructor regarding your request for a week's holiday*

appropriate [ə'prəʊpriət] *adjective* suitable; required, necessary or needed; **appropriate action** = suitable action demanded by the situation

appropriately [ə'prəʊpriətli] *adverb* as appropriate; as needed, as necessary; **to adjust the mixture appropriately** = to adjust the mixture to suit the conditions

approval [ə'pruːvl] *noun* permission; agreement; **with the captain's approval** = with the permission of the captain; **to meet**

with the approval = to have the approval; *plans for restructuring the airline met with the approval of the management*

approve [ə'pruːv] *verb* **(a)** to allow; to agree; *the air traffic controller approved the emergency landing* **(b)** to approve of = to believe something to be right or good; *nearly everybody approved of the new colour scheme for the furnishings*; he **doesn't approve of women being airline pilots** = he believes that it is wrong for women to be airline pilots

approx = APPROXIMATE; APPROXIMATELY (NOTE: the ICAO abbreviation is **APRX**)

approximate 1 [ə'prɒksɪmət] *adjective* not exact; around, about; **an approximate distance of 60 nm** = about 60 nautical miles **2** [ə'prɒksɪmeɪt] *verb* to be close to; to be nearly correct; *the number of people working in the airport approximates 2,000* (NOTE: can be shortened to: **approx** or **APRX** (ICAO))

approximately [ə'prɒksɪmətli] *adverb* not exactly; around, about; *approximately 2,000 people work in the airport* (NOTE: can be shortened to: **approx** or **APRX** (ICAO))

approximation [əprɒksɪ'meɪʃn] *noun* calculation which is not exact but near enough; rough estimate; **an approximation of aircraft height** = a rough estimate of aircraft height

apron ['eɪprən] *noun (at an airport)* area of tarmac, concrete, etc., outside a hangar for parking aircraft (NOTE: also called **ramp** in American English)

APRX *(ICAO)* = APPROXIMATE; APPROXIMATELY

APU = AUXILIARY POWER UNIT

aquaplaning ['ækwəpleɪnɪŋ] *noun* sliding in an uncontrolled way over a thin layer of water on the runway; *aquaplaning is caused by a layer of water between the tyre and the runway*

arbitrary ['ɑːbɪtrəri] *adjective* decided by chance rather than by careful logical thought; done without planning, done at random; *the statute mile is an arbitrary unit of measurement* (NOTE: the nautical mile is not an arbitrary unit: the nautical mile is based on calculations which have a wider use: see **arc**)

arc [ɑːk] **1** *noun* part of the circumference of a circle; *a nautical mile is the length of an arc on the earth's surface subtended by an angle of one minute at the centre of the earth* **2** *verb (of spark, especially produced by an electric current)* to jump across a gap; *the spark arcs from one electrode to another*; *the condenser prevents spark plugs from arcing*; *see also* GAP

Arctic ['ɑːktɪk] **1** *adjective* referring to the area around the North Pole; **cold Arctic air** = cold air from the Arctic; **the Arctic Circle** = parallel running round the earth at latitude 66°32N to the north of which lies the Arctic region **2** *noun* **the Arctic** = area of the earth's surface around the North Pole, north of the Arctic Circle; *the aircraft flew over the Arctic*

area ['eərɪə] *noun* **(a)** defined part of a surface; **restricted area** = airspace of a particular length, width and depth, within which the flight of an aircraft must be carried out in accordance with particular conditions; **manoeuvring area** = part of an aerodrome for the take-off and landing of aircraft, and movement of aircraft on the ground but not including the apron; **danger area** = airspace of a particular length, width and depth, within which activities dangerous to the flight of the aircraft may exist at particular times **(b)** region; **area forecasts** = weather forecast for a region rather than, for example, an aerodrome

argument ['ɑːgjumənt] *noun* **(a)** factor; *QNH is the pressure at station level reduced to sea level using arguments of station height and an international standard atmosphere* **(b)** verbal disagreement; **to have an argument** = to disagree openly and verbally with somebody; *the investigation revealed that there had been an argument between the commander and the copilot about the advisability of continuing with the final* *approach to land* (**c**) reason; *one of the arguments in favour of building the new terminal is the increase in opportunities of employment for the local residents*

arid ['ærɪd] *adjective* very dry; **arid terrain** = desert; **an arid, sub-tropical climate** = a hot, dry climate

arise [ə'raɪz] *verb* to come into being, to happen; to show up, to appear; *an emergency situation may arise which requires the passengers and crew to evacuate the aircraft* (NOTE: **arising - arose - arisen**)

arm ['ɑːm] **1** *noun* (**a**) device similar in function to a human arm; a lever; **rocker arm** = part of the valve mechanism in an internal combustion engine, which transmits the movement of the pushrod to the valve (**b**) horizontal distance from a reference point to the centre of gravity; *the principle of the arm is used in weight and balance calculations for an aircraft* **2** *verb* to make ready for action or use; *door-mounted escape slides are armed before flight*

armature ['ɑːmətʃə] *noun* rotating coils of an electric motor or dynamo; *secondary windings are wound over the primary windings and the whole assembly is known as an armature*

ARR = ARRIVAL

arrange [ə'reɪnʒ] *verb* (**a**) to organize, to plan and prepare; *to arrange a meeting* (**b**) to put in special position; *charts should be numbered and arranged in order of use*; *a series of dipoles are arranged in a circle*

arrangement [ə'reɪnʒmənt] *noun* (**a**) plan; *the arrangements for the VIPs are being handled by the public relations department* (**b**) position of a number of different parts; *the diagram shows a simple arrangement of pistons, cylinders and pipes*

array [ə'reɪ] *noun* arrangement of antennas; *the localizer antenna array is very wide*

arrest [ə'rest] *verb* (**a**) to stop or to prevent something from happening; **to arrest the spread of a fire** = to stop the fire spreading (**b**) to hold someone for breaking the law; *he was arrested at the airport* **2** *noun* holding someone for breaking the law; *his arrest was unexpected*

arrester [ə'restə] *noun* device or substance which prevents or stops something from happening; **flame arrester** = device to prevent flame from an external source from entering a fuel tank

arrival (ARR) [ə'raɪvl] *noun* (**a**) action of reaching somewhere; *Gulf Air announce the arrival of flight GF147 from Abu Dhabi* = flight GF147 has just landed at for example, Bahrain, after leaving from Abu Dhabi (**b**) **arrivals** = part of an airport that deals with passengers who are arriving (NOTE: the opposite for (a) is **departure**, and for (b) **departures**)

arrive [ə'raɪv] *verb* to reach somewhere; *the flight from Tokyo arrived at 8.30* = the flight from Tokyo landed at 8.30

arrow ['ærəʊ] *noun* printed sign which points to something; *non-return valves are marked with an arrow which shows the direction of flow*; **arrow convention** = agreed method of using arrows when drawing wind triangles; *see also* MNEMONIC

article ['ɑːtɪkl] *noun* object, item; **loose articles** = things which may move during flight and cause problems

artificial [ɑːtɪ'fɪʃl] *adjective* not real; man-made; *the small needle indicates cabin altitude or the artificial altitude created by the pressurization system*; **artificial horizon**; *see* ATTITUDE INDICATOR

ascend [ə'send] *verb* to rise, to go up, to move slowly upward; *hot air ascends*; **in ascending order** = in order of number or rank with the smallest or less important at the bottom and the largest or more important at the top (NOTE: the opposite is **descend**)

ascent [ə'sent] *noun* rise, going up, slow upward movement; *the forced ascent of air over high ground*; *in a stable atmosphere where the ascent of air is forced, precipitation is mostly light and occasionally moderate*; mass ascent = slow rise of a very large body of air (NOTE: the opposite is **descent**)

ascertain [æsə'tem] *verb* to make sure, to make certain; *during pre-flight checks, control surfaces should be moved by hand to ascertain that they have full and free movement*

ASI = AIRSPEED INDICATOR

aspect ['æspekt] *noun* **(a)** part of a problem or subject; *vertical motion is an important aspect of meteorology*; safety aspects = matters related to safety **(b)** view from a particular position; *the aspect of the runway on final approach helps the pilot to judge height and progress*

asphyxiation [æsfiksi'eɪʃn] *noun* unconsciousness or death caused by lack of oxygen; *fire may result in the cabin being filled by smoke causing asphyxiation*

assemble [ə'sembl] *verb* **(a)** to put a number of parts together; *parts are made in different countries but the aeroplane is assembled in France* **(b)** to gather together; *passengers should assemble in the departure lounge where an airline representative will meet them*

assembly [ə'sembli] *noun* **(a)** part of the airframe or other part of a larger whole which is, itself, made up of smaller parts; tail assembly = the aft part of the fuselage with the fin and rudder, tailplane and elevators attached; **undercarriage assembly** = wheels, struts and linkages which make up the complete unit **(b)** putting together of parts to make a whole; *final assembly of the A320 takes place in France*

assess [ə'ses] *verb* to check; to estimate; to find out; *cabin crew must assess if their exits are usable*; to assess a situation = to consider all aspects of a situation

assessment [ə'sesmənt] *noun* careful consideration of something; estimate; *the captain's assessment of factors such as aircraft damage, passenger-load, fire, etc., will affect his decision on whether to evacuate the aeroplane or not*

assign [ə'sam] *verb* to set apart beforehand or allocate for a specific purpose; **assigned seats** = seats selected beforehand for particular people; *crew sit in their assigned seats*; *individual carriers assign codes to aircraft*

assist [ə'sɪst] *verb* to help; *if you have any difficulty, cabin staff will assist you*; *when evacuating the aircraft, hand signals by cabin staff assist in directing passengers to the exits*; *see also* POWER

assistance [ə'sɪstəns] *noun* help; to require assistance = to need help; *if a pilot requires assistance, he should contact ATC*; to provide assistance = to give help

associate [ə'səusieɪt] *verb* to come with something else; to be linked to something else; *turbulence is often associated with strong winds*; *the airport authority has to overcome a lot of problems associated with its plans to build a new terminal*

association [əsəusi'eɪʃn] *noun* **(a)** group of people who organize themselves into an official body with common objectives and a code of conduct; *British Air Line Pilots Association* **(b)** in association with = together with; *rain-ice occurs only rarely over the British Isles and is usually found in association with warm fronts*

assume [ə'sjuːm] *verb* **(a)** to take as true before there is proof; *I assume that he's ill because he's not at work today - but I may be wrong* **(b)** to suppose; for our studies we will assume that the earth is a perfect sphere = we know that the earth is not a perfect sphere but it helps if we accept, for the time being, that it is; assuming (that) = accepting or supposing that; *assuming that the return flight from the point of no return to A is made on three engines, calculate the distance from D to the point of no return* **(c)** to take on, to undertake (the duties of somebody); *the copilot assumed control of the aircraft after the captain was taken ill during the flight* (**d**) to take (up) a particular bodily position; *the correct technique of using the escape slides is to assume a sitting position*

assumption [ə'sʌmpʃn] *noun* understanding, belief; *the one-in-sixty rule is based on the assumption that one nautical mile subtends an angle of one (at a distance of 60 nautical miles)*

asymmetric(al) [æsɪ'metrɪk] *adjective* not symmetric(al); not identical or equal on each side of an imaginary central dividing line; **asymmetric flight** = condition in which one engine, displaced from the aircraft's centre-line is not working; **asymmetric power** = power on one side of the aircraft's centre line only

asynchronous [ə'sɪŋkrənəs] *adjective* (**a**) not synchronous; not happening at the same time or rate; *an asynchronous orbit is a 24-hour orbit which enables a satellite to remain overhead one part of the earth's surface* (**b**) not in frequency or phase; **asynchronous computer** = computer which does not process information according to the internal clock

ATC = AIR TRAFFIC CONTROL

ATIS = AUTOMATIC TERMINAL INFORMATION SERVICE

atmosphere ['ætməsfɪə] *noun* (**a**) mass of air (mixture of gases) surrounding the earth; *the surrounding atmosphere moves with the earth*; **lower atmosphere** = layer of the atmosphere in which changes in the weather take place; troposphere (**b**) unit of measurement of pressure; **international standard atmosphere (ISA)** = model atmosphere defined in terms of pressure, density and temperature for all heights, with perfect gases and without any form of water or solid matter; it is used in the calibration of instruments and descriptions of aircraft performance

COMMENT: the main gases found in the atmosphere are: nitrogen and oxygen, less than 1% carbon dioxide and argon, plus traces of hydrogen, helium, krypton, neon, ozone and xenon

atmospheric [ætməs'ferɪk] *adjective* (**a**) referring to the atmosphere; *atmospheric density*; **atmospheric pressure** = normal air pressure on the surface of the earth (**b**) *(of radio signal)* **atmospheric attenuation** = weakening of a radio signal as it passes through the air

atom ['ætəm] *noun* smallest amount of a substance which can take part in a chemical reaction; *an atom consists of a nucleus and electrons*

atomic [ə'tɒmɪk] *adjective* referring to atoms; **atomic structure of matter** = structure of materials and substances at their smallest level

atomization [ætəmaɪ'zeɪʃn] *noun* reduction (of liquids) to a fine spray; *the fuel achieves fine atomization under pressure*

atomize ['ætəmaɪz] *verb* to reduce (liquids) to a fine spray; *the fuel must be atomized or vaporized to combine with the air to permit combustion*

ATPL = AIRLINE TRANSPORT PILOT'S LICENCE

attach [ə'tætʃ] *verb* to join or fix to something; to connect to; *the ice detector is attached to the fuselage*

attachment [ə'tætʃmənt] *noun* (**a**) joining or fixing to something; *the attachment of winglets improved the handling characteristics of the aeroplane*; **attachment point** = place on the airframe where something, such as an engine, is attached by means of bolts; *additional strength is required for the power plant attachment points* (**b**) accessory which can be attached; *the video camera is sold with a number of attachments including a carrying strap and a battery pack*

attain [ə'teɪn] *verb* to reach, to achieve something (often with difficulty); *in order to attain a fuller understanding of gas* *turbines, it is essential to know something about basic gas laws*

attempt [ə'tempt] **1** *noun* try; *any attempt to increase range by applying more power is of little or no benefit* **2** *verb* to try; *he attempted to land despite the poor visibility but then decided to divert to another airfield where he landed safely*

attendant [ə'tendənt] **1** *adjective* which accompanies something else; *fuel spillage and attendant fire risk must be minimized*; **attendant problems** = associated problems **2** *noun* **cabin attendant** *or* **flight attendant** = member of the flight crew who looks after passengers, serves food, etc.; *if you need something, press the call button and a cabin attendant will respond within a few minutes*

attention [ə'tenʃn] *noun* ability or power to concentrate on something; *the crew's attention is alerted by an automatic display*; **attention please** = listen carefully to what will be said; **pay attention** = listen to the speaker and concentrate on what is being said

attenuate [ə'tenjueɪt] *verb* (*of radio signal*) to lose power or strength; *a wave becomes attenuated or loses strength as range increases*

attenuation [ətenju'eɪʃn] *noun* (*of radio signal*) loss of strength; *atmospheric attenuation is negligible until the upper end of the UHF band when it increases rapidly to limit the highest usable frequency to about 10 gHz*

attenuative [ə'tenjuətɪv] *adjective* which weakens (a radio signal); *rain has an attenuative effect*

attitude ['ætɪtjuːd] *noun* (a) position of the aircraft in the air in relation to the horizon; *angle of attack will vary with changes in engine speed and aircraft attitude*; **nose down attitude** = attitude of the aircraft when the nose is at a lower level than the tail; *the aircraft is in a nose down attitude* (b) way of thinking, feeling or behaving about something; *he has an excellent attitude towards his training programme* = he is positive and motivated in his training programme

attitude indicator (AI) ['ætɪtjuːd 'ɪndɪkeɪtə] *noun* flight instrument that gives the pilot pitch and bank information; *in light aircraft, the attitude indicator is situated on the instrument panel, directly in front of the pilot*; *see also* PITCH; BANK

COMMENT: the attitude indicator is sometimes referred to as the 'artificial horizon'. In instrument flight training, the attitude indicator is the primary reference instrument. It is positioned on the instrument panel directly in front of the pilot

attract [ə'trækt] *verb* to cause to draw near; *if two magnets, with unlike poles are brought together, they will attract each other*; **to attract attention** = to behave in such a way that people will notice you

attraction [ə'trækʃn] *noun* act or instance of being drawn near to something; *the attraction of flying was the factor which made him decide to train as a pilot*; *the strength of the magnetic force will depend, amongst other things, on the magnitude of attraction at the magnetic source*

attractive [ə'træktɪv] *adjective* (a) referring to something you feel you would like to have; *after long talks, the prospective buyer made a financially attractive offer for the aircraft* (b) nice to look at; **an attractive design** = an eye-catching, pleasant design (NOTE: although a magnet attracts some metal objects and has magnetic attraction, it cannot be described as **attractive**)

ATZ = AERODROME TRAFFIC ZONE

audible ['ɔːdɪbl] *adjective* which can be heard; *the fire detection system should contain an audible warning device*

audio ['ɔːdiəʊ] *noun* audible sound or sound signal; *the diagram shows an amplitude modulation case where the* *lower frequency of the audio is about 300 Hertz*

augment [ɔːgˈment] *verb* to make larger by adding something; to increase; *the sea breeze may augment the up-slope motion of an anabatic wind*

aural [ˈɔːrəl] *adjective* referring to hearing; *the aural and visual alerts will continue until the crew take action to cancel them* (NOTE: sometimes pronounced [ˈaʊrəl] to show the difference with **oral**)

authoritative [ɔːˈθɒrɪtətɪv] *adjective* in the manner of somebody with authority; in a commanding way; **crew must act in an authoritative manner** = crew must give firm instructions or orders

authority [ɔːˈθrɪti] *noun* **(a)** complete control or power over something; *while boarding, the captain has the authority to ask an unruly passenger to leave the aircraft* **(b)** official or government body with the power to make decisions; **airport authority** = organization responsible for the running of an airport; **Civil Aviation Authority (CAA)** = governing aviation body of the UK; **local authority** = government body responsible for the various services of an area

authorize [ˈɔːθəraɪz] *verb* to allow (officially); to give permission; *a signature is required to authorize the repair* (NOTE: also written **authorise**)

authorized [ˈɔːθəraɪzd] *adjective* (officially) allowed; permitted; *aircraft with a maximum authorized weight of 12,500 lb or less*; **an authorized person** = a person who has been given power to act and perform particular tasks or duties (NOTE: also written **authorised**)

auto- [ˈɔːtəʊ] *prefix meaning* automatic; automated

autoland [ˈɔːtəʊlænd] = AUTOMATIC LANDING

automate [ˈɔːtəmeɪt] *verb* to make (a device or procedure) automatic; **automated systems** = systems which have

been made less dependent on direct human control or management

automatic [ɔːtəˈmætɪk] *adjective* **(a)** *(action)* which is done without needing to think; *in the early stages of training, student pilots have to think about the use of the flying controls but after a while these actions become automatic* **(b)** *(device, system, etc.)* which works by itself without the need of an operator; *the normal activation method is automatic*; **automatic landing (autoland)** = automatic flight control system capable of landing an aircraft 'hands-off'; **automatic mixture control** = subsystem in a piston engine which adjusts the flow of fuel to balance changes in air density; **automatic pilot (autopilot)** = system which automatically stabilizes an aircraft about its three axes, restores original flight path following an upset and, in some systems, causes the aircraft to follow a (pre)selected airspeed, altitude or heading

automatic direction finder (ADF) [ɔːtəmeɪtɪd daɪˈrekʃn ˈfaɪndə] *noun* radio navigation instrument that receives signals from non-directional radio beacons (NDBs); *the needle on the ADF indicator points toward the selected radio signal*

Automatic Terminal Information Service (ATIS) [ˈɔːtəmætɪk ˈtɜːmɪnəl ɪnfəˈmeɪʃn ˈsɜːvɪs] *noun* recording of information played continuously on a specified radio frequency which gives pilots the current weather, runway in use, etc.; *students listen to the ATIS to practise their language skills*

automation [ɔːtəˈmeɪʃn] *noun* **(a)** automatic operation or automatic control of a piece of equipment, a process, or a system; *automation has speeded up baggage handling; automation of throttle control has removed the need for pilots to monitor airspeed so closely; it is possible that the alternate source might provide a reduced level of automation*

autopilot [ˈɔːtəʊpaɪlət] *noun (automatic pilot)* system which automatically stabilizes an aircraft about its three axes, restores original flight path following an upset and, in some systems, causes the aircraft to follow a (pre)selected airspeed, altitude or heading

auxiliary [ɔːgˈzɪlieri] *adjective* secondary system (which is used when necessary); **auxiliary gearbox** = gear box which allows main engine power to be used for secondary systems; **auxiliary power unit (APU)** = a small jet engine used to generate electrical power for air-conditioning, etc., when the aircraft is parked on the ground

availability [əveɪləˈbɪlɪti] *noun* being available; *the status of an airport is determined by the availability of suitable navigation aids*

available [əˈveɪləbl] *adjective* which can be, or is, ready for immediate use; *on a multi-engine aircraft, all the fuel must be available for use by any engine*

average [ˈævrɪdʒ] **1** *adjective* referring to the mathematical mean; *for load sheet purposes, an average weight of the passengers and crew members may be used* **2** *noun* mathematical mean; total divided by the number of items added; *the average of 1, 5, 9, 10 and 15 is 8 (1+5+9+10+15 = 40 ÷ 5 = 8)* **3** *verb* to reach a particular figure as an average; *brake temperatures average around 500ºC during normal operations*

avert [əˈvɜːt] *verb* to avoid; *to avert a collision, he changed direction*

AVGAS = AVIATION GASOLINE

aviation [eɪviˈeɪʃn] *noun* flying an aircraft; *wind speeds in aviation are usually given in knots*; aviation gasoline **(AVGAS)** = fuel used in piston-engined aircraft; **aviation law** = body of laws relating to flying

avionics [eɪviˈɒnɪks] *noun (from 'aviation electronics')* electronic communication, navigation, and flight-control equipment of an aircraft; *the trainee engineer is doing an avionics course*

avoid [əˈvɔɪd] *verb* **(a)** to prevent something from happening; *he just managed to avoid an accident* **(b)** to keep away from something; *avoid flying close to any person or vessel*; *cumulonimbus clouds and thunderstorms should be avoided by as great a distance as possible*

avoidance [əˈvɔɪdəns] *noun* act of avoiding something; **collision avoidance** = the prevention of collisions by taking measures beforehand to ensure that they do not happen; **avoidance of thunderstorms is recommended** = it is recommended to keep away from thunderstorms

await [əˈweɪt] *verb (formal)* to wait for; *await instructions from the flight deck*

aware [əˈweə] *adjective* knowing and being conscious of something; *the pilot should be aware of the positions of all other aircraft in the circuit*

awareness [əˈweənəs] *noun* being aware, being conscious of something; **safety awareness** = being familiar with and prepared for any situation in which safety is important

AWR = AIRBORNE WEATHER RADAR

axial [ˈæksiəl] *adjective* referring to an axis; **axial flow compressor** = compressor in which the flow of air is along the longitudinal axis of the engine; *in spite of the adoption of the axial flow type compressor, some engine retain the centrifugal type*

axis [ˈæksɪs] *noun* **(a)** imaginary line around which a body rotates; *the earth rotates around its own axis*; *an aircraft moves around three axes - vertical, longitudinal and lateral*; *see also* PITCH; ROLL; YAW **(b)** horizontal or vertical scale on a graph, often referred to as the X axis (horizontal axis) and the Y axis (vertical axis); *the plot shows the effect of airspeed on lift with airspeed shown on the horizontal axis and lift on the vertical axis* (NOTE: plural is **axes** [ˈæksiːz])

axle [ˈæksl] *noun* shaft on which a wheel is mounted - either the wheel turns round the axle or is fixed to the axle; *unequal tyre-pressures, where two wheels are mounted on the same axle, will result in one tyre carrying a greater share of the load than the other*

azimuth ['æzıməθ] *noun* the horizontal angle or direction of a compass bearing; *where precision approach radar is installed, the controller can inform the pilot if he departs from either the extended centre-line in azimuth or height or both*

Bb

backup ['bækʌp] *adjective & noun* (second or third) system, instrument, computer disk, etc., available to be used if the first one fails; *the backup system or the backup failed as well*; *backup generators are driven by the engine*

backward ['bækwəd] *adjective* directed towards the back; *a backward movement*

backwards ['bækwədz] *adverb* towards the back; *unlike most aircraft, the C130 can move backwards using its own power* (NOTE: the American English is **backward**)

baffle ['bæfl] *noun* metal plate for preventing the free movement of sound or liquids; *integral fuel tanks can be strengthened by fitting baffle plates*

baggage ['bægɪdʒ] *noun* luggage, cases and bags which you take with you when travelling; *one passenger had a huge amount of baggage*; *she lost one piece of baggage*; *(at airport)* **baggage hall** = area where arriving passengers pick up their baggage; **baggage handling** = process by which passengers' baggage is loaded onto an aircraft, unloaded and moved to the airport terminal; **carry-on baggage** = small bags (of limited size and weight) that passengers are allowed to take with them into the cabin of an aircraft; **excess baggage** = amount, usually expressed as weight, of baggage which exceeds the airline's limit per passenger (NOTE: the word **luggage** is also used in British English)

balance ['bæləns] **1** *noun* **(a)** state in which weight, force or importance are evenly distributed; *the propelling nozzle size is extremely important and must be designed to obtain the correct balance of pressure, temperature and thrust* **(b)** staying steady **2** *verb* **(a)** to provide an opposite and equal weight, force or importance as something else; *the pressure exerted by the weight of the atmosphere above the level of the bowl balances a column of mercury in the tube* **(b)** to stay steady (especially when resting on the centre of gravity)

QUOTE balance refers to the location of the centre of gravity along the longitudinal axis of the aeroplane
Civil Aviation Authority, General Aviation Safety Sense Leaflet

balloon [bə'luːn] *noun* large bag inflatable with hot air or gas to provide lift, but without power; *balloons are sent into the upper atmosphere to collect information useful to meteorologists*

BALPA ['bælpæ] = BRITISH AIR LINE PILOTS ASSOCIATION

band [bænd] *noun* **(a)** (narrow) strip; *a jet stream is a narrow band of high-altitude strong winds* **(b)** range of numbers or frequencies between two limits within a radio system; **high frequency (HF) band** = radio communications range between 3 MHz to 30 MHz; **medium frequency (MF) band** = radio frequency range between 300 kHz and 3000 kHz - often referred to as medium wave (MW); **ultra high frequency (UHF) band** = radio frequency range between 300 MHz and 3000 MHz; **very high frequency (VHF) band** = radio frequency range between 30 MHz and 300 MHz; *see also* FREQUENCY

bandwidth ['bændwɪdθ] *noun* width of a band of radio frequencies; *the sharp setting means the bandwidth is reduced to one kilohertz to minimize noise or interference*

bank [bæŋk] **1** *verb (of aircraft)* to rotate or roll around its longitudinal axis to a particular angle; *stresses are increased when the aircraft banks, turns or pulls out of a dive* **2** *noun (of aircraft)* rotating or rolling movement around its longitudinal axis to a particular angle; *an attitude indicator gives the pilot pitch and bank information*

bar [bɑː] *noun* **(a)** long, straight, rigid piece of metal; *the part is made from a solid bar of aluminium* **(b)** *(meteorology)* unit of atmospheric pressure (1 bar = 1,000 millibars); *see also* MILLIBAR

barograph ['bærəgrɑːf] *noun* instrument for measuring and recording atmospheric pressure; *the most common type of barograph is that which utilizes an aneroid capsule mechanically connected to a pen*

barometer [bə'rɒmɪtə] *noun* instrument for measuring the atmospheric pressure; **aneroid barometer** = barometer which uses an aneroid capsule - a thin flexible cylindrical box, usually made of metal, which has most of the air removed from it an which expands and contracts with changes in atmospheric pressure; **mercury barometer** = type of barometer where the atmospheric pressure is balanced against a column of mercury; *the principle of a mercury barometer has not changed since 1643 when Torricelli demonstrated that the atmosphere can support a column of liquid*

barometric [bærə'metrɪk] *adjective* referring to a barometer; **barometric pressure** = atmospheric pressure as indicated by a barometer; **barometric tendency** = amount of change in pressure with increase in altitude

barrier ['bæriə] *noun* **(a)** something, such as a wall, which prevents the movement of something else; *elevation of the ground over which the aircraft flies can be a dangerous barrier to flight* **(b)** something which prevents a person from making progress; *his medical problems were a barrier to his successful completion of the course*

base [beɪs] **1** *noun* bottom part or lowest part; **cloud base** = bottom part of a layer of cloud; *in general, the lower the cloud base, the less heat is lost by the earth* **2** *verb* to develop (something) from (something else); *the operation of the auxiliary power unit is based on the gas turbine engine; the principle of vapour cycle cooling is based upon the ability of a refrigerant to absorb heat; see also* LEG

basic ['beɪsɪk] *adjective* referring to the most important but often simplest part of something (from which everything else is derived); *this chapter provides a basic understanding from which the study of meteorology can develop*; **basic principle** = central or fundamental idea or theory

basis ['beɪsɪs] *noun* central and most important part of something from which everything else is derived; *the basis of air navigation is the triangle of velocities* (NOTE: the plural is **bases** ['beɪsiːz])

battery ['bætri] *noun* chemical device that produces electrical current; *this piece of equipment is powered by 2 batteries; see also* CHARGER

bay [beɪ] *noun* **(a)** space or area in the structure of an aeroplane where equipment can be located; *to avoid damage to the wheel bay, the nose wheel must be aligned in a fore and aft direction during retraction* **(b)** part of the coast that curves inwards; *the Bay of Bengal*

bayonet ['beɪənɪt] *noun* **bayonet fitting** = type of attachment with two side pins (as found on some light-bulbs); *magnetic chip detectors are of the bayonet type fitting and can be removed and replaced very quickly*

beacon ['biːkən] *noun* light or radio signal for navigational purposes; *if the aircraft turns towards the beacon, signal* *strength will increase*; **ILS locator beacon** = non-directional beacon used for final approach; **non-directional beacon** = radio beacon transmitting a signal by which the pilot can determine his bearing; **survival beacon** = beacon which transmits a signal which enables search aircraft to locate survivors in the water; *VHF and/or UHF survival beacons are carried on all jet transports*

beam [biːm] *noun* **(a)** long thick metal bar used as a support; *a beam is designed with a breaking load of 12 tons but when a three ton load is applied repeatedly, the beam may fail* **(b)** shaft of light or radiation travelling in one direction (such as from a car's headlights); *the electron gun produces a stream of fast-moving electrons and focuses them into a narrow beam*

bear [beə] *verb* **(a)** to carry or to hold; *the undercarriage has to bear the weight of the aircraft on the ground*; **load-bearing structure** = structure which supports the weight of the aircraft in flight or on the ground; **rain-bearing cloud** = cloud carrying moisture which can fall as rain **(b)** **to bear something in mind** = to keep in mind; **it should be borne in mind** = it should be remembered; **bearing in mind** = considering; *bearing in mind that she hadn't flown for three weeks, the student pilot's landings were very good*; **this bears no relation to that** = this is not connected with that **(c)** to tolerate; **he can't bear the heat** = the heat is too much for him (NOTE: **bearing - bore - borne**)

bearing ['beərɪŋ] *noun* **(a)** angle, measured in a clockwise direction, of a distant point, relative to a reference - called compass, magnetic or true (bearing); *to plot a position line from the non-directional radio beacon, it is first necessary to convert the relative bearing to a true bearing and then calculate the reciprocal*; **compass bearing** = direction or position relative to a fixed point measured in degrees on a compass; **magnetic bearing** = bearing with reference to the magnetic north (not true north); **true bearing** = bearing with reference to true north (not magnetic north); **VOR bearing** = direction of the VOR transmitter relative to the aircraft measured in degrees **(b)** device containing steel balls or needles which allows free rotation of one component around another; **wheel bearing** = device which allows the wheel to rotate freely around the axle

belt [belt] *noun* **(a)** long, relatively narrow area; **high-pressure belt** = long narrow area of high pressure; **precipitation belt** = long narrow area of rain, snow or hail; **rain belt** = long narrow area where rain falls; *the cirrus cloud can be 900 miles ahead of the surface front with a rain belt as wide as 200 miles* **(b)** loop of strong material connecting two pulleys or wheels, one driving the other; **belt-driven** = (wheel) moved by a belt linked to another wheel which, in turn, is moved by a motor or an engine; *aircraft generators are belt-driven or shaft-driven*

bend [bend] **1** *noun* curve **2** *verb* to curve from a straight shape; **to bend downwards** = to curve down from a horizontal position; **to bend upwards** = to curve up from a horizontal position; *the wings support the weight of the aircraft and they bend upwards in flight*; *see also* LOAD (NOTE: bending - bent)

Bernoulli's principle *see* LIFT

beware [bɪ'weə] *verb* to be careful; to watch out for; *beware of carburettor icing*; *beware of other aircraft in the circuit*

beyond [bɪ'jɒnd] *preposition* further away than; *the radio horizon extends beyond the visible horizon*; **it is beyond his understanding** = he cannot understand it at all; it is too difficult for him to understand

bi- [baɪ] *prefix meaning* (i) two; (ii) twice

biannual [baɪ'ænjuəl] *adjective* twice yearly, two times a year; **biannual inspection** = inspection done twice every year

bimetallic [baɪme'tælɪk] *adjective* made of two metals; **bimetallic strip** = strip of two separate metals with different rates of expansion, joined together side by side so that when the strip is heated, it bends and makes, or breaks, electrical contact; *circuit breakers use a bimetallic strip as the sensing element*

binary ['baɪnəri] *adjective* number system used in computers that only uses the digits 0 and 1; *logic gates work with binary data*; *computers only process binary information*

biplane *or* **bi-plane** ['baɪpleɪn] *noun* (old) aeroplane design with two pairs of wings, one above the other; *most of the aircraft used in the 1914-1918 war were biplanes*

blade [bleɪd] *noun* flattened (aerofoil) part of a propeller or rotor; **blade angle** = angle between the blade axis and the axis of rotation; *with a variable pitch propeller, the blade angle may be changed in flight*; **blade tip** = the end of the blade furthest from the centre of rotation; **rotor blade** = long thin aerofoil on a helicopter rotor; **turbine blade** = part of a turbine which has an aerodynamic effect on the air

blank ['blæŋk] *adjective* (a) (paper) with nothing written, printed or drawn on it; *a blank sheet of paper*; **a blank form** = a form without the details filled in (b) (TV, computer or video screen) with nothing appearing on it; *when he returned to his computer, the screen was blank*

bleed [bliːd] *adjective* **bleed air** = compressed air from the engine compressor used for cabin pressurization or to drive other services; *bleed air from the right engine can power items normally powered by the left engine*

block [blɒk] **1** *noun* large mass of something; **cylinder block** = casing containing the cylinders in an internal combustion engine; **engine block** = cylinder block with integral crankcase **2** *verb* (a) to prevent (a fluid, etc.) from passing freely (through a pipe or channel); *at high altitude, any water condensing out of the fuel could freeze and block the filters* (b) to prevent a course of action; *the*

government blocked attempts to prevent the building of the new airport

blockage ['blɒkɪdʒ] *noun* (a) collection of something blocking a pipe, narrow channel, filter, etc.; *ice crystals may form to cause a blockage of the fuel filter* (b) being blocked; *the blockage was caused by ice*

blow [bləʊ] **1** *noun* (a) impact; *a blow on the head* (b) disappointment; *the news of his failure in the examination was a severe blow* **2** *verb* (a) (of air, wind) to move; *the sea breeze may blow almost parallel to the coast* (b) (of a fuse) to break, as it should, when the circuit is overloaded (NOTE: **blowing - blew - blown**)

blow-back ['bləʊbæk] *noun* sudden movement of fluid in the opposite direction to the general flow; *a sudden release of pressure may cause a blow-back*

blower ['bləʊə] *noun* device for blowing air; *air for combustion is obtained from a blower*; **cabin compressor and blower system** = part of the air conditioning system for the cabin

board [bɔːd] **1** *noun* (a) flat (square or rectangular) piece of wood or other material; **circuit board** = insulating board which holds components connected into an electrical circuit (b) **on board** = on an aircraft; *the flight plan records the callsign and the number of people on board* **2** *verb* to get on to an aircraft; *in an emergency, many passengers only remember the entrance by which they boarded the aircraft*

boarding gate ['bɔːdɪŋ 'geɪt] *noun* door through which passengers leave the terminal building to get on to an aircraft; *boarding gates 1 - 10 are on the left*

boarding pass ['bɔːdɪŋ 'pɑːs] *noun* temporary pass, issued at the check-in desk, which allows the holder to board the aircraft; *boarding passes must be shown at the gate* (NOTE: the plural is **boarding passes**)

boarding steps ['bɔːdɪŋ 'steps] *noun* stairs used by passengers and crew to get on board an aircraft; *passengers had to wait in the aircraft for 15 minutes before the boarding steps were put in position*

boarding time ['bɔːdɪŋ 'taɪm] *noun* time when passenger are due to board the aircraft; *boarding time is at 13.30 hrs*

body ['bɒdi] *noun* **(a)** (i) the whole of a person or an animal; (ii) the main part of a person, but not the arms or legs **(b)** main part of an aeroplane, system, text, etc.; *the body of an aircraft is also called 'airframe'; a flow-control valve consists of a body and a floating valve* **(c)** large mass of liquid or gas; **body of air** = large quantity of air behaving in a certain way **(d)** *(in scientific description)* an object; *acceleration is the rate of change of velocity of a body*

boil [bɔɪl] *verb* to heat a liquid until it reaches a temperature at which it changes into gas; *water boils at 100°C*; **boiling point** = temperature at which a liquid changes into gas; *the boiling point of water is 100°C*

bolt [bəʊlt] **1** *noun* **(a)** metal rod, with a head, which screws into a nut; *titanium is used for making bolts; the two halves of the wheel are held by bolts* **(b)** **bolt of lightning** = one electrical discharge of lightning **2** *verb* to attach with a bolt; *aircraft wheels are constructed in two halves which are bolted together*

bond [bɒnd] **1** *noun* joining together of surfaces, normally using heat, cold, chemicals or glue; *the de-icing boot breaks the bond between the ice and the outer skin* **2** *verb* to join surfaces together normally using heat, cold, chemicals or glue; *the skin is bonded to the internal members by the redux process*

boost [buːst] **1** *noun* increase; help; improvement; *the improvement in a country's economy often gives a boost to the airline industry* **2** *verb* **(a)** to make or to help something increase; *an oil pump boosts engine oil pressure* **(b)** *(confidence, etc.)* to increase; *the instructor's comments boosted the student pilot's confidence*

booster ['buːstə] *noun* something which gives extra help or support; device which increases the force or amount of something; **booster pump** = centrifugal pump often positioned at the lowest point of a liquid fuel tank to ensure positive pressure in the supply lines to the engine; *fuel is fed through a filter and a booster pump; the purpose of the booster pump is to prevent fuel aeration*

boot [buːt] *noun* flat, flexible tubes bonded to the leading edge or wings and other surfaces which, when pressurized with fluid, break up ice; *the boots on the leading edge of the wings were damaged by hail*

bound [baʊnd] *adjective* **bound for** = on the way to; *an aircraft bound for Paris*; **the Copenhagen-bound flight** = the flight on the way to Copenhagen; **homeward bound** = heading towards home; **outward bound** = leaving home (especially for another country)

boundary ['baʊndri] *noun* (physical or imaginary) limit between two areas; *the boundary between two air masses is called the frontal surface*; **aerodrome boundaries** = the physical or geographical limits of an aerodrome; **boundary layer** = layer of fluid next to the surface over which it is flowing and, because of friction, travels more slowly than layers further from the surface

bowser ['baʊzə] *noun* mobile fuel tank for refuelling aircraft; *it is important to prevent the possibility of an electric spark by earthing the aircraft and the bowser*

brace [breɪs] *verb* **(a)** to strengthen (a construction) using cross-members and/or wires; *early aircraft were of the braced type of construction* **(b)** to take a protective body position in preparation for a crash landing; *the cabin-crew will repeat the 'brace' order and brace themselves*; **to brace yourself** = to quickly prepare yourself mentally and physically for what is shortly to happen

bracket ['brækɪt] *noun* **(a)** metal support often triangular or L-shaped; **component**

bracket = metal device to attach and support a component **(b)** range of frequencies within a band of radio frequencies; *terminal VOR is in the frequency bracket 108-112 MHz* **(c)** *(parentheses)* **round brackets** = printing symbol ( ) used to separate words in a sentence, or within a text; **square brackets** = printing symbol [ ] used to enclose certain types of text

brake [breɪk] **1** *noun* device for stopping a vehicle or a machine; **parking brake** = brake used to prevent the aircraft moving after it has come to a stop; **speed brakes;** *see* SPOILER **2** *verb* to slow down or to stop by pressing the brakes; *he had to brake hard after landing in order to turn off at the correct taxiway* (NOTE: **braking - braked;** do not confuse with **breaking - broke - broken)**

braking [ˈbreɪkɪŋ] **1** *noun* putting on the brakes (to slow down or to stop) **2** *adjective* which brakes or slows down; *the braking effect of drag; see also* ACTION

breather [ˈbriːðə] *noun* **(a)** pipe connecting the crankshaft to the atmosphere to prevent build-up of crankcase pressure **(b)** *(colloquial)* short rest; **to take a breather** = to have a short break, to relax before starting again

breeze [briːz] *noun* gentle wind especially near the coast; *there's no wind, not even a breeze*; **land breeze** = light wind which blows from the land towards the sea; *land and sea breezes occur in coastal areas*; **sea breeze** = gentle wind which blows from the sea towards the land; *the strength of the sea breeze decreases with height*

brief [briːf] **1** *adjective* short; **brief visit** = visit of short duration; **brief letter** = letter containing only a few words **2** *noun* general instructions to enable somebody to perform his duties; *the inspector's brief is to find out as much as possible about the causes of accidents* **3** *verb* to give basic information to somebody; *before take-off, cabin crew must brief passengers on the*

location and use of emergency exits and life jackets

briefing [ˈbriːfɪŋ] *noun* (short) meeting to enable instructions and basic information to be given; **flight-briefing room** = room where instructors talk to trainees immediately before a training flight or where a pilot talks to his crew immediately before boarding the aircraft

British Isles [ˈbrɪtɪʃ ˈaɪlz] *noun* the islands which make up Great Britain and Ireland; *the climate of the British Isles is affected by the Atlantic Ocean*

brittle [ˈbrɪtl] *adjective* which breaks easily (like thin glass); *absorption of oxygen and nitrogen from the air at temperatures above 1,000° F makes titanium brittle*

broad [brɔːd] *adjective* **(a)** very wide; *a broad river* **(b)** wide; general; *three broad categories of aircraft are considered - rotary wing aircraft, light single-engine aircraft and twin-engine aircraft* (NOTE: the opposite is **narrow**)

broadcast [ˈbrɔːdkɑːst] *verb* to transmit, often to a large number of people, a radio signal or message which requires no answer; *the cabin crew can use the public address system to broadcast messages to passengers only* (NOTE: **broadcasting - broadcast)**

broadly [ˈbrɔːdli] *adverb* widely; generally; **broadly speaking** = generally speaking

brush [brʌʃ] *noun* **(a)** a tool with hair or wire, used to paint, clean, etc. **(b)** small, replaceable block of carbon which rubs against the surface of a commutator in a generator or electric motor; *at high altitude, the air becomes drier and this causes a greatly increased rate of wear on the brushes*

buckle [ˈbʌkl] **1** *noun* (metal, etc.) part of a belt used for joining the two ends together **2** *verb* to bend out of shape because of heat or force; *overheating will make the battery plates buckle*

buffet ['bʌfɪt] **1** *noun* shaking of the aircraft caused by the breakdown of the airflow over the upper surface of the wing; *large aircraft use a stick shaker to supplement the natural stall warning of buffet*; **buffet speed** = speed at which buffet is first noticed **2** *verb* to push around with great force (as by water or wind); *the storm buffeted the coast*; *the aircraft was buffeted by strong crosswinds as it made its final approach to land*

COMMENT: buffet is a warning to the pilot that the smooth airflow over the wing is breaking down and that he should take corrective action to prevent a stall

bug [bʌg] *noun* **(a)** fault in computer software which causes the program to operate incorrectly **(b)** **(heading) bug** = moveable plastic pointer on a gyroscopic compass or heading indicator

build up ['bɪld 'ʌp] *verb* to form by accumulation; *in icing conditions, ice builds up on the leading edges*; *see also* BUILT-UP (NOTE: **building up - built up**)

build-up ['bɪldʌp] *noun* gradual accumulation; *a build-up of static electricity*

built-up ['bɪltʌp] *adjective* **built-up area** = area which is full of houses, shops, offices, and other buildings, and with very little open space

bulb [bʌlb] *noun* **(a)** glass ball that gives electric light - the removable element of a lamp; *if a lamp does not work, the bulb may need replacing* **(b)** something shaped like a lamp bulb; *the most common type of hygrometer is the wet and dry bulb thermometer arrangement*

bulkhead ['bʌlkhed] *noun* dividing partition across the structure of the fuselage separating one compartment from another for reasons of safety or strength; *a*

fireproof bulkhead is provided to separate the cool area of the engine from the hot area

bulletin ['bʊlətɪn] *noun* short report or information on a situation; *news bulletin*; *weather bulletin*; *a terminal aerodrome forecast bulletin may consist of forecasts for one or more aerodromes*

BUMF *see* MNEMONIC

burst [bɜːst] **1** *noun* **(a)** minor explosion caused by increased pressure; *the risk of tyre burst through overheating is increased by hard application of the brakes* **(b)** very short period of activity followed by no activity; *the ground installation transmits a code in two short bursts*; **burst of energy** = very short period of energy **2** *verb* to explode because of increased pressure or puncture; *metal debris on the runway may cause a tyre to burst* (NOTE: **bursting - burst**)

busbar ['bʌsbɑː] *noun* electrical conductor used to carry a particular power supply to various pieces of equipment; *complex busbars are thick metal strips or rods to which input and output connections are made*

button ['bʌtən] *noun* little round disc which you push to ring a bell, etc.; **call button** = button (often on the arm of a passenger seat) which can be pushed when you need help from an attendant

Buys Ballot's Law ['baɪz 'bæləts 'lɔː] *noun* rules for identifying low pressure areas (based on the Coriolis effect)

COMMENT: in the northern hemisphere, if the wind is blowing from behind you, the low pressure area is to the left, while in the southern hemisphere it is to the right

bypass ['baɪpɑːs] *noun* alternative pipe, channel, etc.; *a turbine bypass in the form of an alternative exhaust duct is fitted with a valve*

Cc

C = CELSIUS *or* CENTIGRADE

CAA = CIVIL AVIATION AUTHORITY (UK)

cabin ['kæbɪn] *noun* passenger compartment in an aircraft; *air enters at the front of the cabin and leaves at the rear*; **cabin altitude** = artificial altitude inside the cabin created by cabin pressurization; **cabin attendant** = member of the cabin crew; **cabin crew** = airline staff who are in direct contact with the passengers and whose in-flight responsibilities include: ensuring correct seating arrangements, serving food and attending to the general well-being of passengers, etc.; **cabin pressure** = pressure of air inside the cabin which allows people to breathe normally at high altitudes; *see also* STAFF

cable ['keɪbl] *noun* **(a)** thick metal wire; **control cables** = thick metal wire linking the pilot's cockpit controls to control surfaces such as the elevators and ailerons **(b)** thick metal wire used for electrical connections; *earth return is by cable to the negative pole of the battery*

calculate ['kælkjuleɪt] *verb* to find out an answer to a problem by working with numbers; *the total flight fuel can be calculated by multiplying the time of the flight by kilograms of fuel per hour*

calculation [kælkju'leɪʃn] *noun* act or instance of finding out an answer to a problem by working with numbers; **calculation of fuel required** = arithmetic estimation of fuel needed by using time, distance and fuel-consumption factors

calculator ['kælkjuleɪtə] *noun* electronic machine for making calculations; *students are not allowed to use calculators in the examination*

calibrate ['kælɪbreɪt] *verb* to adjust the scale or graduations on a measuring instrument or gauge; *the international standard atmosphere is used to calibrate pressure altimeters*; *see also* AIRSPEED

calibration [kælɪ'breɪʃn] *noun* adjusting of the scale or graduations on a measuring instrument or gauge; *the international standard atmosphere is used for the calibration of instruments*

callsign ['kɔːlsaɪn] *noun* series of words and/or letters and/or numbers to identify an aircraft or station, etc.; *the aircraft's callsign is 'College 23'*; *VOR stations transmit a two or three letter aural Morse callsign*

calorie ['kælərɪ] *noun* amount of heat required to raise the temperature of 1 gramme of water by 1°C (4.186 joules); *after 2 calories have been released the temperature will have risen 2 degrees i.e. to 0°C, and so the freezing process ceases temporarily*

calorific [kælə'rɪfɪk] *adjective* referring calories; *(of fuel)* **calorific value** = the heat produced by the complete burning of a given amount of fuel; *the calorific value of a fuel is an expression of the heat or energy content released during combustion*

cam [kæm] *noun* oval or egg-shaped wheel which, when rotating, converts circular motion into reciprocating motion; *in a piston engine, the shape of each cam is designed to give the correct amount of opening to the valve*

camplate ['kæmpleɪt] *noun* rotating or non-rotating plate with cams on it; *the fuel pump consists of a rotor assembly fitted with several plungers, the ends of which bear onto a non-rotating camplate*

camshaft ['kæmʃɑːft] *noun* rotating shaft carrying cams, which opens and closes valves in a piston engine; *as the camshaft rotates, the cam will transmit a lifting force*

candela [kæn'diːlə] *noun* SI unit of brightness of a light; *the red and green wing tip navigation lights must be at least 5 candela* (NOTE: usually written **cd** with figures)

candle power ['kændl 'paʊə] *noun* unit of brightness of a light; *estimation of visibility is achieved by noting the distances at which lights of a known candle power can be observed*

canopy ['kænəpi] *noun* **(a)** transparent cover, typically on some fighters, light aircraft and gliders, designed to slide backwards and forwards or hinge upwards to allow pilots to enter or leave an aircraft **(b)** covering to protect people in a life raft; *the canopy should be erected to provide protection from the weather*

cantilever ['kæntɪliːvə] *noun* beam fixed and supported at one end only; *the mainplanes or wings are of cantilever design*

cap [kæp] *noun* **(a)** top or lid; *the exhaust valve cap* **(b)** polar ice cap = permanent area of ice at north and south poles

CAP = CIVIL AVIATION PUBLICATION

capability [keɪpə'bɪlɪti] *noun* having the capacity or ability to do something; **the flare has a day and night capability** = the flare is effective in daylight and in the dark

QUOTE France has a large capability in the areas of commercial aviation training and simulation
Civil Aviation Training

capable ['keɪpəbl] *adjective* competent; having an ability; *aircraft used in aerobatics must be capable of withstanding the extra loads imposed on the airframe by the manoeuvres; in most modern multi-engine jet transport aircraft, each fuel tank is capable of feeding any engine*; a capable person = a person who works well

capacitance [kə'pæsɪtəns] *noun* ability of a system of conductors and insulators to store an electrical charge when there is a positive discharge between the conductors; *capacitance is measured in farads and can either be a fixed amount or variable amount; if the supply frequency is low, the voltage has more time to build up a larger charge, or capacitance*

capacitive [kə'pæsɪtɪv] *adjective* referring to the ability of a system of conductors and insulators to store an electrical charge; *overspeed is usually a fault in the constant speed drive unit which causes the generator to overspeed and damage the capacitive loads on the aircraft*

capacitor [kə'pæsɪtə] *noun* (*electronic component*) system of conductors and insulators which store electrical charge; *a capacitor is used in a circuit to store energy for a short while*

capacity [kə'pæsɪti] *noun* **(a)** ability to do something easily; *energy is the capacity for performing work* **(b)** amount (of something) which a container can hold; *each cylinder has a capacity of 0.5 litres*; **battery capacity** = amount of electrical energy a battery can store and deliver expressed in ampere hours; **seating capacity** = the maximum number of people an aircraft, bus, etc., can seat **(c)** ability of an ATC system, in a given area, to provide a normal service, expressed in numbers of aircraft

capillary [kə'pɪləri] *noun* very fine or narrow tube; **capillary action** *or* **capillary flow** = action by which a liquid rises up a narrow tube

capsule ['kæpsjuːl] *noun* small closed container; **aneroid capsule** = thin flexible cylindrical box, usually made of metal, which has most of the air removed from it and which expands and contracts with changes in atmospheric pressure; *a decrease in atmospheric pressure will allow the aneroid capsule to expand*

captain ['kəptɪn] *noun* person in charge of an aircraft; *the captain asked all passengers to remain seated until the aircraft had come to a stop*

captive ['kæptɪv] *adjective* which is not free to move; **captive balloon** = balloon which, when in flight, is attached to the ground by a long cable

carbon ['kɑːbən] *noun* **(a)** non-metallic element - a component of living matter and organic chemical compound - as found in diamonds, charcoal, etc.; **carbon fibre** = thin, light and very strong strand of pure carbon which can be combined with other materials to make them stronger **(b)** black material with good electrical properties; **carbon brush** = small, replaceable, carbon blocks in electric motors, generators and alternators which provides the passage of electric current **(c)** **carbon deposits** = residues of burnt oil deposited in the combustion chamber, etc., in the course of the combustion process; *carbon deposits on a spark-plug electrode may cause misfiring*

carbon dioxide (CO₂) ['kɑːbən daɪ'ɒksaɪd] *noun* colourless, odourless, non-toxic gas found in the atmosphere - it is used in fire extinguishers, fizzy drinks, etc.; *carbon dioxide can be solidified at low temperature to produce dry ice*

carbon monoxide (CO) ['kɑːbən mən'ɒksaɪd] *noun* colourless, but poisonous gas from incomplete combustion found in the exhausts of spark ignition engines

carburation [kɑːbjuˈreɪʃn] *noun* process of mixing fuel with air in a carburettor; *carburation must ensure that rapid and complete burning will take place within the cylinder*

carburettor [kɑːbjuˈretə] *noun* device for mixing air with fuel in the right quantities before combustion; *most carburettors are installed so that they are in a warm position*; **carburettor heat** = system for keeping the carburettor and associated components free of ice; **carburettor icing** = process by which, under certain conditions, ice forms in the venturi tube of the carburettor

cardioid ['kɑːdiɔɪd] *adjective* shaped like a heart; *the cardioid polar diagram of the magnetic field around a bar-magnet*

carousel [kæruːˈsel] *noun* (*at an airport*) turning platform from where arriving passengers can pick up their baggage; *baggage from flight AC123 is on carousel No 4* (NOTE: also written **carrousel** in American English)

carriage ['kærɪdʒ] *noun* action of carrying; *regulations require the carriage of life-rafts when flying over water*

carrier ['kæriə] *noun* **(a)** person or organization that carries people or goods from one place to another; *individual carriers assign codes to aircraft* **(b)** **carrier wave** = continuous transmission of a radio signal of constant amplitude and frequency; *amplitude modulation has only one pair of usable sidebands each at about one sixth of the signal strength of the carrier*

carry ['kæri] *verb* to take someone or something from one place to another; *the aircraft was carrying 120 passengers* (NOTE: **carrying - carried**)

cartridge ['kɑːtrɪdʒ] *noun* removable unit for an air filter; *cabin air filters* normally consist of a casing, housing a replaceable filter cartridge

CAS = CALIBRATED AIRSPEED

case ['keɪs] *noun* **(a)** housing or jacket; outer covering; *cooling air is directed through passages in the engine case to control engine case temperature* **(b)** example, situation or circumstance; *in some special cases, eg for landing and take-off, wind directions are measured from magnetic north*

casing ['keɪsɪŋ] *noun* cover that encloses a piece of equipment, etc.; *annular and outer air casing form a tunnel around the spine of the engine*

CAT = CLEAR AIR TURBULENCE

catastrophe [kə'tæstrəfi] *noun* very bad accident; disaster; *the recent air catastrophe off the Nova Scotia coast; although the family were not at home when it happened, the crash which destroyed their house was a catastrophe for them*

catastrophic [kætə'strɒfɪk] *adjective* terrible, disastrous; *in a catastrophic accident where many persons may be disabled, those who show signs of life should be rescued first*

categorize ['kætɪgəraɪz] *verb* to put into groups, classes or categories; *figure 2 categorizes the types of wave by frequency band; aircraft can be categorized by weight, number of engines, role, etc.* (NOTE: also written **categorise**)

category ['kætɪgəri] *noun* official class or group; *load factors vary depending on the category of aircraft*

cathode ['kæθəʊd] *noun* negative electrode or terminal; *the cathode is a metal cylinder fitted with an internal heater*

cathode ray tube (CRT) ['kæθəʊd 'reɪ 'tjuːb] *noun* high-vacuum tube in which cathode rays produce an image on a screen such as a TV screen; *electronic indicating systems show engine indications, systems monitoring and crew alerting functions on one or more cathode ray tubes or liquid crystal displays mounted in the instrument panel*

cause ['kɔːz] **1** *noun* reason for something happening; *if the ammeter shows a high state of charge after start up, it is quite normal and no cause for alarm* **2** *verb* to make something happen; *air in the fuel line can cause an engine to flame-out or stop*

caution ['kɔːʃn] *noun* **(a)** advice to be careful; warning; *if a problem occurs in the spoiler system, a master caution light illuminates* **(b)** care; *proceed with caution*

cavitation [kævɪ'teɪʃn] *noun* formation of vapour-filled cavities or holes in liquids and gases - caused by low-pressure/high speed (Bernoulli's principle); *most reservoirs are pressurized to provide a positive fluid pressure at the pump inlet and thus prevent cavitation and the formation of bubbles*

cavity ['kævɪti] *noun* hole; *de-icing fluid flows into the cavity in the distributor panels before passing through the porous steel outer skin*

CB = CUMULONIMBUS

cc = CUBIC CENTIMETRES

cease [siːs] *verb* to stop (happening or doing); *if fuel, oxygen or heat is removed from the fire triangle, combustion will cease*

ceiling ['siːlɪŋ] *noun* highest point; *(of aircraft)* the greatest pressure height that can be reached; *the aircraft has a ceiling of 50,000 ft*; **cloud ceiling** = height above the ground or water of the base of the lowest layer of cloud

celestial [sə'lestɪəl] *adjective* referring to the sky; **celestial navigation** = navigation by using the stars in the sky

cell [sel] *noun* **(a)** system of positive and negative plates for storage of electricity that form a battery; *a battery is a device which converts chemical energy into* *electrical energy and is made up of a number of cells* (b) central part of a thunder cloud; *the life cycle of the thunderstorm cell ends when the downdraughts have spread throughout the cloud*

Celsius (C) ['selsiəs] *noun* scale for measuring temperature in which water freezes at 0° and boils at 100°; *compare* FARENHEIT

center ['sentə] *noun* US = CENTRE

centerline ['sentəlaın] *noun* US = CENTRELINE

centigrade (C) ['sentıgreıd] *noun* scale for measuring temperature in which water freezes at 0° and boils at 100°

centimetre ['sentımi:tə] *noun* measure of length that is equal to one hundredth of a metre (NOTE: 2.54 cm = 1 inch)

central ['sentrl] *adjective* which is in the centre or in the middle; *the control knob is moved from the central position*

centralize ['sentrəlaız] *verb* to put into the centre or into the middle position; *the operating jack centralizes the control surface after the turn* (NOTE: also written **centralise**)

Central Standard Time ['sentrəl 'stændəd 'taım] *noun* time zone of the east-central part of the USA and Canada, 6 hours behind GMT (Greenwich Mean Time)

centre ['sentə] **1** *noun* (a) middle; *the plane of the great circle passes through the centre of a sphere*; **centre of a circle** = mid-point of a circle, point in the middle of a circle (b) main building or office; *Area Forecasting Centre* **2** *verb* to move to a central position; *center the control column* (NOTE: **centred - centring** but American spelling **centered - centering**)

centreline ['sentəlaın] *noun* painted or imaginary line running along the centre of the runway (NOTE: also written **centre line**; written **centerline** in American English)

centre of gravity (CG) ['sentə əv 'græviti] *noun* point at which a body can be balanced; point through which the resultant force of gravity acts; *distribution of the tanks and the fuel in the tanks is vital in maintaining the aircraft centre of gravity and trim*

> COMMENT: if the centre of gravity is outside the limits, the aircraft may be difficult or impossible to control

centrifugal [sentrı'fju:gəl   US sen'trıfəgl] *adjective* which goes away from the centre; **centrifugal force** = outward force caused by turning motion; *the blades must be strong enough to carry the centrifugal loads due to rotation at high speed*; *see also* SWITCH

centrifuge ['sentrıfju:dʒ] **1** *noun* device which uses centrifugal force to separate or remove liquids **2** *verb* to separate liquids by using centrifugal force; *the rotating vanes of the breather centrifuge the oil from the mist*

centripetal [sen'trıpıtl] *adjective* which goes towards the centre; **centripetal force** = inward, centre-seeking force working in opposition to centrifugal force; *the magnitude of the centripetal force varies with the square of the wind speed*; *in a turn, lift provides the centripetal force*

certain ['sɜ:tən] *adjective* (a) particular; *in certain areas, at certain times, under certain circumstances, with certain exceptions* (b) sure; *there are certain to be horizontal differences in the mean temperature of a layer*; **to make certain** = to make sure; *make certain that the parking brake is on before doing engine run-up checks*

certificate 1 [sə'tıfıkət] *noun* official document which states that particular facts are true; **medical certificate** = document which confirms that the named person has been medically examined and declared to be in good physical condition **2** [sə'tıfıkeıt] *verb (of an official body)* to award or give a certificate; **aircraft which are certificated for flight** = aircraft which have the necessary paperwork to be authorized to fly

certificate of airworthiness (C of A) [sə'tɪfɪkət əv 'eəwɜːðɪnəs] noun

document issued by an aviation authority stating that an aircraft meets specific safety, etc., requirements that allow it to be used in service; *an authorized person may require production of the Certificate of Airworthiness*

certification [sɜːtɪfɪ'keɪʃn] noun

process of giving certificates; *the inferential method of ice detection is used on flight trials for certification of aircraft*

certify ['sɜːtɪfaɪ] verb

to authorize or permit the use of something; *the aircraft is certified for aerobatic flight*

CFI = CHIEF FLYING INSTRUCTOR

chalk [tʃɔːk] noun

soft white limestone rock that may be used in powder; *oil, which is trapped in the defects is absorbed by the chalk thus indicating their positions*

chamber ['tʃeɪmbə] noun

small enclosed compartment; **combustion chamber** = part of the cylinder in a piston engine where the ignition of the fuel/air mixture takes place; **expansion chamber** = container which allows for expansion of a fluid caused by increase in temperature, etc.; **float chamber** = part of a carburettor which houses the float

channel ['tʃænəl] noun

special frequency band for the transmission of radio signals; *the system operates on VHF communications between 118 and 135.95 MHz giving 360 channels at 50 kHz spacing*

character ['kærəktə] noun

(a) quality or set of qualities which make something different and separate from something else; *the circulation of the atmosphere is zonal in character* (b) individual letter, number or symbol used in printing and writing; **a six character group** = a group of six letters and/or numbers

characteristic [kærəktə'rɪstɪk] 1

adjective typical of a class or group of things; **a characteristic feature** = a normal feature of the thing in question 2 *noun* feature or quality making something different or separate from something else; *air masses have distinct characteristics which can be used to separate air masses on a chart*; **handling characteristics** = features of an aircraft that make it different from other aircraft when handling it; **summer characteristics** = climatic conditions which are typical of summertime

characterize ['kærəktəraɪz] verb

to have qualities or features which make something different and separate; *the stratosphere is characterized by a temperature structure which is steady or increases with height* (NOTE: also written **characterise**)

charge [tʃɑːdʒ] 1 noun

(a) amount of electricity; *friction causes a charge of static electricity; the battery was so old, it would not take a charge*; **a high level of charge** = a high amount of electricity (b) money demanded for the providing of a service; money paid for a service; **landing charges** = money paid to an airport authority by an operator or private pilot for landing an aircraft; **overnight parking is free of charge** = it costs nothing to park overnight 2 *verb* (a) to pass electrical current through something and thereby make it electrically active; *an installed battery becomes fully charged by the aircraft generator*; **charged particles** = atmospheric particles which have either a positive or negative electrical charge (b) to take money for a service; *we do not charge for overnight parking*

charger ['tʃɑːdʒə] noun

battery charger = device for putting an electrical charge into a battery; *see also* **TURBOCHARGER**

chart [tʃɑːt] noun

map for navigational purposes; **aeronautical chart** = map used in air navigation which may include topographic features, hazards and obstructions, navigational aids and routes, designated airspace and airports; **contour chart** = chart which shows high and low grounds; **meteorological chart** = chart of part of the earth's surface with information about weather conditions; **significant weather chart** = weather chart with important weather information marked on it; **surface synoptic chart** = chart of a geographical area with symbols, fronts and isobars giving a representation of the weather over the area at a particular time; information for synoptic chart is provided by observing stations; **upper air chart** = chart showing airflow pattern and distribution of temperatures at specific altitudes above about 10,000 feet

check [tʃek] 1 *noun* examination to make certain that something is as it should be; *safety check*; *a check was made on the undercarriage and airframe after the pilot reported a heavy landing*; *see also* RUN 2 *verb* to examine something in order to find out if it is correct; *it is the pilot's responsibility to check that his aircraft is airworthy*

QUOTE European Union (EU) airports may be empowered to carry out safety checks on foreign airlines
Flight International 1-7 May 1996

check in ['tʃek 'ɪn] *verb* to register by giving in your ticket, showing your passport and giving your baggage at an airline desk before a flight; *passengers should check in two hours before departure*

check-in ['tʃekɪn] *noun* airline desk where passengers register before a flight; *the check-in is on the first floor*; **check-in counter** *or* **desk** = counter where passengers check in; **check-in time** = time at which passengers should check in

checklist ['tʃeklɪst] *noun* list of items, often in booklet form, to be checked in a given sequence; *before every flight, the pilot should perform his pre-flight checks using a checklist*

chemical ['kemɪkl] 1 *adjective* referring to chemistry; *a chemical reaction* 2 *noun*

substance used in or made by a chemical process; *a chemical such as anti-ice for propellers*

chemistry ['kemɪstri] *noun* (a) science of chemical substances and their reactions; the study of substances, what they are made of, how they are made and how they react (b) the nature of something; *the basic chemistry of fire can be illustrated by the three sides of a triangle representing fuel, oxygen and heat*

chief [tʃiːf] *adjective* most important; main; **the chief factors** = the most important factors; **chief flying instructor (CFI)** = senior rank of flying instructor

chock [tʃɒk] *noun* wooden or metal device placed in front of the wheels of a parked aircraft to prevent it from moving; *the accident happened because the chocks had been removed before the engine was started*

choke [tʃəʊk] 1 *noun* valve in a carburettor which controls the amount of air combining with fuel; **choke tube** = carburettor venturi; *increase in rpm increases the speed of air passing through the choke tube or venturi* 2 *verb* (a) to block (a tube, etc.) making a liquid unable to move; **a choked nozzle** = a (partly) blocked nozzle (b) to stop breathing by inhalation of water, smoke, etc.

circle ['sɜːkl] *noun* line forming a round shape; round shape formed by objects or people; *they stood in circle on the tarmac*; **great circle direction** = imaginary circle on the surface of the earth which lies in a plane passing through the centre of the earth

circuit ['sɜːkɪt] *noun* (a) *(electricity)* complete route around which an electrical current can flow; **circuit-breaker** = small protective device in the circuit which blows or breaks before a dangerous overload of current arises (b) pattern of take-off, climb-out, turn onto crosswind leg, turn onto downwind leg, turn onto base leg, turn onto final approach and landing; *when carrying out practice landings at an* *aerodrome, the pilot should keep a sharp lookout for other aircraft in the circuit*

circuitry ['sɜːkɪtri] *noun* system of electrical circuits; *in an anti-skid braking system, circuitry is employed which can detect individual wheel deceleration*

circular ['sɜːkjʊlə] **(a)** *adjective* shaped like a circle; *anodes are circular plates with centre holes*; **semi-circular** = shaped like a half-circle **(b)** document distributed to a large number of people; *an aeronautical information circular*

circulate ['sɜːkjuleɪt] *verb (of fluid)* to move round in such a way as to arrive at the point of departure; *water circulates via the radiator and pump through to the engine block itself*

circulation [sɜːkju'leɪʃn] *noun (of fluids)* moving round in such a way as to arrive at the point of departure; *the general circulation is indicated by the arrows*; **cyclonic circulation** = circulation of air which, if viewed from above, is anticlockwise in the northern hemisphere and clockwise in the southern hemisphere

circulatory [sɜːkju'leɪtəri] *adjective* which circulates; *a self-contained re-circulatory oil system*

circumference [sɜː'kʌmfrəns] *noun* distance around the edge of a circle; *the angle subtended by an arc equal to one 360th part of the circumference of a circle is called one degree*

circumstance ['sɜːkʌmstns] *noun* condition which affects something in a given situation; **in some circumstances** *or* **under certain circumstances** = in some particular situations

cirro- ['sɪrəʊ] *prefix meaning* high altitude (20,000+ feet)

cirrocumulus [sɪrəʊ'kjuːmjʊləs] *noun* layer of broken cloud at about 20,000 feet

cirrostratus [sɪrəʊ'strɑːtəs] *noun* layer cloud at about 20,000 feet

cirrus ['sɪrəs] *noun* high cloud in a mass of separate clouds which are formed of ice crystals

Civil Aviation Authority (CAA) ['sɪvɪl 'eə ɔː'θɒrəti] *noun* the organization which licences operators, aircraft and personnel for non-military, especially commercial aviation; *the CAA issues licences*

Civil Aviation Publication (CAP) ['sɪvɪl 'eə pʌblɪ'keɪʃn] *noun* book, etc., published by the Civil Aviation Authority, each publication having its own reference number; *the procedure for obtaining a bearing can be found in CAP 413*

clad [klæd] *verb* to protect by covering; *alloys can be protected from corrosion by cladding the exposed surface with a thin layer of aluminium*

clamshell door ['klæmʃel 'dɔː] *noun* hinged part of thrust reverser; *clamshell doors are hydraulically or pneumatically opened, and direct the exhaust gases forwards to produce reverse thrust*

classification [klæsɪfɪ'keɪʃn] *noun* putting into groups or classes because of particular common features; *classification of aircraft consists of a multi-level diagram with each category divided into sub-categories*; *a full classification of layer cloud is given in the table*; **aircraft classification number (ACN)** = number expressing the relative effect of an aircraft on a pavement for a specified sub-grade strength; **pavement classification number (PCN)** = number expressing the bearing strength of a pavement for unrestricted operations

classify ['klæsɪfaɪ] *verb* to group items, etc., so that those with similar characteristics are in the same group; *precipitation is classified as light, moderate or heavy according to its rate of fall*; *the weather associated with visibility reductions by particles suspended in the atmosphere is classified either as fog, mist, haze, or smoke*

clear ['klıə] **1** *adjective* **(a)** referring to conditions in which it is easy to see - with no cloud, no fog, etc.; **a clear sky** = a sky with no cloud; **a clear winter night** = a night with no fog, mist or other conditions which might impair visibility; **clear air turbulence (CAT)** = turbulence encountered in air where no cloud is present (CAT is often associated with the jet stream) **(b)** which you can easily see through; **clear ice** = ice which is glass-like rather than white **(c)** with nothing in the way; unblocked; **clear runway** *or* **the runway is clear** = nothing is on the runway *(see also Comment below)* **keep the exits clear** = do not put anything and do not stand in front of the exits; **a clear pass** = exam result which is in no doubt **(d)** *(official permission to do something)* **clear to land** = air traffic control permission to land **(e)** away from; **clear of cloud** = either above or below cloud; out of cloud; **keep clear (of)** = keep away (from) **(f)** easy to hear; **a clear voice** = a voice which is easy to hear and understand **(g)** (i) easy to understand; (ii) understood; *the explanation is very clear*; **is it clear?** = do you understand? **2** *verb* **(a)** to remove a blockage or some other unwanted effect which prevents a system from working correctly; *a heater element is fitted to clear the detector of ice* **(b)** to disappear; *in winter frost and fog are slow to clear* **(c)** to make sure that it is all right to do something; **clear it with the CFI** = make sure that the CFI agrees with the request **(d)** to officially ask people to quickly leave a given area or place; **to clear the building** = to quickly leave the building

COMMENT on 27th March 1977 two Boeing 747s collided on the runway at Los Rodeos airport Tenerife in poor visibility, resulting in 575 deaths. A KLM 747 commenced take-off while a Pan Am 747 was still taxiing towards it on the same runway. There was clearly a breakdown in communications, perhaps a misunderstood radio call. The Pan Am aircraft had been asked by the controller, who was unable to see either aircraft due to low cloud, 'Are you clear of the runway?' The KLM aircraft had already commenced the take-off roll without clearance. It is possible that the KLM pilot mistook the call to the other aircraft thinking that he was 'clear to take off'

QUOTE the principles of weight and balance should have been learned by all pilots during their initial training, but it is clear that, afterwards, some forget
Civil Aviation Authority, General Aviation Safety Sense Leaflet

clearance ['klıərəns] *noun* **(a)** space made to allow for the movement of hardware relative to other hardware; *clearance between rocker arm and valve tip* **(b)** official permission; *obtain clearance for IFR flight* **(c)** disappearance of something unwanted - often rain, fog, snow, etc.; *low temperatures caused a delay in the clearance of fog*

climate ['klaımət] *noun* weather conditions particular to a given area; *Mediterranean climate; Tropical climate*; **continental climate** = type of climate found in, or originating in areas where there is no effect from the sea; **temperate climate** = type of climate which is neither very hot in summer nor very cold in winter; *see also* CONTINENTAL

climatic [klaı'mætık] *adjective* referring to climate; referring to weather conditions particular to a given area; *the aircraft forward speed and altitude as well as climatic conditions will influence the value of thrust*; **climatic zone** = one of the eight areas of the earth which have distinct climates

COMMENT: the climatic zones are: the two polar regions (Arctic and Antarctic); the boreal in the northern hemisphere, south of the Arctic; two temperate zones, one in the northern hemisphere and one in the southern hemisphere; two subtropical zones, including the deserts; the equatorial zone which has a damp tropical climate

climatology [klaɪmə'tɒlədʒi] *noun* science of the study of climate; *although pilots do not need to be experts in climatology, they should have a good understanding of the factors which produce changes in the weather*

climb [klaɪm] 1 *noun (of aircraft)* action of increasing altitude by use of power; *fine pitch enables full engine speed to be used during take-off and climb* (NOTE: the opposite is **descent**) 2 *verb (of aircraft)* to increase altitude by use of power; *after take-off, the aircraft climbed to 5,000 ft* (NOTE: the opposite is **descend**)

climb-out ['klaɪm'aʊt] *noun* flight after take-off from 35 feet to 1,500 feet during which undercarriage and flaps are retracted (not for light aircraft); *turn right after climb-out*

clockwise ['klɒkwaɪz] *adjective* & *adverb* describing a circular movement in the same direction as the hands of a clock (from left to right); *a clockwise direction*; *the relative bearing indicated is measured clockwise from the nose of the aircraft* (NOTE: the opposite is **anticlockwise**)

clog [klɒg] *verb* to block; to prevent movement of fluid through a pipe, etc., because of a build-up of solid matter; *most filters allow unfiltered fluid to pass to the system when the filter becomes clogged*

close [kləʊz] *verb (a window, a system, a book)* to shut; *close the door*

closure ['kləʊʒə] *noun* closing, shutting; *the voltage regulator is turned on by the closure of the generator control relay*

cloud [klaʊd] *noun* mass of water vapour or ice particles in the sky that can produce rain; **cloud base** = bottom part of a layer of cloud; **cloud ceiling** = height above the ground or water of the base of the lowest layer of cloud; **convective clouds** = clouds formed as a result of warm moist air rising and condensing at altitude; **layer cloud** = cloud formed as a continuous horizontal sheet; *layer cloud names consist of a prefix according to height of base, and a* suffix *according to shape*; *see also* COMPLEX

> COMMENT: **altocumulus:** cloud formed at about 12,000 feet as a layer of rounded mass with a level base; **altostratus:** cloud formed as a continuous layer between 6,000 and 20,000 feet usually allowing the sun or moon to be seen from the surface; **cirrocumulus:** layer of broken cloud at about 20,000 feet; **cirrostratus:** layer cloud at about 20,000 feet; **cirrus:** cloud made of ice crystals at 25,000 - 40,000 feet appearing as hair-like formations; **cumulonimbus:** cloud formed as a towering mass and often associated with thunderstorms; **cumulus:** cloud formed in rounded masses with a flat base at low altitude, resulting from up currents of air; **nimbostratus:** thick dark layer cloud at low altitude from which rain or snow often falls (nimbus = rain cloud); **stratocirrus:** cloud like cirrostratus but more compact; **stratocumulus:** layer of connected small clouds at low altitude

cm = CENTIMETRE

co- [kəʊ] *prefix meaning* together; **co-axial** = having the same axis; **co-located** = having the same location

coalesce [kəʊə'les] *verb* to join together to form a large mass or number; *the moisture in the air coalesces into large water droplets*

coalescence [kəʊə'lesəns] *noun* joining together to form a larger mass or number; *coalescence of water vapour in the atmosphere forms larger droplets of water*

coast [kəʊst] *noun* area where the land meets the sea; *Valentia is situated on the coast of south west Ireland*

coastal ['kəʊpstl] *adjective* referring to the coast; **coastal area** = area near a coast; *land and sea breezes occur in coastal areas*

coastline ['kəʊslaɪn] *noun* outline of a coast; coast seen from a distance (altitude) or on a map; *it is normally easy to identify a coastline or island*

coat [kəʊt] **1** *noun* thin covering of a substance such as paint; *the coats of paint on a large aircraft significantly increase its weight* **2** *verb* to cover with a thin layer of a substance (such as paint); *metals are coated for protection against corrosion*

coating ['kəʊtɪŋ] *noun* **(a)** thin layer of a substance; *there are two coatings on the inside of CRT screens* **(b)** act of covering with a thin layer of a substance

cock [kɒk] *noun* manually controlled valve or tap to control the flow of a liquid; *it is necessary to have a master cock for each engine*

cockpit ['kɒkpɪt] *noun* forward area in an aircraft from where the aircraft is controlled by the pilot; *in the case of an in-flight oil loss, a warning indicator will light in the cockpit*

QUOTE in the cockpit of the future there will be two animals, a pilot and a dog. The pilot will be there to feed the dog, and the dog will be there to bite the pilot if he tries to touch anything
NYT News Service

code [kəʊd] *noun* **(a)** system of numbers, letters or symbols used to represent language which has to be learned and decoded in order for the receiver to understand the meaning; *(still used for identifying some radio beacons)* **Morse code** = largely outdated code used for transmitting messages in which letters of the alphabet and numbers are represented by dots and dashes or short and long signals; **a two-figure code** = a code of 2 digits **(b)** series of pulses by which an aircraft transponder replies to a signal from the ground

codeshare ['kəʊdʃeə] *noun* **codeshare deal** = agreement between airlines regarding connecting flights; *the two airlines have entered into a codeshare deal for flights between Dubai and Bangkok*; **codeshare partner** = airline

which has an agreement with another airline regarding connecting flights

codesharing ['kəʊdʃeərɪŋ] *noun* procedure which allows travellers to use connecting flights between one airline and another partner airline for worldwide destinations

coefficient [kəʊɪ'fɪʃənt] *noun (algebra)* mathematical quantity placed before and multiplying another; **in 4xy, 4 is the coefficient of xy**

C of A = CERTIFICATE OF AIRWORTHINESS

C of G = CENTRE OF GRAVITY

coil [kɔɪl] *noun* device consisting of coiled wire for converting low voltage to high voltage; *a voltage coil is connected across the generator*

coiled wire ['kɔɪld 'waɪə] *noun* length of wire twisted round and round; *a coiled wire connects the terminal to earth*

coincide [kəʊɪn'saɪd] *verb* to happen at the same time and/or in the same place; *when the aircraft heading is directly into wind or down wind, track and heading coincide*

coincident [kəʊ'ɪnsɪdənt] *adjective (two or more things)* happening at the same place or at the same time; *the earth's true north and magnetic north poles are not coincident*

col [kɒl] *noun* an area of slack pressure gradient between two centres of high or low pressure; *the persistence and movement of cols are governed by the movement of the adjacent pressure systems*

collapse [kə'læps] **1** *noun* sudden and complete fall; **the collapse of a company** = the end of the existence of the company **2** *verb* **(a)** to fall suddenly and completely; *the magnetic field will reach a maximum in one direction, collapse to zero and reach a maximum in the opposite direction* **(b)** *(of apparatus)* to fold or to close suddenly and unintentionally; **the**

undercarriage collapsed = the undercarriage could not support the aircraft and broke or retracted on its own **(c)** *(of person)* to faint; **the passenger collapsed** = the passenger fell and became semi- or fully unconscious because of some medical problem

QUOTE as the aeroplane slid off the runway, the left landing gear collapsed

Pilot

collect [kə'lekt] *verb* **(a)** to gather over a period of time; *any given object will usually collect ice more quickly at high speed* **(b)** to take or to pick up (goods, documents, etc.) from a place

collection [kə'lekʃn] *noun* **(a)** number of things brought together; *a collection of vintage aircraft* **(b)** act or instance of being collected by somebody; *the documents are in the office waiting for collection*

collide [kə'laɪd] *verb* to bump or to crash into something; *the aircraft left the runway and collided with a fire truck*

collision [kə'lɪʒn] *noun* crash between two objects, two vehicles, etc.; *if there is a risk of collision, alter course to the right*; **collision avoidance** = the prevention of collisions by taking measures beforehand to ensure that they do not happen; **anti-collision light** = flashing white light on an aircraft; *see also* ANTI-COLLISION

column ['kɒləm] *noun* **(a)** *(stick)* **control column** = main hand control used by the pilot to control the aircraft in roll and pitch **(b)** body of fluid or solid with a tall, narrow shape; *Torricelli first demonstrated that the atmosphere has weight by showing that it can support a column of liquid* **(c)** vertical section of a table in a document; *column four of the table shows the totals of the other three*

combat ['kɒmbæt] *verb* to fight against; *fire extinguishers are provided to combat fire*; **combat aircraft** = aircraft designed for warfare

combination [kɒmbɪ'neɪʃn] *noun* two or more things brought together to form

one; *the combination of wind direction and wind speed is called velocity*; **multi-wheel combinations** = undercarriages consisting of a number of wheels on each unit

combine [kəm'baɪn] *verb* to bring two or more things together to make one; *the stabilizing channels for ailerons and elevators are combined*; *thrust and lift combine to overcome drag and gravity*

combustible [kəm'bʌstəbl] *adjective* which burns easily; **combustible materials** = materials which will catch fire easily (such as wood, paper, etc.)

combustion [kəm'bʌstʃən] *noun* burning, especially that which takes place in an engine; *the heat generated by combustion is considerable*; **internal combustion engine** = type of engine in which the fuel is burnt within the cylinders of the engine (as opposed to the steam engine); *an internal combustion engine or piston engine*

command [kə'mɑːnd] **1** *noun* order; **the command to evacuate** = the order to leave the aircraft in an emergency; **in command** = having responsibility for and authority over; **pilot in command** = pilot responsible for the operation and safety of the aircraft during flight time; *see also* STROKE **2** *verb* to order something to be done; *the captain commanded the evacuation of the aircraft*

commander [kə'mɑːndə] *noun* pilot in control of, and responsible for, the aircraft and its contents during flight time; **the commander of an aircraft** = the member of the flight crew specified by the operator as being the commander

commence [kə'mens] *verb* to start (to do something); **commence the evacuation** = start getting people out of the aircraft

commercial [kə'mɜːʃl] *adjective* referring to a business activity; **commercial aviation** = flying as a business enterprise; **commercial aircraft** = aircraft used to carry cargo or passengers for payment; *see also* LICENCE

common sense ['kɒmən 'sens] *noun*
ordinary good sense; *use your common
sense as well as follow the rules if a
passenger feels unwell*

comms = COMMUNICATIONS

communicate [kə'mjuːnɪkeɪt] *verb* to
make contact with somebody in order to
pass information; *the cabin attendants
should communicate with the captain*

communication [kəmjuːnɪ'keɪʃn]
noun passing information to somebody
usually but not always, by using language;
*two methods of communication are
available to crew members - language and
hand signals*

communications [kəmjuːnɪ'keɪʃnz]
noun system of passing information;
satellite communications; *VHF
communications are allocated the
frequency bracket 118-137 MHz*

commutator ['kɒmjuteɪtə] *noun* device
containing metal bars connected to the
coils of a generator to produce electrical
current; *as the power output required is
DC not AC, a commutator is fixed at one
end of the armature*

compact [kəm'pækt] **1** *adjective* small;
close together; not taking much space; *the
annular system, as used on modern
aircraft, provides a compact system, and,
for the same output and mass flow, a
shorter system* **2** *verb* to make smaller or
more dense by pressing; *(of ground)* to
compress, by driving over with heavy
machinery; *when taxiing on grass,
aircraft wheels compact the earth as the
aircraft moves over it*

compaction [kəm'pækʃn] *noun*
pressing together to form one; compressing
(something) and making it hard; *the speed
of impact when the aircraft passes
through a snowstorm causes compaction
of snowflakes into a solid mass on leading
edges and air-intakes*

comparable ['kɒmprəbl] *adjective*
which can be compared equally with
something else; *titanium is non-magnetic
and has an electrical resistance
comparable to that of stainless steel*

comparator [kəm'pærətə] *noun* device
to compare two things; *the autopilot
comparator monitors the operation of the
elevator and aileron channels*

compare [kəm'peə] *verb* to find the
similarities and dissimilarities between two
or more things; *when the chart is properly
orientated, it is easier to compare the
distance between landmarks on the
ground with their corresponding
distances on the chart*; *an aneroid
barometer is small compared with a
mercury barometer* (NOTE: compare with is
regarded by some as better usage than
compare to)

comparison [kəm'pærɪsən] *noun* the
bringing out of differences and similarities
between two or more things; *a table
showing a comparison of fixed points on
various temperature scales is given on
page three*

compartment [kəm'pɑːtmənt] *noun*
small space or area in a structure for a
particular purpose; *engine compartment*;
crew compartment = area reserved for
crew

compass ['kʌmpəs] *noun* instrument
(usually) with a magnetic needle which
always points to the magnetic north;
gyroscopic compass *or* **gyrocompass** =
compass which uses gyroscopic directional
stability rather than magnetism to indicate
directions

compatible [kəm'pætɪbl] *adjective*
referring to a component or system which
can be used with a different component or
system without causing any problems;
*computer software designed for one
particular system may not be compatible
with other systems*

compatibility [kəmpætə'bɪlɪti] *noun*
(of systems) ability of a component to
operate successfully with other
components; *problems of compatibility
caused the computerized system to
malfunction*

compensate ['kɒmpənseɪt] *verb* **(a)** to make up for the loss of something; *the floor covering may be designed to compensate for temperature, pressurization and bending loads*; *the fall in air temperature increases the air density and so compensates to some extent for the loss of the thrust due to atmospheric pressure* **(b)** to give money to a person or organization to make up for a physical or financial loss; to pay for damage; *the money offered by the company did not compensate for the injuries he received in the accident*

compensation [kɒmpə'seɪʃn] *noun* money paid to an individual or organization to replace or make up for physical or financial loss; *the company paid out $2 million in compensation to the families of those who lost their lives in the tragedy*

compilation [kɒmpɪ'leɪʃn] *noun* putting together of suitable information; *the manual is a compilation of materials used by each of the instructors*

compile [kəm'paɪl] *verb* to put together a number of pieces of information; *aviation routine weather reports are compiled half-hourly or hourly at fixed times*

complement ['kɒmplɪmənt] *verb* to fit in with and improve the performance of something; *ultra-sonic detection is used to complement other methods of flaw detection*

complementary [kɒmplɪ'mentri] *adjective* which fits in with and improves the performance of something; *SSR is complementary to the primary radars used by ATC*

complete [kəm'pliːt] **1** *adjective* **(a)** which contains all the parts it should contain; *the centre section can be constructed either as a complete unit or as two separate units* **(b)** absolute and total; a **complete failure** = a total failure **2** *verb* **(a)** to finish; make whole; *the number of revolutions for the crankshaft to complete a full cycle is always two*; complete the

work = continue until the work is finished **(b)** to fill in information; **complete the flight plan** = fill in the required information in the flight plan

completion [kəm'pliːʃn] *noun* (satisfactory) finishing of a task; *it is important to carry out an inspection of an aircraft after completion of de-icing operations*

complex ['kɒmpleks] **1** *adjective* complicated and therefore possibly difficult to understand; *of all the pre-departure activities, route planning is one of the most complex* **2** *noun* **(a)** whole made up of many different parts; a **cumulonimbus cloud complex** = a collection of cumulonimbus clouds forming a system **(b)** building made up of many different parts; **the terminal three complex** = the main building and associated buildings which together make up terminal three

complexity [kəm'pleksɪti] *noun* being complex; complication; *up-to-date design does not necessarily mean structural complexity*

complicate ['kɒmplɪkeɪt] *verb* to make more difficult; *map reading is often complicated by seasonal variations*

complicated ['kɒmplɪkeɪtɪd] *adjective* not easy to understand; a **complicated problem** = a problem which takes a long time and a lot of work to solve

complication [kɒmplɪ'keɪʃn] *noun* difficulty or problem; *the complication with the Mercator's projection is that great circle directions must be converted to rhumb line directions by the application of conversion angle before they can be plotted*

comply [kəm'plaɪ] *verb* to be or do what is required by an instruction or law; *equipment and furnishings of modern jet transports must comply with safety regulations; passengers must comply with the no-smoking signs* (NOTE: **complying - complied**)

component [kəmˈpəʊnənt] *noun* **(a)** part (of an aircraft, aircraft system or piece of equipment); *the undercarriage is made up of a number of different components* **(b)** one part of a force which consists of a number of different parts, such as wind; **crosswind component** = that part of the wind force acting at an angle to the direction of flight **(c)** substance which forms part of a compound

compose [kəmˈpəʊz] *verb* to make something from a number of parts; *the atmosphere is composed of a mixture of gases*

composite [ˈkɒmpəzɪt] **1** *adjective* referring to something made up of a number of different parts; *composite material*; *the flight crew route flight plan is a composite document which serves as a navigation log* **2** *noun* lightweight but very strong man-made material used in aircraft manufacturing; *to make a composite it is necessary to combine the reinforcing glass fibres with special glue or resin* (NOTE: the word composite is an adjective but through frequent usage the term 'composite material' has been shortened to 'composite')

COMMENT: composites are used in the construction of many modern aircraft, from gliders to aircraft such as the Airbus A320 because they are strong and lighter than metals

QUOTE Canadian Aerospace Group (CAG) is working with Pratt & Whitney Canada on a turboprop-powered version of its Windeagle all-composite light aircraft
Flight International 16-22 July 1997

composition [kɒmpəˈzɪʃn] *noun* make-up, structure of something; **composition of the atmosphere** = combination of gases which make up the atmosphere

compound [ˈkɒmpaʊnd] **1** *adjective* referring to something made up of two or more parts or substances; **a compound wound generator** = generator which consists of a number of windings **2** *noun*

substance made up of two or more components; *a chemical compound has qualities that are different from those of the substances from which it is made*; *advances in sealing compounds have now made fuel tanks less liable to leaks*

compress [kəmˈpres] *verb* to put under pressure thereby reducing volume; *pressure is created when a fluid is compressed*

compressibility [kəmpresəˈbɪlɪti] *noun* natural ability of a substance to change volume when under varying pressures; *in systems using very high pressure, the compressibility of the liquid becomes important*

compressible [kəmˈpresəbl] *adjective* referring to something that can be compressed; *air is compressible but water is not*

compression [kəmˈpreʃn] *noun* act or instance of putting pressure on something; **compression stroke** = the stage of an internal combustion cycle when the fuel/air mixture comes under pressure from the upward-moving piston; *see also* ADIABATIC

compressive [kəmˈpresɪv] *adjective* referring to forces caused by pressure on a surface; *a strut is designed to withstand compressive loads*; **compressive stress** = the resistance of a body to crushing by two forces acting towards each other along the same straight line

compressor [kəmˈpresə] *noun* device (such as a pump) to compress air (to increase pressure); *a shaft connects the turbine to the compressor*; *see also* AXIAL

comprise [kəmˈpraɪz] *verb* to be made of; *a simple fuel system comprises a tank, pipes, a filter, a pump and a carburettor* (NOTE: The correct use of comprise is often disputed. Some people regard it as a synonym for the verb **consist of**, while others believe it should be used in an opposite sense: **a tank, pipes, a filter, a pump and a carburettor comprise the fuel system**. It is sometimes used in its passive form: **the fuel system is comprised of a number of different parts.**)

concentrate ['kɒnsəntreɪt] *verb* **(a)** to collect in a particular place rather than spread around; *most of the mass of air is concentrated at the lowest levels of the atmosphere* **(b)** to give attention and thought to something in particular; *this chapter concentrates on charts*; **to concentrate hard** = to give all one's thought and attention to something

concentration [kɒnsən'treɪʃn] *noun* **(a)** collection in a particular place rather than spread around; *the maximum concentration of ozone is between 20 and 25 km above the earth's surface* **(b)** giving attention and thought to something; *in the early stages of training, instrument flying requires great concentration on the part of the student pilot*

concentric [kən'sentrɪk] *adjective* having the same centre; **concentric circles** = circles of different diameters but with the same centre point

concept ['kɒnsept] *noun* idea or abstract principle; **a complicated concept** = an idea or series of ideas or principles which are difficult to understand; *the concept of open skies is not one with which everybody agrees*

concern [kən'sɜːn] **1** *noun* **(a)** serious interest; **a matter for concern** = something which must be taken very seriously **(b)** responsibility; *attention to the welfare of passengers is the concern of the cabin crew*; *safety is everybody's concern*; **this is no concern of ours** = this is nothing to do with us **2** *verb* **(a)** to worry; **this report concerns me enormously** = I am not at all happy about this report **(b)** to be about or to be the subject of; *if there is serious vibration, the crew should shut down the engine concerned*; **this report concerns me** = this report is about me **(c)** to be of interest and relevance to; **the regulations concern all employees** = the regulations apply to all employees; all employees must obey these regulations (NOTE: the difference in meaning (a) and (b) of

the verb would be made clear by the stress: (a) this report **concerns** me; (b) this report concerns **me**)

QUOTE the correct storage and handling of cargo and especially dangerous goods is an area which is of considerable concern to the Federation

INTER PILOT

concrete ['kɒŋkriːt] *noun* substance made of cement, sand and water used in the construction of buildings, roads, etc.; *rock, sand and concrete reflect only 10-20% of radiation*

condensation [kɒndən'seɪʃn] *noun* process by which vapour changes into liquid; *if the air becomes saturated, further cooling results in condensation* (NOTE: the opposite is **evaporation**)

condense [kən'sens] *verb* **(a)** to change from vapour to liquid form; *the most common type of hygrometer is one in which a surface in contact with the atmosphere is cooled until moisture begins to condense on the surface* (NOTE: the opposite is **evaporate**) **(b)** to remove unnecessary parts from a text to make it shorter; *the synoptic code condenses information without loss of sense*

condenser [kən'densə] *noun* electrical capacitor; *the condenser prevents spark plugs from arcing*

condition [kən'dɪʃn] *noun* **(a)** present state of something; **although the aircraft is old, it is in good condition** = the aircraft is old but well-cared for **(b)** state of the surrounding; *in a high relative humidity condition, the evaporation rate is low*; **abnormal weather conditions** = unusual or unfavourable weather; **adverse weather conditions** = bad weather; **flying conditions** = the weather and its suitability for flying **(c)** circumstances; **working conditions** = those aspects of working lives which affect the way people feel about their work **(d)** something on which another thing depends; **on condition that** = only if; **the flight will depart on condition**

that the weather improves = the flight will depart only if the weather improves

conducive [kən'djuːsɪv] *adjective* (circumstance, etc.) which allows something to happen more easily; *atmospheric conditions conducive to the formation of ice are detected and these operate a warning system*

QUOTE when refuelling, ensure the aircraft is properly earthed. The very low humidity on a crisp, cold day can be conducive to a build-up of static electricity
Civil Aviation Authority, General Aviation Safety Sense Leaflet

conduct 1 ['kɒndʌkt] *noun* (a) manner or way of doing something; *the captain is responsible for the safe conduct of the flight* (b) behaviour; *the investigation found that the flight attendant's conduct was unacceptable* **2** [kən'dʌkt] *verb* (a) to organize and do something; to carry out; *crew will conduct area checks; security personnel conducted a search of the building* (b) to allow (electricity, heat) to pass through; *water conducts electricity*

conduction [kən'dʌkʃn] *noun* process by which heat or electricity passes through a substance; *heat is transferred to the layer of air next to the earth's surface by conduction*

conductive [kən'dʌktɪv] *adjective* referring to the ability of a substance to allow heat or electricity to pass through; *steel is a conductive material; land masses are less conductive than water*

conductivity [kɒndʌk'tɪvɪti] *noun* ability of a material to allow heat or electricity to pass through; *because of the poor conductivity of air, heat is transferred from the earth's surface upwards by convection*

conductor [kən'dʌktə] *noun* substance through which heat or electricity can pass; *water and steel are good conductors*

cone [kəun] *noun* solid body with a base in the shape of a circle, and with sides which narrow to a point (or any object which has that shape); **nose cone of an aircraft** = the foremost part of the nose of a multi-engine aircraft which may house electronic equipment, but not an engine

configuration [kənfɪgə'reɪʃn] *noun* pattern or way in which things are arranged; **configuration of an aircraft's fuel tank system** = the way in which the tanks are laid out; **aircraft configuration** = particular combination of moveable parts such as flaps, landing gear, etc., which affect the aerodynamics of the aircraft

confine [kən'faɪn] *verb* (a) to limit to a particular area; *cooling is confined to the air in contact with the ground; the damage was confined to a small area* (b) to limit to a given subject; **the report confines itself to the incident of 3rd January** = the report deliberately does not mention anything other than the incident of the 3rd January

confined [kən'faɪnd] *adjective* limited; small; **a confined space** = a small defined space which does not allow free movement

confirm [kən'fɜːm] *verb* to agree that something is correct; to repeat something to remove any uncertainty; *the attitude indicator shows that the aircraft is in a nose down attitude and the increasing airspeed confirms that the aircraft is not in level flight; can you confirm that the instructor was flying the aircraft at the time of the collision?; VHF and/or UHF radio aids confirm ADF bearings*

COMMENT: cross-checking of certain flight instruments is used to confirm readings from other instruments, eg the airspeed indicator and vertical speed indicator confirm pitch information from the attitude indicator

conform [kən'fɔːm] *verb* **to conform to regulations** = to do what is required by rules and regulations; *(of product)* to correspond to required standards; *fuels must conform to strict requirements*

conformal [kən'fɔːml] *adjective (of charts)* which represent angles, bearings, etc., correctly; *Lambert's conformal projection*

congestion [kən'dʒestʃn] *noun* blocking of free movement of people, traffic, etc.; *when leaving the aircraft in an emergency, to avoid congestion, passengers should be directed to move away from exits quickly*

conic ['kɒnɪk] *adjective* which is based on the shape of a cone; **conic projection** = the standard two-dimensional representation of the earth

conical ['kɒnɪkl] *adjective* shaped like a cone; *the nose of Concorde has a conical shape*

conjunction [kən'dʒʌŋkʃn] *noun* togetherness; **in conjunction with** = (working or operating) together with; *built-up areas, used in conjunction with other features such as rivers, railways and coastlines which are near them are more easily identified*

connect [kə'nekt] *verb* to join; *batteries are sometimes connected in series*; *a cockpit lever is connected to a needle valve in the float chamber*

connecting flight [kə'nektɪŋ 'flaɪt] *noun* second aircraft which a passenger should arrive on time to catch, and which will take him to his final destination; *instead of flying direct to London, take the flight to Amsterdam and then take a connecting flight to London Heathrow*

connection [kə'nekʃn] *noun* (a) point at which things are joined; *there is an electrical connection to the battery* (b) link or feature which makes things interdependent; *there is a connection between temperature change and altitude* (c) catching a second aircraft to arrive to a final destination; *follow the 'Flight Connection' signs*

connector [kə'nektə] *noun* device which connects two or more things; *a connector is used to connect two lengths of wire together*; *standard connectors*

consist of a metal coupling with a rubber sandwich joint

consecutive [kən'sekjjʊtɪv] *adjective* which follow one another without a break; *4, 5 and 6 are three consecutive numbers*; **a period of 28 consecutive days** = 28 days following immediately one after the other

consequence ['kɒnsɪkwəns] *noun* result of an action; *the accident was a consequence of the pilot's actions*; **as a consequence** = as a result

consequent ['kɒnsɪkwənt] *adjective* resulting; *as temperature rises, there will be a consequent increase in the volume of the gas*

consequently ['kɒnsɪkwentli] *adverb* therefore, as a result; *he was late, consequently he missed the start of the examination*

conserve [kən'sɜːv] *verb* to avoid using unnecessarily; to keep; *release the brakes when necessary and conserve main system pressure*; **to conserve energy** = to use only as much energy as you really need; **to conserve fuel** = to use as little fuel as possible

consider [kən'sɪdə] *verb* to think carefully about something; *if the aircraft is low on fuel, the commander should consider diverting to the nearest suitable airport*

QUOTE many purchasers of flight simulators would argue that, when considering the major manufacturers, there is little to choose between them
Civil Aviation Training

considerable [kən'sɪdrəbl] *adjective* a lot of; quite large; *the required range of trim change is considerable*; **a considerable amount of fuel** = a lot of fuel, a large amount of fuel; **a considerable distance** = a long distance; **a considerable force** = a lot of force (NOTE: considerable does not mean that something should be thought about, as the meaning for the verb 'consider' might suggest)

consideration [kənsɪdə'reɪʃn] *noun*
(a) something important to remember and
to think carefully about; **safety**
consideration = matter related to safety;
an important safety consideration; **to take**
into consideration = to remember to
include when thinking about something,
solving a problem or making a calculation
(b) thoughtfulness; respect; **to show**
consideration for other people and
property = to show respect for what
belongs to other people

consist [kən'sɪst] *verb* **to consist of** = to
be made up of; *layer cloud names consist*
of a prefix, according to height of base,
and a suffix according to shape; *a TAF*
(terminal area forecast) bulletin may
consist of forecasts for one or more
aerodromes; **to consist in** = to mean, to be

consistent [kən'sɪstənt] *adjective*
which always reacts or behaves in the same
way; *human hair responds in a consistent*
manner to changes in the relative
humidity; **consistent performance** =
performance which maintains a particular
standard

consolidate [kən'sɒlɪdeɪt] *verb* to
make more solid or strong; **revision of the**
subject helps to consolidate it = revision
of the subject helps to set it more firmly in
the memory

consolidation [kɒnsɒlɪ'deɪʃn] *noun*
(a) process by which something is made
more solid or strong; **revision aids**
consolidation of the subject matter =
revision of the subject helps to set it more
firmly in the memory **(b)** grouping goods
together for shipping

constant ['kɒnstənt] *adjective* which
does not change; **the temperature of the**
gas remains constant = the temperature of
the gas stays the same; **constant pressure**
= pressure which stays the same; **constant**
speed drive unit (CSDU) = device fitted
to aircraft with constant speed propellers;
constant speed propeller = propeller with
a control system which automatically
adjusts pitch to maintain selected rpm

constituent [kən'stɪtjuənt] *noun* part
which makes up a whole; *water, whether*
in the form of vapour, liquid or ice, is a
very important constituent of the
atmosphere

constitute ['kɒnstɪtjuːt] *verb* to make
up, to form; *oxygen and nitrogen together*
constitute most of the atmosphere

constrain [kən'streɪn] *verb* **(a)** to
prevent someone from being completely
free, from doing something they want to
do; *the airline was constrained in its*
purchase of new aircraft by lack of
financial resources **(b)** to force someone
to do something; *lack of financial*
resources constrained the airline to
cancel the purchase of new aircraft

constraint [kən'streɪnt] *noun* something
which prevents something else; *the*
number of landings per 24-hour period is
subject to constraint

constrict [kən'strɪkt] *verb* to make
something narrower; to make the flow of
gas or liquid more difficult by narrowing
the passage through which it flows; *in the*
carburettor venturi, the flow of air is
constricted

constriction [kən'strɪkʃn] *noun*
narrowing; *a thermometer has a*
constriction in the base of the tube
between the bulb and the beginning of the
scale

construct [kən'strʌkt] *verb* **(a)** to put
together; *the table on page 4 can be used*
to construct the low level forecast for the
route **(b)** to build; **to construct an aircraft**
= to manufacture or build an aircraft; *wings*
are constructed of light alloy pressed ribs
and an outer skin

construction [kən'strʌkʃn] *noun* **(a)**
way of putting things together; *the basic*
construction of the lead-acid cell consists
of a positive electrode and negative
electrode **(b)** building; *the construction of*
the home-built aircraft took two years

consume [kən'sjuːm] *verb* **(a)** *(fuel,*
electricity, etc.) to use up in a given time;
drag must be overcome with thrust, which

requires engines, which in turn consume fuel **(b)** *(food)* to eat

consumption [kən'sʌmpʃn] *noun* **(a)** *(of fuel, electricity, etc.)* (i) amount used up in a given time; (ii) process of using up fuel, etc.; *fuel consumption is higher in bigger, more powerful engines* **(b)** *(of food)* (i) amount eaten; (ii) the act of eating

contact ['kɒntækt] **1** *noun* **(a)** touch; in contact with = touching; *the air in contact with the earth's surface cools* **(b)** to be in contact with = to communicate with (by telephone, radio, etc.); **to be in visual contact** = to see; *(by radio)* **to make contact** = to communicate; **to lose contact** = to stop communicating; *ATC lost contact with the aircraft* **(c)** person who can be contacted in order to get something done; **I have a contact in Madrid who can help** = I know somebody in Madrid who can help; **contact number** = telephone number where information can be obtained **(d)** electrical connection; *dirty contacts were the cause of the problem*; **contact breaker** = the contact breaker assembly is a mechanically operated switch which is timed to break the primary circuit when maximum current is flowing **2** *verb* to get in touch with someone (by radio, telephone, etc.); *the captain couldn't contact ATC*

contain [kən'teɪn] *verb* to hold, to have inside; *most clouds contain some super-cooled water droplets*; *the booklet contains details of the airline's flight schedule*

container [kən'teɪnə] *noun* box, bottle, etc., which holds something else; *a smouldering fire in a waste container could become very active due to pressure changes during ascent*

contaminate [kən'tæmɪneɪt] *verb* to make something impure, harmful or dangerous; *if contaminated air enters the cabin, the dump valve can be opened*; **contaminated fuel** = fuel which contains an unwanted substance, such as water - such fuel shouldn't be used

contamination [kəmtæmɪ'neɪʃn] *noun* process by which a liquid, gas or object is made unusable because impurities or foreign matter are allowed into or onto it; **contamination of air** = pollution; **fuel contamination** = act or instance of fuel being made unusable because of an unwanted substance such as water in it; **nuclear contamination** = damage done to an object, person or substance because of contact with nuclear radiation

content ['kɒntənt] *noun* whatever is contained within something - often expressed as a percentage; *the stratosphere is a layer in which the water vapour content is low*; **the moisture content of the atmosphere** = the amount of water vapour in the air

continent ['kɒntɪnənt] *noun* one of the seven great land masses; *the continent of Europe*

COMMENT: the seven continents are: Asia, Africa, North America, South America, Australia, Europe and Antarctica

continental [kɒntɪ'nentl] *adjective* referring to a continent; **continental climate** = type of climate found in, or originating in areas where there is no effect from the sea; *a tropical continental climate*; *a polar continental climate*

contingency [kən'tɪnʒənsi] *noun* something which might happen in the future and therefore must be planned for; **contingency reserve fuel** = fuel which would only be used in an unusual situation such as a diversion

continuity [kɒntɪ'njuːəti] *noun* continuing; **continuity of precipitation** = continuing rain, snow or hail

contour ['kɒntʊə] *noun* shape of something; **contour charts** = charts which show high and low ground; **contour gradient** = steepness of change in elevation; **contour line** = line on a map or chart joining points of equal elevation

contract [kən'trækt] *verb* to become smaller in volume; *liquids will expand or* *contract as a result of temperature changes* (NOTE: the opposite is expand)

contraction [kən'trækʃn] *noun* the decrease in volume of a substance brought about by cooling; *due to contraction, the length of a mercury column shortens* (NOTE: the opposite is expansion)

contrast ['kɒntrɑːst] *noun* (a) the amount of light and dark in something seen; *contrast and colour enable a pilot to identify ground features* (b) difference between two things; *there is an enormous contrast between the performance of the two aircraft*; in contrast to = when compared with; *air at altitude is cold in contrast to air at the surface*

contribute [kən'trɪbjuːt] *verb* to give or provide as part of the whole; *exhaust gases contribute to engine power*; although the weather was bad, pilot error contributed to the accident = pilot error was partly responsible for the accident

contribution [kɒntrɪ'bjuːʃn] *noun* giving or providing as part of the whole; *the differences in the effect of solar radiation on land and sea make the biggest contribution to weather and climate*

contributor [kən'trɪbjutə] *noun* person or thing that contributes to something; *there are other factors which cause the division of the lower atmosphere into two layers but the ozone effect is a major contributor*

control [kən'trəʊl] 1 *noun* (a) directing; working in the correct way; control surfaces = moveable aerofoils, usually on the wings and tailplane, which can be operated from the cockpit by the pilot thus changing aircraft attitude; control zone = designated ATC area (b) controls = manual or automatic devices that are used to control a machine, a system, etc., or to make a machine, a system, etc., work in a correct way; *the pilot at the controls of the aircraft* = the pilot who is operating the flying controls; flying controls = the yoke or control column, rudder pedals, etc., used by the pilot in order to manoeuvre the

aircraft (c) crowd control = management of the movements of large numbers of people (d) checking or examining; passport control = (i) action of checking passports of people arriving in or leaving a country; (ii) place where passports are checked when people arrive in or leave a country; *we now have to go through passport control; at passport control, a customs official checks passports* 2 *verb* to make a machine, system, procedure, etc., work in the correct way; to direct; to manage; *the purpose of the centrifugal switch is to control the starting and ignition circuits* (NOTE: the word control in English is used in a different way to similar words in other languages. In English, the verb check is more often used to mean 'look at and verify' while control is used in the sense of 'to make something work in a certain way': the yoke and rudder pedals are used to control the movement of the aircraft. Note also: controlling - controlled)

controller [kən'trəʊlə] *noun* (a) device which ensures that something operates in the correct way; *the propeller speed controller*; sidestick controller = small side-mounted control column used on aircraft such as the Airbus A340 (b) person who manages systems to ensure the smooth operation of procedures; air traffic controller = person whose job it is to ensure correct separation of aircraft in all phases of flight; *air arrival controllers; air departure controllers*

convection [kən'vekʃn] *noun* process by which hot air rises and cool air descends; *heat is transferred from the earth's surface upwards largely by convection*

convective [kən'vektɪv] *adjective* referring to convection; (something) which is affected by the vertical circulation of air; convective clouds = clouds formed as a result of warm moist air rising and condensing at altitude; convective movement = movement caused by warm air rising and cool air descending

convenience [kən'viːnɪəns] *noun* (a) easily used service; *reading lights are* provided for passengers' convenience; **at your convenience** = when it is least troublesome for you **(b)** ease of understanding; *for convenience we will assume that the earth is round*

convenient [kən'viːniənt] *adjective* **(a)** useful; *the circular slide rule has a convenient scale for converting weights and volumes* **(b)** which is suitably free from problems; *we must arrange a convenient time and place for the meeting*

convention [kən'venʃn] *noun* **(a)** idea which, because of long usage has become normal and accepted; *by convention, wind direction is the direction from which the wind blows* **(b)** meeting involving large numbers of people and long discussions in order to arrive at an agreed course of action often outlined in a public statement; *the Tokyo Convention*

conventional [kən'venʃnl] *adjective* *(behaviour or practices)* which are familiar to most people; usual; *every pilot must know the conventional symbols used for depicting the various ground features on charts*

converge [kən'vɜːdʒ] *verb* to move towards a point; to come together at a certain point; to meet; *meridians converge towards the poles*; **aircraft on converging courses** = aircraft on courses which may eventually be too close to each other if no corrective action is taken (NOTE: the opposite is **diverge**)

convergence [kən'vɜːdʒəns] *noun* moving towards a point; coming together at a certain point; meeting; *the inter-tropical convergence zone is the zone in which the trade winds from the two hemispheres approach each other; there is convergence of meridians of longitude at the north and south poles* (NOTE: the opposite is **divergence**)

converse ['kɒnvɜːs] *noun* the opposite; *the converse of port is starboard*; *warm air rises - the converse is also true* = in other words, cool air descends

conversion [kən'vɜːʃn] *noun* **(a)** change to a different system or set of rules; *the conversion of km into nm is not difficult* **(b)** conversion course = flying training which enables and qualifies a pilot to fly a different aircraft type; **a tailwheel conversion course** = course which familiarizes qualified pilots with the differences in handling characteristics between nosewheel and tailwheel aircraft

convert [kən'vɜːt] *verb* to change to a different system or set of rules; *to convert km into nm; how do you convert degrees C into degrees F?*

converter [kən'vɜːtə] *noun* device which alters the form of something; *a backup converter converts the alternating current power into direct current*

convertible [kən'vɜːtəbl] *adjective* which can easily be changed (to fit in with a new system or set of standards); *the statute mile, unlike the nautical mile is not readily convertible into terms of angular measurements*

convey [kən'veɪ] *verb* to pass, to carry or to move from one place to another; *a large number of tubes convey the cooling medium through the matrix*; *buses are used to convey passengers from the aircraft to the terminal building*; **to convey information** = to pass information from one person to another, or from one place to another

cool [kuːl] **1** *adjective* a little cold; **cool weather** = weather which is not hot, warm nor very cold **2** *verb* to make less hot; to become less hot; *the airflow is used to cool the oil*; *see also* AIR

coolant ['kuːlənt] *noun* substance, usually liquid, used to cool something such as an engine; *radiator coolant*; *the coolant is sprayed into the combustion chamber inlet*

cooler ['kuːlə] *noun* device for cooling; *a self-contained system, consisting of an oil tank, pump, filter, cooler, and oil jets, lubricates the auxiliary power unit*

cooling ['ku:lɪŋ] **1** *noun* the action of making something cool; *the cooling of the oil by the airflow* **2** *adjective* which cools; **cooling medium** = substance which reduces the temperature of another substance or material

coordinate 1 [kəʊ'ɔ:dɪnət] *noun* **coordinates** = values used to locate a point on a graph or a map; *the airfield can be seen on the map at coordinates B:12* **2** [kəʊ'ɔ:dɪneɪt] *verb* **(a)** to bring together the various parts of a procedure or plan to ensure that the operation works correctly; *it is the task of air traffic controllers to coordinate the movement of traffic in and out of a terminal* **(b)** to make different parts of the body work well together; *during a hover, helicopter pilots must be able to coordinate movements of both hands and feet*

coordinated flight [kəʊ'ɔ:dɪneɪtɪd 'flaɪt] *noun* flight, especially during turns, in which the horizontal and vertical forces acting on the aircraft are in balance; *in coordinated flight, the ball in the turn coordinator will be in the centre*

COMMENT: the ball in the balance indicator of the turn coordinator shows the pilot if the aircraft is in coordinated flight or if it is slipping or skidding. When the ball moves to the left the pilot should apply left rudder pedal pressure, if the ball moves to the right, the pilot should apply right rudder pedal pressure

Coordinated Universal Time (UTC) [kəʊ'ɔ:dɪneɪtɪd ju:nɪ'vɜːsəl 'taɪm] *noun* time used in aviation based on the 24-hour clock format; GMT or Zulu time; *see also* GMT; ZULU TIME (NOTE: 7 P.M. is **1900 hours**, say: 'nineteen hundred hours')

coordination [kəʊɔ:dɪ'neɪʃn] *noun* **(a)** bringing together the various parts of a procedure or plan to ensure that the operation works correctly; *a rescue coordination centre is set up to control the emergency* **(b)** ability to use different parts of the body together well; *a pilot must have good hand/eye coordination*

cope [kəʊp] *verb* to manage to do something, often with some difficulty; *in heavy rainstorms, the windscreen wipers may not be able to cope; the aircraft structure must be able to cope with increased loads caused by turning movement*

copilot *or* **co-pilot** ['kəʊpaɪlət] *noun* licensed pilot (not a pilot under instruction) who is second in command to the captain of an aircraft; *the copilot landed the aircraft*

cord [kɔ:d] *noun (to reinforce tyres)* strong thread, usually of nylon; *tyres are of pure rubber and are either cord-strengthened or reinforced*

core [kɔ:] *noun* central part, heart of something; *the primary windings consist of heavy gauge wire mounted on a soft iron core*; **the core of a problem** = the central, most fundamental part of a problem

Coriolis force [kɒri'əʊlɪs 'fɔ:s] *noun* force which accelerates movement of a rotating mass perpendicular to its motion and towards the axis of rotation; *the Coriolis force explains why wind patterns are clockwise in the northern hemisphere and anti-clockwise in the southern hemisphere*

COMMENT: the Coriolis force acts at a right angle to wind direction and is directly proportional to wind speed. Named after G. G. Coriolis, a French engineer who died in 1843

correct [kə'rekt] **1** *adjective* which is right; **correct tyre pressure** = pressure at which the tyres should be maintained **2** *verb* **(a)** to adjust in order to make right; *a servo-motor fitted in the elevator trim system will automatically correct for loads; calibrated airspeed or rectified airspeed is indicated airspeed corrected for instrumentation and installation error* **(b)** to mark answers right or wrong as in an examination; *the instructor has corrected the students' examination papers*

correction [kə'rekʃn] *noun* **(a)** making something correct; adjusting or changing something to make it correct; **course or heading correction** = changing of heading in order to deal with a new situation **(b)** use of a mathematical formula for adjusting a known inaccuracy of calculation; *in applying this correction the reading is converted to that which would occur at mean sea level* **(c)** alteration on, for example a test answer, which provides the right answer in place of the wrong answer given; showing a mistake in a text, etc.; *I made several corrections to the text*

corrective [kə'rektɪv] *adjective* referring to something designed to correct; **corrective action** = action taken to put a situation right; *if the pilot realizes that he is too high on the approach, he should take corrective action immediately*

correlate ['kɒrəleɪt] *verb* to measure something against something else in order to form a relationship between the two; *power is measured not by the amount of work done, but by units of accomplishment correlated with time*

correlation [kɒrə'leɪʃn] *noun* measurable and predictable relationship; *at a given speed, there is a correlation between time and distance*

correspond [kɒrɪs'pɒnd] *verb* **(a)** to have a direct relationship with; to fit with; *movements of the control surfaces correspond to movements of the pilots flying controls* **(b)** to be similar to; *in the interests of passenger comfort, the ideal cabin conditions to maintain would be those corresponding to sea level*

corrode [kə'rəud] *verb* **(a)** to destroy by a slow chemical process such as rust; to cause corrosion; *turbine fuels tend to corrode the components of the fuel and combustion systems mainly as a result of the sulphur and water content of the fuel* **(b)** to suffer corrosion; *aluminium will not corrode easily*

corrosion [kə'rəuʒn] *noun* destruction of material by chemical process; *aluminium has a high resistance to corrosion*; *see also* ANTI-CORROSION

corrosive [kə'rəusɪv] *adjective* causing corrosion; *sulphuric acid is very corrosive*

counter- ['kauntə] *prefix meaning* against; **counter-rotating propellers** = propellers which turn in opposite directions (NOTE: also called **contra-rotating propellers**)

counter ['kauntə] *verb* to work against something to remove the effect of it; *for level flight, lift must counter the force of gravity*; *some people find that swallowing hard counters the effects of changes in pressure*

couple ['kʌpl] **1** *noun* two of something; **a couple of minutes** = two or three minutes **2** *verb* **(a)** to connect or to join (mechanically); *the auxiliary power unit is a self-contained unit which normally consists of a small gas turbine engine which is coupled to a gearbox* **(b)** to combine; to connect, to link; *pilot error, coupled with poor weather conditions, resulted in an accident*

coupling ['kʌplɪŋ] *noun* joining or connecting component; *when not in use, the coupling is sealed by a dust cap*

course [kɔːs] *noun* **(a)** imaginary line across the surface of the earth which must be followed in order to arrive at the destination; **to alter course** = to change route; to follow a different route **(b)** formal period of study; *a meteorology course* **(c)** continuing time; during; **in the course of the briefing** = during the briefing

cover ['kʌvə] **1** *verb* **(a)** to include (the complete extent of a period of time, a whole area, etc.); *the restriction covers the period from 4th-8th July*; **the area covered by the forecast** = the area which the forecast deals with **(b)** to deal with a subject (as in a text); *the subject of central warning systems is covered in the systems book* **(c)** to be completely over something so as to hide what is underneath; *the area is covered in snow* **2** *noun* something which goes over something else completely; **cloud cover** = the amount of cloud; **snow**

cover = situation in which there is a layer of snow on top of the earth so that the earth cannot be seen

coverage ['kʌvrɪdʒ] *noun* **(a)** amount of space or time given to a subject, an event, etc.; *more complete coverage of the one-in-sixty rule is given in the plotting section of these notes* **(b)** complete given area of radar signal transmission and reception; *glidepath coverage*; *localizer coverage*

cowl [kaʊl] *noun (for an installed engine)* covering usually made up of hinged or removable panels; **cowl flap** = removable or hinged panel of a cowl; *further cooling can be obtained by the use of controllable cowl flaps which regulate the amount of air flowing across the cylinders*

cowling ['kaʊlɪŋ] *noun (for an installed engine)* covering usually made up of hinged or removable panels; cowl; *access to the engine compartment is normally via hinged cowling panels*

CPL = COMMERCIAL PILOT'S LICENCE

craft [krɑːft] *noun* (i) boat, etc., for carrying people or goods on water; (ii) aeroplane (aircraft) or space-shuttle (spacecraft) for carrying people or goods in the air or in space; *an airship is classified as a lighter-than-air craft*

crankcase ['kræŋkkeɪs] *noun* part of the engine that houses the crankshaft and also usually the oil pump; *oil passages in the crankcase allow lubricating oil to pass through*

crankshaft ['kræŋkʃɑːft] *noun* part of a piston engine connecting the pistons, via the connecting rods, to the flywheel and gearbox; *rpm is the number of revolutions per minute that the engine crankshaft is making*

crash [kræʃ] **1** *noun* accident that causes damage **2** *verb* to have an accident that causes damage; to collide; *the aircraft crashed into the sea*

crash-land ['kræʃlænd] *verb (of an aircraft)* to land heavily without using the undercarriage, so that the aircraft is damaged; *the aircraft crash-landed short of the runway*

crash-landing ['kræʃlændɪŋ] *noun* act of landing an aircraft heavily sometimes without the undercarriage; *the crash-landing did not damage the aircraft as much as the pilot expected*

create [kriˈeɪt] *verb* to make, to produce; *the velocity and pressure of the exhaust gas create the thrust in the turbojet engine*

creep [kriːp] *noun* **(a)** weakening and slow damage to something; *creep is a particular feature of components which are subjected to operation at high temperatures* **(b)** slight movement of a tyre on a wheel caused by landing; *aligned white marks on the wheel and tyre indicate that there is no creep*

crest [krest] *noun* top of a mountain or wave; *wind speeds increase with height, the speed of the wind at the crest of a mountain or wave being the greatest*

crew [kruː] *noun* person or (usually) two or more people who have responsibility for flight operations; **cabin crew** = airline staff who are in direct contact with the passengers and whose in-flight responsibilities include: ensuring correct seating arrangements, serving food and attending to the general well-being of passengers, etc.; **flight crew** = crew responsible for flying the aircraft

criteria [kraɪˈtɪrɪə] *see* CRITERION

criterion [kraɪˈtɪrɪən] *noun* standard by which you define, decide or judge something; **the criterion for promotion is seniority** = senior staff will be promoted first; **VMC (visual meteorological conditions) criteria** = all the factors which define the limits of flying in visual meteorological conditions; *see also* VISUAL METEOROLOGICAL CONDITIONS (NOTE: the plural is **criteria**)

critical ['krɪtɪkl] *adjective* **(a)** extremely important, essential; *temperature and oil pressure are critical to any type of system* **(b)** (a point, angle, speed, etc.) at which an important change occurs; *as the angle of attack is increased, it reaches the critical point when the airflow over the upper surface of the wing begins to break down*

cross [krɒs] *verb* **(a)** to get from one side of an area to another; **to cross the Atlantic** = to go from one side of the Atlantic to the other **(b)** *(of lines, roads, etc.)* to go across each other at an angle; *meridians intersect at the poles and cross the equator at right angles*

cross-check ['krɒs'tʃek] **1** *noun (brief message from one crew member to another, as from pilot to cabin staff giving or confirming situation)* verification; making certain **2** *verb* to verify; to make certain; *cross-check doors closed and locked and escape slides armed*

cross-country ['krɒs'kʌntri] *noun* flight during which the student pilot must demonstrate navigation skills; *the qualifying cross-country flight for the PPL*

cross-section ['krɒs'sekʃn] *noun* view of an object seen as if cut through; *the diagram is a cross-section of a turbojet engine*

crosswind ['krɒswɪnd] *noun* surface wind which blows at an angle to the landing or take-off heading; *on some aircraft, crosswind take-offs should be made with full aileron deflection in the direction from which the wind is blowing*; *see also* LEG

COMMENT: a crosswind landing is one of the most difficult exercises for a student pilot. The final approach is usually made with the aircraft yawed into wind, while tracking the extended runway centreline. Just before touchdown, the pilot aligns the aircraft with the direction of flight using the rudder pedals. Correct timing for the alignment and accurate airspeed are required to achieve positive contact

with the runway surface otherwise the aircraft may depart the runway to one side

CRT = CATHODE RAY TUBE

cruise [kruːz] **1** *noun* main part of the flight between top of climb after take-off and descent for landing **2** *verb* to fly the main part of the flight between top of climb after take off and descent for landing; *we are cruising at 500 kt*; *cruising speed, cruising power and cruising altitude are selected to give maximum engine efficiency and prolong engine life*

cruising altitude ['kruːzɪŋ 'æltɪtjuːd] *noun* altitude at which most of a flight is flown en route to a destination, from top of climb to top of descent; *our cruising altitude will be 35,000 feet*

cruising power ['kruːzɪŋ 'paʊə] *noun* engine power used to give required speed from top of climb to top of descent usually giving fuel economy and long engine life; *cruising power is about 2,300 rpm*

cruising speed ['kruːzɪŋ 'spiːd] *noun* speed selected from top of climb to top of descent, usually giving fuel economy and long engine life; *the cruising speed is 110 knots*

crush ['krʌʃ] *verb* to damage by pressure; *excessive load on the beam may crush the core*

crystal ['krɪstl] *noun* regular geometric shape formed by minerals, or as water freezes; **ice crystal** = type of precipitation composed of crystals in the form of needles, plates or columns

CDSU = CONSTANT SPEED DRIVE UNIT

cubic ['kjuːbɪk] *adjective* measured in volume - by multiplying length, depth and width; **cubic centimetres (cc)** = usual unit of measure for the capacity of an engine; *the engine has a capacity of 2,000cc*

cumuliform ['kjuːməlɪfɔːm] *adjective (of clouds)* which develop vertically;

cumuliform clouds such as cumulonimbus

cumulonimbus (CB)
[kjuːmjʊləʊ'nɪmbəs] *noun* dark, low cumulus - type of cloud associated with thunderstorms; *a cumulonimbus has a characteristic anvil shape*

cumulus ['kjuːmjʊləs] *adjective* big, fluffy, white (or grey) cloud heaped or piled up, which develops at low altitude; *cumulus clouds may develop because of thermal activity resulting from the warming of the surface*; *grey cumulus often develop into cumulonimbus*; *see also* ALTOCUMULUS; STRATOCUMULUS

current ['kʌrənt] **1** *adjective* present; actual; happening at the moment; **current weather conditions** = present weather conditions; **current position** = position now **2** *noun* **(a)** electrical supply; *alternating current*; *direct current* **(b)** flow; **wind currents** = movement of air in a particular direction through a mass of air which is not moving so much (NOTE: in general English, the term 'current flow' can be understood in two ways - flow of something such as fuel at the present moment in time, or the flow of electricity in a circuit)

curvature ['kɜːvətʃə] *noun* curved shape; **curvature of the earth** = curving of the earth's surface due to the spherical form of the earth

customary ['kʌstəmri] *adjective* normal or usual (practice, etc.); *it is customary for the senior cabin supervisor to introduce herself to passengers at the start of a flight*

customs ['kʌstəms] *noun* official department of government concerned with movement of people and freight across national borders; **customs aerodrome** = aerodrome, usually near a border and/or coast, with customs facilities

cycle ['saɪkl] *noun* series of actions which end at the same point as they begin; *with the piston engine, the cycle is intermittent, whereas in the gas turbine, each process is continuous*; **life cycle of the thunderstorm cell** = process of formation, development and decay of a thunderstorm; **four-stroke (4-stroke) cycle** = induction, compression, power and exhaust phases in the operation of a four-stroke combustion engine

cyclic(al) ['sɪklɪkl] *adjective* referring to a cycle; which happens in cycle; *off-shore and on-shore wind patterns are cyclic*

cyclone ['saɪkləʊn] *noun* system of winds rotating inwards to an area of low barometric pressure; *these areas of low pressure are called hurricanes in the Atlantic Ocean, cyclones in the Indian Ocean and Bay of Bengal, and typhoons in the China Sea* (NOTE: also called a **low** or **depression**)

cyclonic [saɪ'klɒnɪk] *adjective* referring to air movement, which turns in the same direction as the earth and which, when seen from above, is anticlockwise in the northern hemisphere and clockwise in the southern hemisphere; *in winter the sub-tropical high retreats and gives way to cyclonic pressure patterns which produce cool unsettled conditions with rain at times*

cylinder ['sɪlɪndə] *noun* device shaped like a tube, in which a piston moves; *smaller aircraft have a static hydraulic system similar to a car, with a master cylinder and individual brake cylinders at each wheel*; **cylinder block** = casing containing the cylinders in a internal combustion engine; **cylinder head** = removable top part of a piston engine cylinder containing plugs, inlet and exhaust connections and valves

cylindrical [sɪ'lɪndrɪkl] *adjective* with the shape of a cylinder; like a cylinder; *the modern jet engine is basically cylindrical in shape*

Dd

DALR = DRY ADIABATIC LAPSE RATE

damage ['dæmɪdʒ] **1** *noun* harm that is caused to something; *if the temperature rises it can cause serious damage to the engine* **2** *verb* to cause harm to something; *small stones around the run-up area may damage propellers*

dampen ['dæmpn] *verb* **(a)** to decrease; to reduce; *an accumulator is fitted to store hydraulic fluid under pressure and dampen pressure fluctuations* **(b)** to make slightly wet

damper ['dæmpə] *noun* device to decrease or reduce something; *a yaw damper is used for rudder control*

data ['deɪtə] *noun* **(a)** information made up of numbers, characters and symbols often stored on a computer in such a way that it can be processed; *airspeed information is supplied from an air data computer*; **meteorological data** = information about weather conditions stored on a computer **(b)** information (used for plans, etc.); *see also* RECORDER

datum ['deɪtəm] *noun* reference or base point of a scale or measurement such as mean sea level; **reference datum** = line fixed by the designer from which measurements are made when checking or adjusting wing angles, etc.; **QNH datum** = barometric level from which altitude is measured

DC = DIRECT CURRENT

de- [diː] *prefix meaning* undo; remove; stop; *deactivate; depressurize*

deactivate [diˈæktɪveɪt] *verb* to turn off a system or a piece of equipment thus stopping it being ready to operate; to stop a process; *on some aircraft nose wheel steering must be deactivated prior to retraction*

dead reckoning *or* **ded reckoning** ['ded 'rekənɪŋ] *noun* navigation using calculations based on airspeed, course, heading, wind direction and speed, ground speed, and time; *in the early stages of practical navigation, the student pilot navigates by using dead reckoning* (NOTE: the term comes from 'deduced' reckoning or 'ded' reckoning)

de-aerate [diˈeərəeɪt] *verb* to remove gas - especially carbon dioxide or air - from a liquid such as fuel; *the pump helps to de-aerate the fuel before it enters the engine*

de-aeration [dieəˈreɪʃn] *noun* removing gas from a liquid such as fuel; *partial de-aeration of fuel takes place in the pump*

de-aerator [dieəˈreɪtə] *noun* device to remove gas from a liquid; **de-aerator tray** = device in the lubrication system to remove air bubbles from oil

deal [diːl] **1** *noun* **a great deal** = a large amount of, a lot of; *a great deal of damage was done to the aircraft as a result of the fire* **2** *verb* to handle; to manage; *a computer can deal with the constant inputs required to control an unstable aircraft*

debris ['debriː] *noun* (scattered) broken pieces; *before running up the engine,* *check that the aircraft is on firm ground and that the area is free of stones and other debris; the aircraft exploded in mid-air, spreading debris over a wide area of the countryside*

decal [dɪˈkæl] *noun* peel-off sticker; picture, letters, digits printed on special paper, which is transferred onto a surface; *a red decal with AVGAS 100LL in white letters indicates the type of fuel to be used*

decelerate [diːˈseləreɪt] *verb* to slow down; *reverse thrust and brakes help to decelerate the aircraft after landing* (NOTE: the opposite is **accelerate**)

deceleration [diːseləˈreɪʃn] *noun* slowing down; *anti-skid braking systems units are designed to prevent the brakes locking the wheels during landing, thus reducing the possibility of wheel skid caused by the sudden deceleration of the wheel* (NOTE: the opposite is **acceleration**)

decimal [ˈdesəml] **1** *noun* decimal fraction **2** *adjective* **decimal fraction** = a fraction as expressed in the decimal system; *0.50 is a decimal fraction that is equal to 1/2*; **decimal notation** = method of writing a number in the decimal system; *the fraction 3/4 can be written as 0.75 in decimal notation*; *prices and number are normally written using decimal notation; he finds it difficult to understand how the computer works because it uses binary not decimal notation*; **decimal place** = the position of a number to the right of the decimal point; **correct to three places of decimal** *or* **to three decimal places** = correct to three figures after the decimal point; *2.754 is correct to three decimal places, 2.7 is correct to one decimal place*; **decimal point** = the dot (.) used to separate a whole number from a decimal fraction; **decimal system** = system of counting based on the number 10 and using the digits 0 - 9

COMMENT: the decimal point is used in the USA and Britain. In most European countries a comma (,) is used to show the decimal, so 4,75% in Germany is written 4.75% in Britain

decision [dɪˈsɪʒn] *noun* act of deciding (something) or of making up one's mind (about something); **to make a decision** = to choose a course of action; *the decision to evacuate the aircraft was made by the captain*

decision height (DH) [dɪˈsɪʒn ˈhaɪt] *noun* altitude at which, during an ILS landing approach, a pilot must decide whether to land or carry out a missed approach; *the pilot waited until he was at decision height before initiating the missed approach procedure*

COMMENT: a ILS approach generally has a DH of 200 ft (60 m) above ground level

deck [dek] *noun* floor of a ship or aircraft; **flight deck** = place where the flight crew of an airliner sit while flying the aircraft; **flight deck instruments** = instruments used by the flight crew when flying an aircraft

decode [diːˈkəʊd] *verb* to change coded information into readable form; *incorrectly spaced information pulses can result in failure by the ground station to decode the aircraft information*

decoder [diːˈkəʊdə] *noun* device used to decode signals from the air traffic control radar beacon system; *the aircraft receiver is set to the required frequency and linked to a selective call system decoder which has a 4-letter code*

decrease 1 [ˈdiːkriːs] *noun* lessening; reduction; *a decrease in power results in the aircraft descending* **2** [diːˈkriːs] *verb* to become less, to fall; *air density and pressure decrease with an increase in altitude* (NOTE: the opposite is **increase**)

deduce [dɪˈdjuːs] *verb* to reach a conclusion, to work out in the mind using information provided; *sometimes, it is possible to estimate the depth of the layer of mist or fog from the ground observations and hence to deduce the ground range from any height*

defect [ˈdiːfekt] *noun* fault, error; *low oil pressure or excessive temperature* *indicate the development of a possible defect*

defective [dɪ'fektɪv] *adjective* referring to something which is faulty or which is not operating correctly; *loss of supply pressure is caused by either a defective booster pump or lack of fuel*

define [dɪ'faɪn] *verb* (a) to give an exact explanation (as in a dictionary); **it is not easy to define the word** = it is difficult to say exactly what the word means (b) to set the limits of (something); *cloud tops are very difficult to define*

definite ['defɪnət] *adjective* referring to something which is not in doubt, which is certain; *using a time scale on the track, the pilot should be prepared to look for a definite feature at a definite time* (NOTE: the opposite is **indefinite**)

definition [defɪ'nɪʃn] *noun* exact explanation of what a word or expression means; *the definition of a year is the time taken for a planet to describe one orbit around the sun*; **by definition** = understood by the use of the word itself; *a sphere is, by definition, round*

deflate [diː'fleɪt] *verb (tyre, balloon, etc.)* to allow air to escape from something (so that it becomes smaller or collapses); **to deflate a tyre** = to remove or depress the valve and allow the air to escape (NOTE: the opposite is **inflate**)

deflation [diː'fleɪʃn] *noun (of tyre, balloon, etc.)* the act or instance of allowing air to escape from something (so that it becomes smaller or collapses); *deflation of a tyre is done by depressing the valve*

deflect [dɪ'flekt] *verb* (a) to move away from a neutral or central position; *during an out-of-balance turn, the ball in the slip indicator will be deflected to the left or right* (b) to move a moving object, gas or liquid away from its intended path; *in an open-cockpit aircraft, the windshield deflects the airflow over the pilot's head*

deflection [dɪ'flekʃn] *noun* (a) movement away from a central or neutral position; *full deflection of the ailerons is sometimes needed on take-off to counteract a crosswind* (b) movement of a moving object, gas or liquid away from its intended path; *in the southern hemisphere the deflection of wind at the equator is to the left*

deformation [diːfɔː'meɪʃn] *noun* change of correct shape caused by stress; *deformation of wing panels may be an indication of serious structural damage*

deg = DEGREE

degradation [degrə'deɪʃn] *noun (of radio signal, radar return, etc.)* decrease in quality; *degradation of the radio signal sometimes makes it impossible to understand the message*

degrade [dɪ'greɪd] *verb* to decrease the quality of something; *interfering signals degrade VOR performance*

degree [dɪ'griː] *noun* (a) level; amount, quantity; **the degree of compression** = the amount of compression; **a high degree of safety** = a high level of safety; **to a greater degree** = more than; **to a lesser degree** = less than (b) unit of temperature (°); *twenty degrees Celsius (20°C); twenty degrees Centigrade (20°C); seventy degrees Fahrenheit (70°F)* (c) unit of measurement of an angle equal to 1/360th of a circle (°) - each degree is divided into 60 minutes and each minutes into 60 seconds; *make a turn to the right at a bank angle of 30°*; **an angle of 90°** = a right angle (d) unit of direction as measured on a compass (°); *east = 090°; west = 270°; see also* TRUE

DH = DECISION HEIGHT

dehydration [diːhaɪ'dreɪʃn] *noun* (unwanted and sometimes dangerous) loss of water from the body; *dehydration can be avoided by drinking plenty of water*

de-ice [diː'aɪs] *verb* to remove ice; *the ground crew de-iced the aircraft prior to take-off*

de-icer [diːˈaɪsə] *noun* device or substance used to remove ice; *de-icer spray should be checked to make sure it is not harmful to light aircraft windscreens*

de-icing [diːˈaɪsɪŋ] **1** *noun* removing ice; removal of ice **2** *adjective* referring to the removal of ice; *de-icing fluid*; *see also* ANTI-ICING; ICING

delay [dɪˈleɪ] **1** *noun* being made late; length of time something is late; *by day, the presence of cloud can cause a delay in clearance of fog* **2** *verb* (a) to make late, to cause to be late; *take-off was delayed because of fog* (b) to put something off until later; *he delayed telling her the news until they had landed*

delayed-action [dɪˈleɪd ˈækʃn] *adjective (of some type of device or situation)* in which there is an unusual passing of time between stimulus and response; *the door is fitted with a delayed-action lock which operates one minute after the power has been switched off*

deliver [dɪˈlɪvə] *verb* to provide; to give; *the motor will continue to run but will deliver only one-third the rated power*; *the pump can deliver fuel at the rate of 2,000 gph*

delivery [dɪˈlɪvri] *noun* providing; giving; *on some pumps, a depressurizing valve is used to block delivery to the system*; **delivery pressure** = pressure normally expected when fuel is being pumped

deluge [ˈdeljuːdʒ] *noun see* FIRE

demand [dɪˈmɑːnd] **1** *noun* (a) need or use caused by necessity; **high current demand on a generator** = situation requiring the generator to produce a lot of electricity (b) something which is asked firmly; **on demand** = when asked for or ordered; *a computer will produce, on demand, a flight plan giving the optimum route, levels and fuel* **2** *verb* (a) to require as a necessity; *higher operating weights of modern aircraft demand an increase in the number of wheels fitted to the landing*

gear (b) to ask firmly; *he demanded an explanation*

demonstrate [ˈdemənstreɪt] *verb* to show by clear example or explanation; *Torricelli first demonstrated that the atmosphere has weight*; *it will be demonstrated in chapter 12 that turbulence is associated with strong winds*

demonstration [demənˈstreɪʃn] *noun* clear, often visual, description or explanation; *your instructor will give a demonstration of the stall-recovery technique*

dense [dens] *adjective* (a) referring to a substance which is closely compacted; **dense fog** = thick fog (b) referring to the amount of mass of a substance for a given unit of volume; *air which contains water vapour is less dense than air which does not*

density [ˈdensɪti] *noun* quantity of mass for a given unit of volume; *air density*; **density altitude** = pressure altitude corrected for non-ISA temperature; **relative density** = the ratio of density of a liquid with reference to water, or a gas with reference to air

COMMENT: density altitude is a very important factor in calculating aircraft performance because of the effect on engine performance, time to reach takeoff speed (and therefore length of take-off run) and rate of climb

depart [dɪˈpɑːt] *verb* to leave; *the flight departs at 0200 GMT* (NOTE: the opposite is **arrive**)

department [dɪˈpɑːtmənt] *noun* separate part of a complex whole especially in an organization; **operations department** = that part of an airline or airport organization which deals with flight operations

departure [dɪˈpɑːtʃə] *noun* (a) the act of leaving; **departure time** = time when an aircraft becomes airborne; **departure point** = exact place of departure marked on a chart; *see also* ESTIMATED; LOUNGE (b) **departures** = part of an airport that deals with passengers who are leaving **(c)** distance between two meridians at any given latitude (NOTE: the opposite for (a) is **arrival**, and for (b) is **arrivals**)

depend [dɪˈpend] *verb* **(a)** to be controlled or affected entirely by something; *whether or not an object can be seen by aircrew at a given distance will depend on factors such as size, shape and colour of the object*; *if an aircraft ditches in the sea, early rescue depends on rapid location of survivors* **(b)** to rely on; *pilots depend on air traffic controllers to help them conduct a safe flight*

dependable [dɪˈpendəbl] *adjective* reliable, trustworthy; *mercury barometers have largely been replaced by precision aneroid barometers which are smaller, simpler to use, and more dependable*

dependent [dɪˈpendənt] *adjective* unable to do without; which relies on something; *the height indicated by an altimeter is dependent on the pressure which is set on the sub-scale*

deploy [dɪˈplɔɪ] *verb* to come into action; to become ready to be used; *slide rafts are door-mounted and automatically deploy and inflate when the door is opened in the armed position*

deposit [dɪˈpɒzɪt] *noun* layer of collected matter on a surface; *a deposit of ice crystals causes the aircraft surfaces to change their aerodynamic characteristics*; *wheel brakes should be inspected for snow or ice deposits*

depreciate [dɪˈpriːʃɪeɪt] *verb* to decrease in value; *the aircraft depreciated by 100% over the 5 year period* (NOTE: the opposite is **appreciate**)

depreciation [dɪpriːʃɪˈeɪʃn] *noun* decrease in value; *there was a depreciation of 100% in the value of the aircraft over the 5 year period* (NOTE: the opposite is **appreciation**)

depress [dɪˈpres] *verb (button, switch, etc.)* to push down; *switches on the control columns instantly disengage the autopilot when depressed*

depression [dɪˈpreʃn] *noun* **(a)** area of low atmospheric pressure; *in the northern hemisphere, the wind blows anticlockwise round a depression and clockwise round an anticyclone and vice versa in the southern hemisphere*; **deep depression** = area of very low relative atmospheric pressure; **frontal depression** = area of low pressure found together with a weather front; **shallow depression** = area of slightly low relative atmospheric pressure **(b)** lower area on a surface (often difficult to see); *a depression on the wing surface must be investigated in case it is an indication of more serious structural damage*

depressurization [diːpreʃəraɪˈzeɪʃn] *noun* loss, especially sudden, of cabin pressure; *emergency oxygen must be available in the event of depressurization* (NOTE: also written **depressurisation**)

depressurize [diːˈpreʃəraɪz] *verb* to lose pressure suddenly; to cause to lose pressure; *the aircraft began to depressurize at 20,000 feet* (NOTE: also written **depressurise**)

depth [depθ] *noun* distance from the top surface of a layer, etc., to the bottom; *the troposphere's depth is variable in temperate latitudes*; **depth of the sea** = distance from the surface to the sea bed

derive [dɪˈraɪv] *verb* to get or to obtain; *performance data is derived from flight tests*; *Kepler derived the laws which relate to the motion of planets in their orbits*

descend [dɪˈsend] *verb (of aircraft)* to lose altitude, usually in a planned manoeuvre; **the aircraft descended to 10,000 feet** = the pilot reduced altitude until the aircraft was at 10,000 feet (NOTE: the opposite is **climb** *or* **ascend**)

descent [dɪˈsent] *noun (of aircraft)* planned loss of altitude; *the descent from cruise altitude took 40 minutes*; **in the descent** = during planned loss of altitude, usually in preparation for landing; **rate of descent** = speed of descent measured in feet per minute (NOTE: the opposite is **climb** *or* **ascent**)

QUOTE: a search of radar recordings showed that a DC-10 had tracked within a few hundred metres of the house while passing 9,500 feet in the descent to Gatwick

Pilot

describe [dɪ'skraɪb] *verb* **(a)** to give the particular features of something; **to describe what happened** = put into words exactly what happened **(b)** to draw or to make a geometric figure; *the definition of a year is the time taken for a planet to describe one orbit around the sun*; **to describe an arc** = to draw an arc

description [dɪ'skrɪpʃn] *noun* **(a)** giving the particular features of something; *a detailed description of world climate* **(b)** drawing or making of a geometric figure; **the description of a triangle** = the drawing of a triangle (NOTE: a description of a triangle could be understood in two ways - the drawing of a triangle or the giving of the particular features of a triangle, i.e. a three-sided figure whose angles total 180⁰)

desert ['dezət] *noun* large area of dry often sandy country; *over desert areas the lack of water vapour produces cold nights*

design [dɪ'zaɪn] **1** *noun* plan or drawing of something before it is made; *the design and testing of aircraft are important stages in the development programme* **2** *verb* to draw plans using accurate information in preparation for constructing something; **to design an aircraft** = to have the idea, make drawings, calculate data, etc., with the intention of producing an aircraft

designate ['dezɪgneɪt] *verb* to particularize; to choose for a special purpose; *this region is designated as a fire zone*

designator ['dezɪgneɪtə] *noun* group of letters and/or numbers which identify something; *a runway visual range group always includes the prefix R followed by the runway designator*

designer [dɪ'zaɪnə] *noun* person who has the idea for, and makes plans to produce something; *Rutan is a designer of unusual-looking aircraft*

QUOTE: test-pilot's tip for a safe first flight - take the designer with you

Flight International 9-15 Oct. 1996

desirable [dɪ'zaɪrəbl] *adjective* preferred; wanted; *equalization of the air pressure across the eardrum is more difficult to achieve during descents than ascents, and a minimum rate of pressure change is desirable*

despite [dɪ'spaɪt] *preposition* in spite of; although; *many beacons and aids which are provided for low operations are left out to keep the chart clear - despite this, the charts still look very difficult to understand*; **despite the weather, we took off** = although the weather was bad, we took off

DEST = DESTINATION

destination [destɪ'neɪʃn] *noun* place where someone or something is going; *aerodrome forecasts are normally given in code form for destination and alternates*

destroy [dɪ'strɔɪ] *verb* to damage so much as to make useless; *the aircraft was destroyed in the accident*

destruction [dɪ'strʌkʃn] *noun* act or instance of making completely useless by breaking; *by testing selected parts to destruction, a safe life can be assessed for all structures and components*

destructive [dɪ'strʌktɪv] *adjective* referring to something which destroy; **the winds of a tornado are extremely destructive** = tornadoes cause a lot of serious damage

detach [dɪ'tætʃ] *verb* to unfix and remove; to become unfixed; *a fuselage panel became detached and had to be replaced*; *the parachute flare is a device which is fired to a height of 1,200 ft where a red flare and parachute detach*

detachable [dɪ'tætʃəbl] *adjective* referring to something which can be unfixed and removed; **detachable wheel spats** = streamlined coverings for the wheels of light aircraft which can be taken off to allow inspection and repair of tyres

detail ['diːteɪl] *noun* the important and less important facts about something; *the amount of detail which appears on a topographical chart depends upon the scale*

detect [dɪ'tekt] *verb* to discover the presence of something; *apart from sensing the abnormal rate of descent of a false glide slope, the pilot can detect an error by comparing height with distance to go*

detection [dɪ'tekʃn] *noun* discovery of the presence of something; **fire detection system** = system to detect the presence of fire (in an aircraft); **ultrasonic detection** = method using high frequency sound to check metal components for internal weaknesses; *see also* FLAW

detector [dɪ'tektə] *noun* device for discovering the existence of something; **ice detector** = device for detecting the presence of ice on the airframe; *when ice forms on the vibrating rod ice detector head, the probe frequency decreases*

deteriorate [dɪ'tɪəriəreɪt] *verb* to make bad or worse; to become bad or worse; *the electrolyte in the cells of a nickel-cadmium battery does not chemically react with the plates and so the plates do not deteriorate*; **deteriorating weather** = worsening weather

deterioration [dɪtɪəriə'reɪʃn] *noun* worsening; **a deterioration in the situation** = a worsening of the situation

determination [dɪtɜːmɪ'neɪʃn] *noun* (a) finding out by calculation; *structure design for a given safe life has led to the determination of the minimum number of flying hours which should pass before major failure occurs* (b) strength of mind to do what is required; *determination was a major factor in the trainee passing his exams*

determine [dɪ'tɜːmɪn] *verb* (a) to find out by calculation; *to determine the average age, divide the total number of years by the number of people; when we wish to fly from one place to another, it is first necessary to determine the direction of the destination from the departure point* (b) to set or to fix precisely; *on a large transport aircraft, the safety of hundreds of passengers is involved, and regulations determine the minimum crew that must be carried*

detonation [detə'neɪʃn] *noun (in the cylinders of a piston engine)* sudden, explosive burning of the air/fuel mixture; *prior to the accident, engine detonation could be heard by people on the ground*

COMMENT: detonation imposes excessive loads on the pistons and other engine components possibly causing engine damage and resulting engine failure

develop [dɪ'veləp] *verb* (a) to come into being; *carburettor icing may develop in any type of carburettor in relatively warm air temperatures; vertical motion and therefore turbulence suggest that thunderstorms may develop* (b) to get bigger; to grow and change; to grow in number; *during the day, light breezes may develop into strong winds*

development [dɪ'veləpmənt] *noun* (a) something new, made as an improvement on something older; *satellite navigation aids for light aircraft are a recent development* (b) growth and change; *to study weather and its development, the meteorologist has to be aware of the horizontal changes in atmospheric pressure both in space and time*

deviate ['diːvieɪt] *verb* to move away from the normal position or path; *if the aircraft deviates beyond the normal ILS glide slope, the flight crew are alerted*

deviation [diːvi'eɪʃn] *noun* (a) moving away from the normal position or path; *on final approach, any deviation from the extended centreline of the runway should be corrected immediately* (b) magnetic compass error in a particular aircraft caused by magnetic influences in the structure and equipment of the aircraft itself; *deviation is not a constant value but varies from one aircraft to another*

device [dɪ'vaɪs] *noun* object, especially mechanical or electrical, which has been made for a particular purpose; *a capacitor is a device with the ability to temporarily store an electric charge*

dew [dju:] *noun* drops of condensed moisture left on the ground, etc., overnight in cool places; **dew point** = temperature at which air is saturated with water vapour and condensation begins

> COMMENT: weather reports usually include the air temperature and dew point temperature. When the difference between temperature and dew point is small, there is a strong possibility of fog, clouds, or precipitation

DFDR = DIGITAL FLIGHT DATA RECORDER

DI = DIRECTION INDICATOR

diagonal [daɪ'ægənl] **1** *adjective* **(a)** *(of a line)* joining two opposite corners of a rectangle **(b)** *(of line, plane or structure)* sloping halfway between the vertical and horizontal; *early aircraft were of the wire braced type of construction, the wire being superseded by tubular diagonal struts* **2** *noun* line joining two opposite corners of a rectangle; line sloping halfway between the vertical and horizontal

diagram ['daɪəgræm] *noun* often simplified drawing showing the structure or workings of something; *the diagram shows a simple open-circuit system*

diagrammatic [daɪəgrə'mætɪk] *adjective* referring to something which is shown as a drawing of a system or structure; **diagrammatic format** = in the form of a diagram

dial ['daɪəl] *noun* face of an instrument showing a scale; *a cup anemometer is connected to an instrument with a dial showing wind speed in knots*

diameter [daɪ'æmɪtə] *noun* distance from one side of a circle to the other and passing through the centre; **equatorial diameter** = distance from the equator, through the centre of the earth to the equator on the opposite side of the globe; **polar diameter** = distance from one pole, passing through the centre of the earth, to the other pole; *the earth's polar diameter is shorter than its average equatorial diameter*

diaphragm ['daɪəfræm] *noun* thin sheet of material used to separate parts or chambers; *some switches are operated by a diaphragm which flexes under fluid or air pressure*

differ ['dɪfə] *verb* to be unlike; *track and heading differ by the amount of drift; because the chart time and the departure/arrival times differ, it is necessary to consider the movement of any weather system which might affect the route*

differential [dɪfə'renʃl] *adjective* referring to things which react differently when measured against a norm or standard; **differential heating of the atmosphere** = heating of the atmosphere to varying temperatures depending on the relative warmth of the land at the equator and the poles

differentiate [dɪfə'renʃɪeɪt] *verb* to recognize the difference between two things; to show two things to be different; *some types of colour blindness make the sufferer unable to differentiate between blue and red*

diffraction [dɪ'frækʃn] *noun* breaking down of a beam of radiation; *diffraction produces a surface wave which follows the curvature of the earth*

diffuse [dɪ'fju:s] **1** *adjective* spread out in every direction; *glare caused by diffuse reflection of sunlight from the top of a layer of fog or haze can seriously reduce air-to-ground visibility* **2** *verb* to spread out in every direction; *light diffuses as it passes through fog*

diffusion [dɪ'fjuːʒn] *noun* spreading out; *gas from the turbine enters the exhaust system at high velocities but, because of high friction losses, the speed of flow is decreased by diffusion*

digit ['dɪdʒɪt] *noun* any number from 0 to 9; *information is provided in a four-digit group*

digital ['dɪdʒɪtl] *adjective* referring to a system or device which uses signals or information in the form of numbers; **digital display** = information shown as numbers; *the clock uses a digital display to show the time of 12:33; a digital display barometer uses solid-state electronics to measure barometric pressure*

diluted [daɪ'luːtɪd] *adjective* referring to the decreased strength or concentration (as of a liquid by adding water, etc.); *spillage from a lead acid battery may be neutralized by washing with a diluted solution of sodium bicarbonate*

diluter [daɪ'luːtə] *noun* device for decreasing the strength or concentration of a liquid or gas; *most flight decks use the diluter demand system in which the oxygen is diluted with cabin air*

dimension [daɪ'menʃn] *noun* measurable distance such as height, length, etc.; measurement of height, length, etc.; *variations of atmospheric pressure produce changes in the dimension of the capsule chamber*

diminish [dɪ'mɪnɪʃ] *verb* to decrease or to reduce in size or importance; *friction is greatest near the ground and diminishes with height; at higher altitudes, ground objects are less easily seen because of diminished size*

diode ['daɪəʊd] *noun* electronic component that allows an electrical current to pass in one direction and not the other

dioxide [daɪ'ɒksaɪd] *noun* oxide containing two atoms of oxygen; *see also* CARBON DIOXIDE

direct ['daɪrekt] **1** *adjective* **(a)** in a straight line; by the shortest route; *air temperatures are taken in such a way as to be representative of the air temperature near the surface but unaffected by direct surface heating or cooling* **(b)** exact; complete; *the direct opposite* **2** *verb* to guide or control the movement of something; *clamshell doors are hydraulically or pneumatically opened, and direct the exhaust gases forwards to produce reverse thrust*

direct current (DC) ['daɪrekt 'kʌrənt] *noun* electric current flowing in one direction only; *an electric starter is usually a direct current electric motor, coupled to the engine which automatically disengages after the engine starts*

direction [dɪ'rekʃn] *noun* course taken by someone or something; *the earth rotates about its own axis in an anticlockwise direction*; **wind direction** = description of where the wind is blowing from, given as north, south, east, west, etc., or a number of degrees, eg a wind coming from the west would be a wind direction of 270°; *wind direction and speed only affect the movement of the aircraft over the ground*

directional [dɪ'rekʃnl] *adjective* referring to the course taken by someone or something; **directional radar beam** = signal from a directional beacon enabling the pilot to determine a bearing from the beacon with a communications receiver; *see also* NON-DIRECTIONAL

directional gyro [dɪ'rekʃnl 'dʒaɪrəʊ] *noun* free gyroscopic instrument which indicates direction; *the directional gyro should be set to correspond with the magnetic compass*; *see also* HEADING INDICATOR

directive [dɪ'rektɪv] **1** *adjective* referring to the ability of a device to send or receive signals in straight lines; *the antenna is highly directive in transmission and reception* **2** *noun* general or detailed instructions from management to staff to guide them in their work;

according to the management directive, all late arrivals should be logged

director [daɪˈrektə] *noun* **(a)** device with a central controlling function; *EFIS is a highly sophisticated type of flight director system* **(b)** person on the board of a company; *managing director*

disadvantage [dɪsədˈvɑːntɪdʒ] *noun* drawback; unwanted situation or condition; factor which makes someone or something less likely to succeed; *the disadvantage of a booster pump is that the output is constant so that when engine demand is high, fuel pressure tends to be low and vice versa*

disadvantaged [dɪsədˈvɑːntɪdʒd] *adjective* physically disadvantaged (person) = such as a person who needs a wheelchair, etc.

COMMENT: the word 'disadvantaged' may be regarded by some people as a 'politically correct' term for 'disabled'. With the help of specially-adapted controls, more and more disabled people are learning to fly

disappear [dɪsəˈpɪə] *verb* **(a)** to vanish; *if air blew at right angles to isobars, the horizontal pressure differences would eventually disappear* **(b)** pass out of sight; *the aircraft took off, climbed out and soon disappeared from view*

disarm [dɪsˈɑːm] *verb* **(a)** to switch off an active or live system; *on the ground approaching the terminal, the flight deck will instruct the cabin crew to disarm the escape devices* **(b)** to forcibly remove a weapon from somebody; *the hijacker was disarmed by security forces*

disc [dɪsk] *noun* circular flat plate; *a turbine consists of a disc on which is mounted a number of blades*

discharge [dɪsˈtʃɑːdʒ] **1** *noun* *(electrical)* release of power from a source (such as a battery); *a lightning flash is a large-scale example of an electrical spark, or discharge*; battery discharge = loss or release of electrical supply from a battery **2** *verb* to release electrical supply from a

source (such as a battery); *the battery discharged overnight*

disconnect [dɪskəˈnekt] *verb* to separate two things attached to one another; to pull a plug out of an electric socket; *the electrical supply can be disconnected by pulling out the plug*

discrimination [dɪskrɪmɪˈneɪʃn] *noun* the ability to know or see the difference between two (similar) things; *targets on the same bearing which are separated radially by less than half a pulse length distance will appear at the receiver as one echo, so good target discrimination requires short pulses*

discuss [dɪsˈkʌs] *verb* to write about or talk about a subject; *this chapter will discuss HF and VHF voice communications*

disembark [dɪsɪmˈbɑːk] *verb* to leave the aircraft after landing; *the passengers finally disembarked at 20.00 hours*

disembarkation [dɪsembɑːˈkeɪʃn] *noun* the act of leaving the aircraft after landing; *the exits are used as conventional doors for disembarkation*

disengage [dɪsɪnˈgeɪdʒ] *verb* to switch off a system or device; *switches on the control columns instantly disengage the autopilot when depressed*

dish [dɪʃ] *noun* shallow container for food; dish antenna = circular antenna with a shape like a shallow bowl

disintegration [dɪsɪntɪˈgreɪʃn] *noun* the falling apart or destruction of something; *electromagnetic radiations resulting from the disintegration of radioactive materials are known as gamma rays*

dismantle [dɪsˈmæntl] *verb* to take apart into single components; *one type of inspection is able to reveal fatigue cracks, corrosion, internal damage, the presence of loose articles and mercury spillage without the need to dismantle the aircraft* (NOTE: the opposite is **assemble** or **reassemble**; the verb 'mantle' is not used)

disorientation [dɪsɔ:rɪən'teɪʃn] *noun* state of confusion in which there is loss of understanding of where one is or which direction one is facing, etc.; *when the cabin is rapidly and completely filled by smoke and fumes passengers will suffer from disorientation*; **spatial disorientation** = situation of bad visibility and/or unusual manoeuvres which result in the pilot not knowing what attitude the aircraft is in

dispensation [dɪspən'seɪʃn] *noun* permission not to have to do something; *at very high altitudes the flying pilot must be on oxygen at all times, unless an aircraft dispensation has been obtained*

dispense (with) [dɪ'spens] *verb* not to include or not to use something; *in some cases the rivets are dispensed with and the skin is fixed to the internal members by the redux process*

dispersal [dɪs'pɜ:səl] *noun* (a) leaving an area and going in different directions; **the dispersal of a crowd** = the disappearance of a crowd (b) clearing away; *the dispersal of hill fog; dispersal of cloud takes place when surface heating lifts the cloud base or drier air is advected*

disperse [dɪs'pɜ:s] *verb* (a) to leave an area going in different directions; **the crowd dispersed** = the people in the crowd left the area, going in different directions, so that eventually the crowd disappeared (b) to clear away; *the fluorescent green dye will disperse slowly in a calm sea but quickly in a moderate to rough sea*

displace [dɪs'pleɪs] *verb* to move out of the normal position; *the atmosphere is said to be stable if, when a parcel of air is displaced vertically, it tends to return to its original level*

displacement [dɪs'pleɪsmənt] *noun* movement away from the normal position; *the ILS is a cross-pointer indicator which shows the aircraft horizontal displacement from the localizer and vertical displacement from the glide path*

display [dɪ'spleɪ] 1 *noun* (a) appearance of information on a monitor screen or on the panel of an instrument or of an indicator; *there are three different types of electronic display systems: EFIS, EICAS and ECAM*; **digital display** = information shown as numbers; *the clock uses a digital display to show the time of 12:33*; **liquid crystal display (LCD)** = liquid crystals that turn black when a voltage is applied, used in many watches, calculators and digital displays (b) show; demonstration; **aerobatic display** = demonstration, often public, of piloting skill and aircraft performance 2 *verb* to show (on a panel, on a screen, etc.); *alerting and warning information is displayed*

disseminate [dɪ'semɪneɪt] *verb* to send out; to spread; *meteorological stations make routine weather observations at fixed intervals and disseminate this information locally*

dissimilar [dɪ'sɪmɪlə] *adjective* referring to something which is not the same as or is unlike something else; *differential expansion switches operate on the principle that the coefficients of expansion of dissimilar metals are different*

dissipate ['dɪsɪpeɪt] *verb* to (cause to) spread out and lose power or strength; *tropical storms often dissipate as they pass from sea to land*

dissipation [dɪsɪ'peɪʃn] *noun* spreading out with loss of power or strength; *the rubber used on nose or tail wheels is usually constructed to form a good electrical conductor for the safe dissipation of static electricity*

dissolve [dɪ'zɒlv] *verb* to become or to cause to become part of a liquid and form a solution; *sugar dissolves in water; there is a possibility that in some types of accumulator, gas may be dissolved into the fluid and thus introduced into the system*

dissolved [dɪ'zɒlvd] *adjective* which has become part of a liquid and forms a solution; **dissolved water** = water in solution in fuel

distance ['dɪstəns] *noun* space between two places, two points, etc.; the measurement of such a space; *the distance from point A to point B is 100 nm; the distance from point A to point B on the diagram is 2 cm; the height of the aircraft is the vertical distance, measured in feet, of the aircraft above the surface of the earth*

distance measuring equipment (DME) ['dɪstəns 'meʒrɪŋ ɪ'kwɪpmənt] *noun* airborne secondary radar whose signal is converted into distance; *it is quite common to find a VOR located together with DME (Distance Measuring Equipment) to give simultaneous range and bearing from the same point on the ground*

COMMENT: DME equipment is usually located in a VOR station. Other equipment in the aircraft transmits a signal to the VOR station, which replies. The equipment in the aircraft converts the signal into distance and also calculates ground speed and time to reach the station

distillation [dɪstɪ'leɪʃn] *noun* the process by which a liquid is heated, the resulting vapour being then condensed and collected; *with kerosine-type fuels, the volatility is controlled by distillation*

distinct [dɪ'stɪŋkt] *adjective* clear and easily seen or understood; *when a lead-acid battery is fully charged, each cell displays three distinct indications*

distinction [dɪ'stɪŋkʃn] *noun* point of difference; something which makes one thing different from another; *a clear distinction is made between showers and general precipitation*

distinctive [dɪ'stɪŋktɪv] *adjective* easily recognized because of particular features or characteristics; *Concorde is a very distinctive-looking aeroplane*

distinguish [dɪ'stɪŋgwɪʃ] *verb* to know or to see the difference between things; *a receiver antenna would be unable to distinguish between signals unless they had some differing characteristics*

distinguishable [dɪ'stɪŋgwɪʃəbl] *adjective* easily recognized as different from; *useful ground features must be easily distinguishable from their surroundings*

distort [dɪ'stɔːt] *verb* (a) to put out of shape; *stress could cause the body to distort or change its shape* (b) to produce a bad radio signal; *the sound of the transmission is distorted if the volume is set too high*

distortion [dɪ'stɔːʃn] *noun* (a) putting out of shape; *difficulty in closing a door may be caused by distortion of the airframe* (b) electrical distortion; *distortion of the signal made it difficult for the controller to understand what the pilot said*

distress [dɪ'stres] *noun* (a) serious danger; serious difficulty; **distress signal** = signal transmitted by an aircraft in imminent danger; **International Distress/Calling frequency** = 2182 kHz or 500 kHz (b) personal worry or anxiety; *some passengers were in distress after the incident*

distribute [dɪ'strɪbjuːt] *verb* (a) to give; to send out; *there are two basic configurations which are used to distribute electrical power, the parallel system and the split bus system* (b) to spread over a wide area; *multiple wheel undercarriage units distribute the weight of the aircraft*

distribution [dɪstrɪ'bjuːʃn] *noun* (a) giving; sending out; *parallel AC and DC power distribution systems are found on commercial aircraft containing three or more engines* (b) act or instance of spreading over a wide area; *there is a high distribution of used and disused airfields in the south of England*

distributor [dɪ'strɪbjʊtə] *noun* device which sends an electrical charge to each spark plug in turn; *the distributor directs* the high voltage impulses to the cylinders in turn as they reach their ignition point

disturb [dɪ'stɜːb] *verb* to upset the normal condition of something; *small hills can disturb the flow of air*

disturbance [dɪ'stɜːbəns] *noun* upsetting the normal condition of something; *in general, the higher the mountain and the faster the air flow the greater is the resulting disturbance*

ditch [dɪtʃ] *verb* to land a plane in the sea, in an emergency; *even though aircraft have ditched successfully, lives have been lost because life rafts were not launched in time*

ditching ['dɪtʃɪŋ] *noun* landing a plane in the sea, in an emergency; *after all four engines stopped, the captain had to seriously consider the possibility of a ditching in the Indian Ocean*

diurnal [daɪ'ɜːnl] *adjective* referring to the 24-hour cycle of day and night; *diurnal changes in surface temperature over the sea are small*

dive [daɪv] **1** *noun* steep nose-down attitude of an aircraft; **to pull out of** *or* **from a dive** = to return the aircraft to level flight after a nose-down flight path; *during manoeuvring of an aircraft, when banking, turning and pulling out from a dive, stresses on the airframe are increased* **2** *verb* to put the aircraft into a steep nose-down attitude; *the aircraft dived to avoid the other aircraft* (NOTE: diving - dived)

diverge [daɪ'vɜːdʒ] *verb* to move further apart from something else; *air diverges at low levels and converges at high levels, causing a sinking or subsiding effect in the atmosphere* (NOTE: the opposite is converge)

divergence [daɪ'vɜːdʒəns] *noun* moving apart; *divergence of air at high levels leads to rising air at low levels with a consequent pressure fall* (NOTE: the opposite is convergence)

divergent [daɪ'vɜːdʒənt] *adjective* referring to something which moves further apart from something else; **divergent duct** = duct which has an inlet area which is smaller than the outlet area (NOTE: the opposite is convergent)

diversion [daɪ'vɜːʃn] *noun* change in route or destination caused by bad weather, technical problem, etc.; *the diversion of the flight to another airport was due to fog*

divert [daɪ'vɜːt] *verb* to turn away from a course or a destination; *an automatic cut-out valve is fitted to divert pump output to the reservoir when pressure has built up to normal operating pressure*; *the aircraft was diverted to Manchester airport because of fog*

divide [dɪ'vaɪd] *verb* **(a)** to separate into parts; *air masses are divided into two types according to source region and these are known as polar and tropical air masses* **(b)** to calculate how many times a number is contained in another number; *eight divided by four equals two (8 ÷ 4 = 2)*

division [dɪ'vɪʒn] *noun* **(a)** separation into parts; **the division of the lower atmosphere** = separation of the atmosphere into its component layers **(b)** calculation of how may times a number is contained in another number; *the division sign is ÷*

DME = DISTANCE MEASURING EQUIPMENT

document ['dɒkjumənt] *noun* memo, letter, report, etc.; *the flight crew route flight plan is a composite document which also serves as a navigation log*

documentation [dɒkjumen'teɪʃn] *noun* collection of letters, memos, reports, etc.; *flight crews are provided with a full meteorological briefing, backed by documentation, a short time before ETD*

domestic [də'mestɪk] *adjective* referring or belonging to inside a country; *domestic flights usually leave from Terminal 1*

dominant ['dɒmɪnənt] *adjective* main or most influential; *both pressure and temperature decrease with height but the pressure change is the dominant one and so, as pressure decreases with height, so does density*

dominate ['dɒmɪneɪt] *verb* to have the most effect or influence on; *because the chart time and the departure/arrival times differ, it is necessary to consider the movement of any weather system which will dominate the route*

Doppler ['dɒplə] *noun* **Doppler radar** = radar which can distinguish between fixed and moving targets or provide groundspeed and track information from an airborne installation; **Doppler VOR** = an adaptation of VOR to reduce errors caused by location (NOTE: named after **C.J. Doppler**, Austrian physicist)

dot [dɒt] *noun* small circular mark on paper; *the highest point in a locality is marked by a dot with the elevation marked alongside*

down draught ['daʊn 'drɑːft] *noun* (i) cool air which flows downwards as a rainstorm approaches; (ii) air which flows rapidly down the lee side of a building, mountain, etc. (NOTE: also written **down draft** in American English)

down draft ['daʊn 'drɑːft] *noun* US = DOWN DRAUGHT

downstream [daʊn'striːm] *adverb (of fuel, etc.)* in the direction of flow; further along the line of flow (than); *internally driven superchargers are generally used on medium and high powered engines and are fitted downstream of the throttle valve*

downward ['daʊnwəd] **1** *adjective* moving to a lower level; *when flying in turbulent air conditions, an aircraft is subjected to upward and downward gust loads* **2** *adverb* US = DOWNWARDS (NOTE: the opposite is **upward**)

downwards ['daʊnwədz] *adverb* to a lower level; towards the bottom; *pull the toggles downwards to inflate the lifejacket*

(NOTE: the opposite is **upwards**; in American English, **downward** is used as an adverb and as an adjective)

downwind [daʊn'wɪnd] **1** *adjective* in the same direction as the wind is blowing; **the downwind leg** = part of the airfield traffic circuit which runs parallel to, but in the opposite direction to, the approach to land which is made into wind **2** *adverb* **turn downwind** = turn the aircraft so that it is flying in the same direction as the wind is blowing (NOTE: the opposite is **upwind**)

draft ['drɑːft] *US noun* local current of air; *a down draft or an updraft* (NOTE: **draught** is preferred in British English)

drag [dræg] *noun* resistance of the air created by moving the aircraft through the air; *to reduce the effect of drag on an aircraft by the fixed undercarriage a retractable type was introduced; if an engine failure occurs, the windmilling propeller may cause considerable drag*

> COMMENT: there are two basic types of drag called parasite drag and induced drag. Parasite drag is caused by friction between the air and the aircraft surface, aerials, landing gear, etc. Induced drag is produced by lift

drain [dreɪn] **1** *noun* device to allow fluid to escape from its container; *when the cabin is pressurized the drains close, preventing loss of pressure* **2** *verb* to allow fluid to escape by providing a hole or tube, etc., through which it can pass; *the moisture drains in the lower skin of the cabin are open when the cabin is unpressurized, allowing moisture to drain*

drainage ['dreɪnɪdʒ] *noun* **(a)** act or instance of allowing a fluid to escape from its container; *drainage of water from the fuel system should be carried out before the first flight of the day* **(b)** system of outlets for fluid, such as water or fuel, to pass out of a closed area

draught [drɑːft] *noun* local current of air; *a down draught or an updraught* (NOTE: written **draft** in American English)

draw [drɔ:] *verb* **(a)** to make a picture (as with a pencil, on paper, etc.); *because there is a temperature gradient across each front it is possible to draw isotherms which reduce in value from warm to cold air* **(b)** to pull; to take; *fluid is drawn into the pump body* **(c)** to pull towards oneself (NOTE: **drawing - drew - drawn**)

drift [drɪft] **1** *noun* movement away from the desired course, created by wind blowing at an angle to the intended direction of flight; *if the wind direction is not the same as the aircraft track or its reciprocal, then the aircraft will experience drift* **2** *verb* to move away from the desired course; *when landing, a cross-wind from the right will cause the aircraft to drift to the left*

drill [drɪl] *noun* **(a)** short series of actions carried out in a particular sequence; *the starting drill varies between different aircraft types and a starting check procedure is normally used* **(b)** tool, often electrically powered, for making holes in metal, wood, etc.

drive [draɪv] **1** *noun* devices concerned with transmitting power to wheels, propellers, etc.; *rotation of the engine for starting is done by an electric starter motor connected to a drive shaft in the accessories gearbox* **2** *verb* **(a)** to make something move or turn; **belt-driven generator** = generator whose pulley is turned by a belt attached to an engine-driven pulley; **shaft-driven** = using a rotating shaft as a means of transmitting power from one part to another (eg from a turbine engine to a helicopter rotor); *see also* **ENGINE (b)** *(vehicle)* to control and guide; *he's learning to drive* (NOTE: **driving - drove - driven**)

driven [ˈdrɪvn] *see* **DRIVE**

drizzle [ˈdrɪzl] *noun* precipitation (often persistent) in the form of very small drops of water; *drizzle is the lightest form of precipitation consisting of fine water droplets*

COMMENT: in weather reports and forecasts, drizzle is abbreviated to DZ

drop [drɒp] **1** *noun* **(a)** a small amount of liquid that falls; *a drop of water; a few drops of rain* **(b)** (sudden) lowering; *the passage of a cold front is usually followed by a drop in temperature; a sudden drop in oil pressure is normally an indication of serious engine trouble* **2** *verb* to become lower; to decrease (suddenly); *the temperature dropped by several degrees*

droplet [ˈdrɒplət] *noun* small drop of liquid; *experiments show that smaller droplets of rain can remain super cooled to much lower temperatures than large droplets*

drove [drəʊv] *see* **DRIVE**

drum [drʌm] *noun* cylindrical device often with closed ends; **brake drum** = part of the brake mechanism attached to the wheel and against which the brake shoes rub, thus preventing the wheel from turning; *see also* **EARDRUM**

dry [draɪ] *adjective* containing no water or no moisture; not wet; not moist; *dry air*; *see also* **LAPSE RATE**

dual [ˈdjuːəl] *adjective* double; in pair; *most light aircraft with side-by-side seating have dual controls*

duct [dʌkt] *noun* channel or tube through which fluids or cables, etc., can pass; *the modern jet engine is basically a duct into which the necessary parts are fitted*

due [djuː] **1** *adjective* **(a)** expected to arrive; **the flight is due at 10 o'clock** = the flight should arrive at 10 o'clock **(b)** **due to** = because of; *due to daytime heating, the stability decreases and the wind speed increases* **2** *adverb* exactly and directly; *the aircraft flew due east*

dump [dʌmp] *verb* to offload quickly; *normal operating cabin pressure can be reduced rapidly in the event of emergency landings, by dumping air; the aircraft flew out to sea in order to dump fuel before landing*

duplication [djuːplɪˈkeɪʃn] *noun* copying; doubling; *control surfaces are divided into sections operated by a separate control unit, thus providing duplication to guard against failure of a unit*

durability [djʊərəˈbɪlɪti] *noun* ability of a substance or device to last a long time; *high quality components have good durability*

duration [djuˈreɪʃn] *noun* length of time for which something continues; *the duration of the examination is two hours*; *the duration of the flight was three hours*

dust [dʌst] *noun* fine powdery substance blown by the wind and found on surfaces; *solid particles in the air include dust, sand, volcanic ash and atmospheric pollution*

duty [ˈdjuːti] *noun* **(a)** period of work; **on duty** = at work; **off duty** = not at work **(b)** customs duty *or* import duty *or* duty = payment made to a government on certain goods imported or exported; *the duty payable on a carton of cigarettes*; *see also* HEAVY-DUTY

dye [daɪ] *noun* colouring used to change the colour of something; *minute surface cracks which are difficult to detect by visual means may be highlighted by using penetrant dyes*

dynamic [daɪˈnæmɪk] *adjective* referring to something in motion; **dynamic pressure** = pressure created by the forward movement of the aircraft; *if the dynamic pressure increases due to an increase in forward speed, the force required to move the control column will increase* (NOTE: the opposite is **static pressure**)

DZ = DRIZZLE

Ee

E = EAST

ear ['ɪə] *noun* hearing organ

ear defenders ['ɪə dɪ'fendəz] *noun see* EAR MUFFS

eardrum ['ɪədrʌm] *noun* membrane inside the ear which vibrates with sound and passes the vibrations to the inner ear; *equalization of the air pressure across the eardrum is more difficult to achieve during descents than ascents*

ear muffs ['ɪəmʌfs] *noun (personal equipment)* acoustic ear muffs = coverings to protect the ears from loud noise; ear protectors, ear defenders

ear protectors ['ɪə prə'tektəz] *noun see* EAR MUFFS

earth [ɜːθ] **1** *noun* **(a) (the planet)** Earth = the planet where we live **(b)** ground; soil **2** *verb* to connect an electrical appliance to a position of zero potential; *when refuelling a light aircraft, ensure that the aircraft is properly earthed* (NOTE: the American expression is **to ground**)

east [iːst] **1** *noun* **(a)** compass point on the mariner's compass 90° clockwise from due north and directly opposite west; *London is east of New York* **(b)** direction in which the earth rotates; direction of the rising sun **2** *adjective* **(a)** referring to areas or regions lying in the east; *the east coast of Canada* **(b)** east wind = wind blowing from or coming from the east (NOTE: a wind is named after the direction it comes from) **(c)** eastern part of a region; *East Africa* **3** *adverb* towards the east; *the aircraft was flying east*

eastbound ['iːstbaʊnd] *adjective* travelling towards the east; *an eastbound flight*

easterly ['iːstəli] **1** *adjective* **(a)** situated towards the east **(b)** *(wind)* easterly component = one part of the wind direction coming from the east; **easterly wind** = wind blowing from or coming from the east **(c)** *(direction)* to move in an easterly direction = to move towards the east **2** *noun* wind which blows from the east

eastern ['iːstən] *adjective* situated in the east; *one of the eastern provinces of Canada*

Eastern Standard Time (EST) ['iːstən 'stændəd 'taɪm] *noun* time zone of the eastern USA and Canada, 5 hours behind GMT (Greenwich Mean Time)

eastward ['iːstwəd] **1** *adjective* going towards the east; **eastward flight** = flight heading towards the east **2** *adverb US* = EASTWARDS

eastwards ['iːstwədz] *adverb* towards the east; *flying eastwards or westwards for long periods of time affects sleep patterns*

ECAM = ELECTRONIC CENTRALIZED AIRCRAFT MONITOR

echo ['ekəʊ] *noun* **(a)** repetition of a sound by reflection of sound waves from a surface **(b)** return of a signal back to the source from which it was transmitted; *the strength of the returning echo from a radar transmission depends on a number of factors*

economic [iːkəˈnɒmɪk] *adjective* financially rewarding; *it was no longer economic to keep the maintenance operation going*

economical [iːkəˈnɒmɪkl] *adjective* referring to a substance or device for which input is minimized and output maximized (thereby saving costs); **economical engine** = engine which uses relatively less fuel to produce the same or more power as comparable engines; *jet engines are more efficient and economical when operated at high altitudes*

ECS = ENVIRONMENTAL CONTROL SYSTEM

eddy [ˈedi] *noun* current of air moving in the opposite direction to the main current, especially in a circular motion; *when wind flows over an obstruction such as a building, an eddy is formed on the lee, or downwind side*

edge [edʒ] *noun* line of intersection or joining of two surfaces; **leading edge** = forward part of an aerofoil which meets the oncoming airstream; **trailing edge** = aft part of an aerofoil

effect [ɪˈfekt] **1** *noun* **(a)** something which happens because of a cause; result; influence; *ultra-violet radiation has the effect of warming the atmosphere*; *pressure patterns have an effect on weather*; **a marked effect** = a noticeable effect **(b)** the condition of being in full force; **in effect** = which is in operation; **to take effect** *or* **to come into effect** = to start to operate; *a new regulation comes into effect tomorrow*; **with effect from** = starting from **2** *verb* to make; to carry out; **to effect a change** = to make a change; **modifications were effected** = modifications were carried out (NOTE: compare **affect**)

effective [ɪˈfektɪv] *adjective* **(a)** having an expected and satisfactory result; **the new cleaning fluid was very effective** = it cleaned well **(b)** operative; in effect; *the regulation is effective immediately*

effectiveness [ɪˈfektɪvnəs] *noun* how well something works; *ice covering reduces the effectiveness of an aerial*

efficiency [ɪˈfɪʃənsi] *noun* **(a)** comparison of the effective or useful output to the total input in any system; being able to act or produce something with a minimum of waste, expense, or unnecessary effort; *efficiency is a key component of a successful business* **(b)** ratio of the energy delivered by a machine to the energy supplied for its operation; *mechanical efficiency; propeller efficiency*; **thermal efficiency** = the efficiency of conversion of fuel energy to kinetic energy

efficient [ɪˈfɪʃnt] *adjective* able to act or produce something with a minimum of waste, expense, or unnecessary effort; *at certain speeds and altitudes the pure jet engine is less efficient than a piston engine*; **efficient combustion** = combustion in which fuel energy is used to its maximum capability

effort [efət] *noun* **(a)** use of physical or mental energy to do something; *in order to qualify for a licence, it is necessary to put some effort into the training course*; *flying a high performance aerobatic light aircraft to its limits requires a lot of physical effort on the part of the pilot* **(b)** *(physics)* force applied against inertia; *actuators are capable of exerting low-speed turning effort*

EFIS = ELECTRONIC FLIGHT INSTRUMENT SYSTEM

EICAS = ENGINE INDICATING AND CREW ALERTING SYSTEM

eject [ɪˈdʒekt] *verb* to throw out forcefully; *on depressurization the oxygen mask is ejected automatically from the service panel*

ejection [ɪˈdʒekʃn] *noun* action of throwing out forcefully; *ejection seat*; *see also* EJECTOR SEAT

ejector [ɪˈdʒektə] *noun* **(a)** device to throw out forcefully; **ejector seat** *or* **ejection seat** = emergency escape seat in military aircraft which is fired out of the aircraft while the crew-member is still in it **(b)** device using a jet of water, air, or steam to withdraw a fluid or gas from a space; *a jet transfer pump or fuel ejector is used to transfer fuel*

elapse [ɪˈlæps] *verb (of time)* to pass; *the radio altimeter works on the principle that, if the path followed by the radio wave is straight down and up, then the elapsed time between the outgoing and incoming signal is a function of the aircraft's height*

elastic [ɪˈlæstɪk] *adjective* easily returning to original shape after being stretched or expanded; flexible; *at low values of stress, if the plot of stress and strain is a straight line, this indicates that the material is elastic within this range*

elasticity [ɪlæˈstɪsɪti] *noun* the property of returning to an original form or state following deformation; *titanium falls between aluminium and stainless steel in terms of elasticity, density and elevated temperature strength*

electric [ɪˈlektrɪk] *adjective* powered or worked by electricity; **electric current =** mass movement of electric charge in a conductor

electrical [ɪˈlektrɪkl] *adjective* **(a)** referring to electricity; *an electrical fault* **(b)** powered or worked by electricity; *activation may be mechanical or electrical*

electricity [ɪlekˈtrɪsɪti] *noun* electric current used to provide light, heat, power

electro- [ɪˈlektrəʊ] *prefix meaning* electricity

electrode [ɪˈlektrəʊd] *noun* solid electrical conductor through which an electric current enters or leaves an electrolytic cell; *a battery has a positive and a negative electrode*

electrolyte [ɪˈlektrəlaɪt] *noun* chemical compound that becomes conductive when dissolved or molten; *the electrolyte in a lead-acid battery consists of sulphuric acid diluted with distilled water*

electrolytic [ɪlektrəˈlɪtɪk] *adjective* **electrolytic cell =** cell consisting of electrodes in an electrolyte solution

electro-magnet [ɪˈlektrəʊˈmægnɪt] *noun* magnet consisting of a coil of insulated wire wrapped around a soft iron core that is magnetized only when current flows through the wire

electro-magnetism [ɪlektrəʊˈmægnɪtɪzm] *noun* force exerted by a magnetic field found around any conductor carrying current, the strength of which will depend on the amount of current flow

electromotive force (emf) [ɪˈlektrəʊˈməʊtɪv ˈfɔːs] *noun* source of electrical energy required to produce an electric current, produced by devices such as batteries or generators and measured in volts

electron [ɪˈlektrɒn] *noun* sub-atomic particle - negative particle of an atom; *electrons in the outer orbits of an atom may not be strongly attracted to the nucleus and may be lost*

electronic [elekˈtrɒnɪk] *adjective* referring to, based on, operated by, or involving the controlled conduction of electrons especially in a vacuum, gas, or semi-conducting material; *lightning does not often seriously damage aircraft but it may affect sensitive electronic equipment*; **electronic centralized aircraft monitor (ECAM) =** display on two CRTs (Cathode Ray Tubes) giving pilots engine and systems information; **electronic flight instrument system (EFIS) =** primary flight and navigation information on a CRT (Cathode Ray Tube)

COMMENT: the EFIS can show basic flight information, engine performance information, moving maps, checklists, etc.

element [ˈelɪmənt] *noun* **(a)** *(chemistry)* a substance composed of atoms with an identical number of protons in each nucleus; *elements cannot be reduced to simpler substances by normal chemical* methods (b) *(electricity)* the resistance wire (coil) in an electrical device such as a heater (c) removable component or removable part (such as in an air filter or oil filter); **filter element** = filter cartridge; removable paper or metal component in a filter housing (which must be replaced periodically); *from time to time the filter element must be removed and cleaned or replaced*

elevate ['elɪveɪt] *verb* to move something to a higher place or position from a lower one; to lift; *in some light aircraft the magnetic compass is elevated to a position as far away from the interfering effect of other components as possible*

elevated ['elɪveɪtɪd] *adjective* **elevated temperature** = increased or raised temperature

elevation [elɪ'veɪʃn] *noun* height at which something is, above a point of reference such as the ground or sea level; *the highest point in a locality is marked by a dot with the elevation marked alongside*; **aerodrome elevation** = distance in feet of the aerodrome above sea level; *elevation is indicated on charts by means of contour lines, spot heights, etc.*

elevator ['elɪveɪtə] *noun* (a) *(aerofoil)* movable control surface, usually attached to the horizontal stabilizer of an aircraft, used to produce the nose up/down motion of an aircraft in level flight known as pitch; *elevators should be checked for full and free movement immediately prior to take-off* (b) *US* = LIFT (b)

COMMENT: some aircraft have an all-moving tailplane called a 'stabilator' (a combination of the words stabilizer and elevator)

eliminate [ɪ'lɪmɪneɪt] *verb* to get rid of; to remove; *air dryers are provided to eliminate the possibility of ice forming*; *to eliminate the need for complex mechanical linkage, the selector is operated electrically*; **to eliminate a danger** = to remove a danger

ellipse [ɪ'lɪps] *noun* oval-shaped line; *each planet moves in an ellipse and the sun is at one of the foci*

elliptical [ɪ'lɪptɪkl] *adjective* having an oval shape; *the elliptical path of the Earth around the sun*

ELR = ENVIRONMENTAL LAPSE RATE; *(ICAO)* EXTRA LONG RANGE

embarkation [embɑː'keɪʃn] *noun* act of going onto an aircraft; *embarkation will start in ten minutes*; **embarkation time** = time when passengers will be asked to go onto the aircraft (NOTE: **boarding** is usually preferred)

embed [ɪm'bed] *verb* to fix firmly in a surrounding mass; *a temperature probe is embedded into the stator of the generator*; *water outlets have heater elements embedded in rubber seals in the outlet pipe*

emergency [ɪ'mɜːdʒənsi] *noun* serious situation that happens unexpectedly and demands immediate action; *to deal with an emergency or to handle an emergency*; **medical emergency** = situation when someone is unwell and quickly needs medical care; **emergency descent** = planned rapid losing of altitude because of a serious situation; **emergency equipment** = devices for use only in serious situations; **emergency frequency** = 121.5 MHz - frequency on which aeronautical emergency radio calls are made; **emergency landing** = landing made as a result of an in-flight emergency; **emergency procedures** = set of actions pre-planned and followed in the event of a serious situation; **emergency services** = fire, ambulance and police services; *the alarm will activate the emergency services*; **emergency situation** = serious situation which demands immediate action

emf = ELECTROMOTIVE FORCE

emission [ɪ'mɪʃn] *noun* (a) sending out (matter, energy, signals, etc.); *light emissions; radio emission*; *one factor on which the operational range of a radio emission depends is the transmitted power*

(b) substance discharged into the air, as by an internal combustion engine; *exhaust emissions contain pollutants*

emit [ɪ'mɪt] *verb* to send out (matter, energy, etc.); *radiation emitted by the sun; an X-ray tube emits radiation; latent heat is emitted when condensation takes place* (NOTE: **emitting - emitted**)

empennage [em'penɪdʒ] *noun* tail assembly of an aircraft; *the empennage usually includes the fin, rudder, horizontal stabilizer (or tailplane), and elevator*

emphasis ['emfəsɪs] *noun* force of expression that gives importance to something; *it is only in recent years that much emphasis has been placed on determining the causes of metal fatigue*

emphasize ['emfəsaɪz] *verb* to give importance to something; *on some maps, different elevations are emphasized by colouring* (NOTE: also written **emphasise**)

employ [ɪm'plɔɪ] *verb* **(a)** to use; *there are two methods employed to cool the cylinders down; in some aircraft, particularly those employing nickel-cadmium batteries, temperature sensing devices are located within the batteries to provide a warning of high battery temperatures* **(b)** to give someone regular paid work

enable [ɪ'neɪbl] *verb* to make it possible (for someone to do something); to facilitate; *isolation valves are fitted to enable servicing and maintenance to be carried out*

enclose [ɪn'kləʊz] *verb* to surround on all sides; to close in; *the housing encloses the various mechanical parts; fuses form a weak link in a circuit and are usually made of a strip of tinned copper enclosed in a glass tube*

encode [ɪn'kəʊd] *verb* to put into code; *weather information is encoded to allow large amounts of information to be given in a short space of time*

encounter [ɪn'kaʊntə] *verb* to meet something unexpected or unwanted; *severe icing can be encountered in wave cloud*

endurance [ɪn'djʊrəns] *noun* length of time an aircraft can stay in the air without refuelling; *the flight time to the PNR and back will equal the endurance of the aircraft*

energy ['enədʒi] *noun* **(a)** the ability of a physical system to do work **(b)** power from electricity, petrol, heat, etc.; *the engine converts heat energy into mechanical energy; the generator converts mechanical energy into electrical energy;* **kinetic energy** = energy of motion

engage [ɪn'geɪdʒ] *verb* **(a)** to switch on and use; *the autopilot may be engaged during climb or descent* (NOTE: the opposite is **disengage**) **(b) engaged in** = working on a particular job or task; *personnel engaged in ground running must ensure that any detachable clothing is securely fastened and they should wear acoustic ear muffs*

engine *noun* ['endʒɪn] machine that converts energy into mechanical force or motion, different from an electric or hydraulic motor because of its use of a fuel; *jet engine; piston engine; internal combustion engine; see* COMBUSTION; JET; PISTON; **aero-engine** = engine designed for use in an aircraft; **engine capacity** = swept volume of engine; **engine compartment** = space in the airframe where the engine is located; **engine-driven** = referring to equipment and devices which take their power from the engine when it is running; *engine-driven generator; engine-driven pump;* **engine failure** = unwanted stoppage of an engine during running; **engine indicating and crew alerting system (EICAS)** = cockpit display for monitoring engines and warning of malfunction; **engine instruments** = instruments which give the pilot information about engine temperature, speed, etc.; **engine intake** = front part of the engine where air enters the engine; **engine malfunction** = situation in which the engine does not work as it should;

engine performance = description of how well the engine works or detailed statistical information about the capabilities of the engine; **engine running** = engine operating or working; **the engine is running** = the engine is working; *the accident investigation demonstrated that the engine was running at full power when the aircraft hit the ground*; **engine run-up** = testing of piston engine at high power, in a light aircraft, just before take-off

COMMENT: in British usage, there is a clear distinction between the terms engine and motor, the term motor only being used for electric power units, but in American usage it is used for all types of power unit including internal combustion

engineer [endʒɪ'nɪə] *noun* **aeronautical engineer** = engineer who specializes in the design of aircraft; **aircraft engineer** = engineer who specializes in the maintenance and repair of aircraft

engineering [endʒɪ'nɪərɪŋ] *noun* use of scientific and mathematical principles to practical ends such as the design, manufacture, and operation of machines, and systems, etc.; **aeronautical engineering** = science or study of the design of aircraft; **aircraft engineering** = branch of aviation concerned with the maintenance and repair of aircraft; *reinforced plastics or composites are being used in aircraft engineering instead of metals because they are much lighter*

enhance [ɪn'hɑːns] *verb* to make greater or better or clearer; *chances of survival are enhanced if passengers know where the emergency exits are*

QUOTE any automation must be designed to enhance the decision making abilities of the crew, not replace them
INTER PILOT

enhancement [ɪn'hɑːnsmənt] *noun* making greater or better or clearer; **enhancement of an image on a screen** = improvement of an image on a screen

enlarge [ɪn'lɑːdʒ] *verb* to make bigger or larger; **enlarge the hole** = make the hole bigger

en route ['ɒn 'ruːt] *adverb & adjective* on or along the way; **en route from New York to London** = on the way from New York to London; **en route alternate** = an airfield where it is possible to land if there is an in-flight problem; **en route weather conditions** = description of the weather along the path of flight

ensure [en'ʃɔː] *verb* to make certain, to make sure; *the generator cut-out ensures that the battery cannot discharge*; *before the engine is stopped, it should normally be allowed to run for a short period at idling speed, to ensure gradual cooling*

enter ['entə] *verb* **(a)** to come into; to go into; *air enters at the front of the cabin and leaves at the rear* **(b)** to write down (information, etc.); *enter the rectified airspeed in the log*; *enter your name in the correct place in the form* **(c)** *(into a computer)* to type in, to keyboard (data); *enter the data into the computer*

entire [ɪn'taɪə] *adjective* whole; having no part excluded or left out; **the entire life of a thunderstorm** = the complete life of a thunderstorm; **Arctic air can sweep over the entire continent** = Arctic air can move over the whole continent

entry ['entri] *noun* **(a)** the act or instance of going in; *the flow of traffic from entry/exit points to the airfield*; *see also* POINT **(b)** the writing in of an item, as in a record or log; *an entry should be made in the technical log*

envelop [ɪn'veləp] *verb* to surround and cover; *the atmosphere envelops the earth*

envelope ['envələʊp] *noun* **(a)** the set of limitations within which a technological system, especially an aircraft, can perform safely and effectively; *the boundaries of flight envelopes vary between aircraft categories and performance groups but in each case, there is a speed which must not be exceeded which is called the Vne (never-exceed speed)* **(b)** cover; *the* *atmosphere is the gaseous envelope surrounding the earth*

environment [ɪn'vaɪrənmənt] *noun* (a) nearby conditions or circumstances; *a body of air warmer that its environment will rise*; **cabin environment** = small-scale conditions of the aircraft cabin including the temperature, the space, the colour scheme, the seating arrangements, etc.; **a non-computer environment** = a computer-free working situation (b) large-scale circumstances and conditions of the planet Earth; *people are interested in issues to do with the environment, such as global warming*

environmental [ɪnvaɪrən'mentl] *adjective* referring to the immediate surroundings; *environmental conditions*

environmental control system (ECS) [ɪnvaɪrən'mentl kən'trəʊl 'sɪstəm] *noun* air-conditioning system for the aircraft

environmental lapse rate (ELR) [ɪnvaɪrən'mentl 'læps 'reɪt] *noun* rate at which the temperature of the air falls as one rises above the earth; *although there is an average ELR of 1.98°C per 1,000 feet, in practice the ELR varies considerably with space and time*

epoxy [ɪ'pɒksi] *noun* **epoxy-based primer** = primer containing epoxy resin, a substance which, with the addition of hardeners, becomes very strong and hard after a time at normal temperatures

equal ['iːkwəl] **1** *adjective* having the same quantity, measure, or value as another; *for every action, there is an equal and opposite reaction* **2** *verb* (*mathematics*) to be the same in value as; *two plus two equals four (2 + 2 = 4)*

equalize ['iːkwəlaɪz] *verb* to make the same quantity, measure or value as another; to make equal; *fluid pressure and gas pressure equalize at normal system pressure* (NOTE: also written **equalise**)

equate [ɪ'kweɪt] *verb* to be the same as; *in an electrical circuit, an increase in length equates to an increase in resistance*

equation [ɪ'kweɪʒn] *noun* statement, usually in symbols, that two quantities or mathematical expressions are equal; $X2 + Y2 = Z2$; *the equation Vg = P can be used to find the geostrophic wind*

equator [ɪ'kweɪtə] *noun* the imaginary great circle around the earth's surface, equidistant from the poles and perpendicular to the earth's axis of rotation which divides the earth into the northern hemisphere and the southern hemisphere; *every point on the equator is equidistant from the poles*

equatorial [ekwə'tɔːriəl] *adjective* referring to the equator or to conditions that exist at the earth's equator; *equatorial heat; equatorial climate*

equilibrium [iːkwɪ'lɪbriəm] *noun* a state of physical balance; *when an aircraft is in unaccelerated straight and level flight at a constant speed, the forces of lift, thrust, weight and drag are in equilibrium*

equipment [ɪ'kwɪpmənt] *noun* devices, systems, machines, etc., that are needed for a particular purpose; **distance measuring equipment (DME)** = airborne secondary radar whose signal is converted into distance; **electrical equipment** = devices, components, systems, etc., which use electricity (NOTE: no plural; for one item say: **a piece of equipment**)

equivalent [ɪ'kwɪvələnt] *adjective* being equal, all things considered; *the function of a logic gate is equivalent to that of a switch; a metal part could be as much as 25 times heavier than an equivalent plastic part*

error ['erə] *noun* (a) mistake; incorrect calculation; *an error in somebody's work; errors caused by location* (b) known inaccuracy of an instrument or system which has to be corrected by calculating the true value; **instrument error** = difference between indicated instrument value and true value (c) **density error** = correction to airspeed to give TAS (true airspeed)

escape [ɪ'skeɪp] **1** *noun* getting away from or out of a place after being held; **escape of fuel or oil** = unwanted loss of fuel or oil; **escape from danger** = getting to a safe place; **escape hatch** = small doorway only used in emergencies; **escape routes** = path from the passengers seat to the emergency exit; **escape slide** = device which allows passengers to exit the aircraft safely in an emergency, when no steps are available **2** *verb* to get away from or out of after being held; *if there is a hole in the fuselage of a pressurized aircraft, air escapes from the cabin to the atmosphere*

essential [ɪ'senʃl] *adjective* absolutely necessary; *teamwork within the crew is essential*; *a knowledge of the tropopause is essential*; **non-essential** = not necessary

EST = EASTERN STANDARD TIME; *(ICAO)* ESTIMATE(D)

establish [ɪ'stæblɪʃ] *verb* **(a)** to be confirmed as stable in a particular flight condition, such as a flight level or glideslope, etc.; *once established on the downwind leg, the pilot should perform his checks* **(b)** to work out or to calculate; **establish your position** = find out where you are **(c)** to position; *low-power NDBs (Non-Directional Radio Beacons) are often established at the outer or middle marker sites* **(d)** to **establish communication** = to make contact with; **to establish control** = to get control

estimate ['estɪmeɪt] *verb* **(a)** to calculate approximately the cost, value or size of something; *I estimate that it will take about two hours for us to reach our destination*; *cloud heights may be measured or estimated* **(b)** to form a judgement about; **to estimate the chances of something** = to weigh the possibilities and form an opinion

estimated ['estɪmeɪtɪd] *adjective (of aircraft)* **estimated time of arrival (ETA)** = time when an aircraft is expected to land; **estimated time of departure (ETD)** = time when an aircraft is expected to take off

estimation [estɪ'meɪʃn] *noun* **(a)** approximate calculation; *an estimation of groundspeed*; *estimation of visibility is achieved by noting the distances at which lights of known candle power can be observed and relating these distances to visibility-by-day values* **(b)** opinion; **in my estimation** = in my opinion

ETA = ESTIMATED TIME OF ARRIVAL

ETD = ESTIMATED TIME OF DEPARTURE

evacuate [ɪ'vækjueɪt] *verb* **(a)** *(people)* to remove all people (from somewhere) in the event of an emergency; *to evacuate all passengers from the airport* **(b)** *(building, aircraft, village, etc.)* to empty (of all people) in the event of an emergency; *to evacuate the aircraft* **(c)** to create a vacuum; **evacuate a glass jar** = remove all the air from a glass jar

evacuation [ɪvækju'eɪʃn] *noun* **(a)** *(of people)* act of removing all people (from somewhere) in the event of an emergency; *the evacuation of the passengers from the airport was not ordered* **(b)** *(of building, aircraft, village, etc.)* act of emptying (of all people) in the event of an emergency; *the evacuation of the aircraft did not take long*; **evacuation command** = evacuation order from the captain; **ditching evacuation** = evacuation after the aircraft has force-landed on water

evaluate [ɪ'væljueɪt] *verb* to examine and judge carefully; *deposits of ice are detected and continuously evaluated to operate a warning system*

evaluation [ɪvælju'eɪʃn] *noun* examination and judgement of something; *the ice detector system provides continuous evaluation of conditions conducive to the formation of ice*

evaporate [ɪ'væpəreɪt] *verb* to convert or change a liquid into a vapour; *in the heat of the day, water evaporates from the surface of the earth* (NOTE: the opposite is **condense**)

evaporation [ɪvæpə'reɪʃn] *noun* changing of a liquid into vapour; vaporization; *carburettor icing can be caused by the expansion of gases in the carburettor and the evaporation of liquid fuel*; **evaporation fog** = steam fog; *evaporation fog is usually confined to water surfaces and adjacent areas of land* (NOTE: the opposite is **condensation**)

even ['iːvn] **1** *adjective* **(a)** *(surface)* flat or smooth, with no bumps or dents **(b)** *(distribution)* uniform; *an even distribution of passengers*; *an even application of paint* **(c)** *(mathematics)* **even numbers** = exactly divisible by 2 - such as 4, 6, 20 **2** *adverb* **(a)** invites comparison with something stated; *it will be even higher than the new building*; **even faster** = not just as fast as, but more **(b) even if** = whether or not; *stop at the holding point even if there are no other aircraft on the approach*; **even though** = although; in spite of the fact (that); *he gained his private pilot's licence even though he was 73 years old*

event [ɪ'vent] *noun* happening; *the Paris air show is a major event*; **in the event of** = if it should happen; *passengers should fasten their seat belts in the event of turbulence*; **in the event of main pump failure** = if there should be a failure of the main pump; **in the event of fire** = if there should be a fire

eventual [ɪ'venʃul] *adjective* happening at an unspecified time in the future; *water in the fuel may lead to eventual engine stoppage*

eventually [ɪ'venʃuli] *adverb* at an unspecified time in the future; *vapour cools and eventually condenses*; *when water vapour cools adiabatically and eventually condenses, the cloud droplets formed are small and light at first and thus tend to be held in the atmosphere by up-currents*

evidence ['evɪdəns] *noun* an outward sign; **external evidence of cracks** = something which can be seen on the surface which suggests that there is a deeper structural problem; *deformed wing panels may be evidence of an over-stressed airframe*

evident ['evɪdənt] *adjective* easily seen or understood; obvious; *it is evident from the information available that language problems played a part in the cause of the accident*; **self-evident** = clear in itself, without further explanation

exact [ɪg'zækt] *adjective* completely accurate or correct; *the exact fuel flow and pressure is adjusted*; **the calculation is not exact** = the calculation is not 100% correct

exactly [ɪg'zæktli] *adverb* **(a)** accurately; correctly; *measure the quantity exactly* **(b)** absolutely; completely; *a fuel injection system performs exactly the same function as a carburettor*

examination [ɪgzæmɪ'neɪʃn] *noun* **(a)** set of questions or exercises testing knowledge or skill; *the examination includes a flight plan* **(b)** **medical examination** = medical check-up **(c)** careful observation or inspection; *the examination of a faulty component*; **visual examination** = close observation or inspection with the eyes

QUOTE the pilot of a Grumman Cheetah refused to be breathalysed, and was taken to a police station for examination by a police surgeon, who confirmed that he had been drinking
Pilot

examine [ɪg'zæmɪn] *verb* **(a)** to find out the qualifications or skills by means of questions or exercises; *students will be examined in four subjects* **(b)** to test or check the condition or health of (someone); *to examine a patient* **(c)** to study or analyse; *to examine charts*

exceed [ɪk'siːd] *verb* to be greater than; *vertical velocity of updrafts can exceed 50 kt*

exception [ɪk'sepʃn] *noun* something or someone not included; **an exception to the**

rule = an example which does not conform to a generality; **with the exception of** = not including; **with the exception of Smith, all the students passed their exams** = Smith did not pass, but the other students did

exceptional [ɪkˈsepʃnl] *adjective* **(a)** being an exception; uncommon; **in exceptional circumstances** = in unusual circumstances **(b)** well above average; extraordinary; **an exceptional pilot** = a very good pilot

excess [ɪkˈses] *noun* amount or quantity beyond what is normal or sufficient; a surplus; **excess baggage** = amount of baggage (in weight) over what is accepted (free) by an airline; *the passenger had to pay $100 for excess baggage*; **excess power** = difference between horsepower available and horsepower required; **in excess of** = more than; **a height in excess of 50,000 feet** = higher than 50,000 feet

excessive [ɪkˈsesɪv] *adjective* more than normal, usual, reasonable, or proper limit; *excessive use of power when taxiing will require excessive use of brakes*

exchange [ɪksˈtʃeɪnʒ] *verb* to give in return for something received; *meteorological stations exchange information with other meteorological stations*

excitation [eksɪˈteɪʃn] *noun* act of supplying a small current to the windings of larger electrical motors, etc.; *pilot excitation consists of a pilot exciter and a main exciter, to provide the direct current for the motor of the alternating current generator*

exciter [ɪkˈsaɪtə] *noun* source of a small current such as a battery, to supply electrical current to the windings of larger electrical motors, etc.; *pilot excitation consists of a pilot exciter and a main exciter, to provide the direct current for the motor of the alternating current generator*

exclude [ɪksˈkluːd] *verb* to keep out; to prevent from entering; *joints and*

interfaces should exclude moisture and improve fatigue life

exercise [ˈeksəsaɪz] **1** *noun* activity that requires physical or mental effort or practice; *a classroom exercise*; *swimming is good physical exercise for people, such as pilots, who spend a lot of time sitting down* **2** *verb* to put into play or operation; to employ, to use; *student pilots must exercise special care when landing in a strong crosswind*

exert [ɪgˈzɜːt] *verb* **to exert a force** = to put a force on(to); *pressure is the force per unit area exerted by the atmosphere on a given surface area*; **to exert an influence** = to have an influence; **to exert pressure** = to put pressure onto something

exhaust [ɪgˈzɔːst] **1** *noun* **(a)** *(from an engine)* the escape or release of vaporous waste material; **exhaust gas** = gas which is the product of the combustion process and which is passed out through the exhaust system; *exhaust gases contain carbon monoxide* **(b)** pipe through which waste gases pass out of the engine; *the exhaust valve opens to allow for the exit of exhaust gases*; **exhaust system** = system of pipes, silencers, etc., which carry exhaust gases from the engine to a point where they are released into the atmosphere **2** *verb* to consume or use up all of something; *supplies of fuel are exhausted* (NOTE: **to run out** is less formal)

exhibit [ɪgˈzɪbɪt] *verb* to show; to have; to display; *composites, due to their construction, exhibit good fatigue behaviour*; *altocumulus are (usually) white layers or patches of cloud frequently exhibiting a waved appearance*

exist [ɪgˈzɪst] *verb* to be present under certain circumstances or in a specified place; to occur; *water can exist in the atmosphere in three forms; a fire risk may exist following failure or leakage of any component*

existence [ɪgˈzɪstəns] *noun* the fact or state of being; *warning systems are provided to give an indication of a* *possible failure or the existence of a dangerous condition*

exit ['eksɪt] *noun* (**a**) the act of going out of a place; *the exhaust valve opens to allow for the exit of exhaust gases*; **exit velocity** = the velocity of exhaust gases from a jet engine (**b**) *(passage, door, etc.)* way out; **emergency exit** = way out only to be used in case of an emergency; *how many emergency exits are there in the aircraft?* (NOTE: the opposite for (a) is **entry**, and for (b) is **entrance**)

expand [ɪk'spænd] *verb* to increase in size, volume, quantity; to enlarge; *air expands when heated and contracts when cooled* (NOTE: the opposite is to **contract**; but when it is caused by factors other than temperature it is to **compress**)

expansion [ɪk'spænʃn] *noun* increase in size, volume, quantity; enlargement; *there is an expansion of the gas when it is heated*; **differential expansion switches** = switches which operate on the principle that the coefficients of expansion of dissimilar metals are different (NOTE: the opposite is **contraction**; but when it is caused by factors other than temperature it is **compression**)

expect [ɪk'spekt] *verb* to hope or to assume that something is going to happen; *the weather to be expected along a route*; *we expect flight AC 309 within ten minutes*; **as might be expected** = as one thinks would happen

QUOTE by 1959 there were some 40 pilots past age 60 flying the line with the number expected to rise to 250 within the next few years
INTER PILOT

expected [ɪk'spektɪd] *adjective* which is thought or hoped will happen; estimated; *the expected number of passengers*; *see also* ESTIMATED

expedite ['ekspədaɪt] *verb* to speed up the progress of; **to expedite the evacuation** = to speed up the evacuation; **to expedite the disembarkation** = to get the passengers off the aircraft quickly

expel [ɪk'spel] *verb* to force out, to drive out; *exhaust gases are expelled from the cylinder by the upward movement of the piston*; *the piston draws fluid into the cylinders on the outward stroke and expels fluid into the system on the inward stroke*

experience [ɪk'spɪəriəns] **1** *noun* (**a**) building up of knowledge or skill over a period of time by an active participation in events or activities; *a pilot with 20 years' experience* (**b**) event; incident; *the first solo is an experience most pilots never forget* **2** *verb* to participate in (something); to find oneself in a particular situation; to undergo; *it is not unusual to experience traffic delays on the ground prior to departure*; *turbulence can be experienced when flying through a trough*

experiment 1 [ɪk'sperɪmənt] *noun* scientific test - under controlled conditions - that is made to demonstrate or discover something; *experiments have shown that left-handed people often have better hand/eye co-ordination than right handed people*; **to conduct an experiment** = to perform an experiment **2** [ɪk'sperɪment] *verb* **to experiment (with)** = to carry out a scientific test - under controlled condition - in order to demonstrate or discover something

experimental [ɪksperɪ'mentl] *adjective* referring to something still at an early stage of development; not tried and tested; *the experimental and testing stages of a new type of aircraft*; **an experimental aircraft** = an aircraft designed to be used for experimental purposes; *the experimental aircraft were used to investigate high-speed flight*

explanatory [ɪk'splænætri] *adjective* referring to something which explains; **explanatory paragraph** = paragraph of text which explains something; **self-explanatory** = something which does not need any further explanation

explosion [ɪk'spləuʒn] *noun* (**a**) release of energy in a sudden and often violent way; *an explosion caused by a bomb* (**b**) bursting as a result of internal pressure; *tyre explosion due to overheating* **(c)** the loud sound made as a result of an explosion; *the passengers heard an explosion*

explosive [ɪk'spləʊzɪv] **1** *adjective* referring to something having the nature of an explosion; **an explosive effect** = having the effect of an explosion **2** *noun* substance, especially a prepared chemical, that explodes or causes explosions (such as Semtex)

expose [ɪk'spəʊz] *verb* to subject or allow to be subjected to an action or an effect; *when the slope of a hill is exposed to solar radiation, wind currents are set up*; **exposed to the sun** = in sunlight without covering; **exposed surface** = surface without paint or covering of any sort

exposure [ɪk'spəʊʒə] *noun* **(a)** being exposed, especially to severe weather or other forces of nature; *after 24 hours in the sea, he was suffering from the effects of exposure and was taken to hospital* **(b)** being subjected to something; *exposure to radio-active substances may cause cancer*

express [ɪk'spres] *verb* to put into words, symbols or signs; *bearings may be expressed as true or relative*; *an angle may be expressed in degrees, minutes and seconds*; *pressure altitudes are expressed in hundreds of feet*

extend [ɪk'stend] *verb* to stretch or spread from one point to another in space or time; *air from the Gulf of Mexico can extend into Canada*; *cumulonimbus clouds may extend to over 50,000 ft*; **to extend the duration of something** = to prolong the time; *the visit was extended to allow time for more discussions*

extensive [ɪk'stensɪv] *adjective* large in range or amount; **an extensive area** = a large area; **extensive cloud** = a lot of cloud; **extensive use is made of** = much use is made of

extent [ɪk'stent] *noun* range or amount of something; *the horizontal extent of the*

cloud averages about 50 km; *clouds of great vertical extent are not uncommon*; **to a certain extent or to some extent** = partly; *the accident was caused, to a certain extent, by the poor weather*; **to a lesser extent** = not as much as something previously stated; *the cloud types which are most likely to affect flying conditions in terms of icing, precipitation and turbulence are cumulus, cumulonimbus and, to a lesser extent, nimbostratus*

external [ɪk'stɜːnl] *adjective* referring to, existing on, or connected with the outside or an outer part; exterior; *the only external force acting on air is gravity*; **external ambient pressure** = pressure outside the aircraft; **external appearance** = the appearance of something from the outside (NOTE: the opposite is **internal**)

extinguish [ɪk'stɪŋgwɪʃ] *verb (fire)* to put out; *the fire services extinguished the fire*

extinguisher [ɪk'stɪŋgwɪʃə] *noun* portable mechanical device for spraying and putting out a fire with chemicals; *hand-operated fire extinguishers are provided to combat any outbreaks of fire in the flight crew compartment and passengers' cabins*

extinguishing [ɪk'stɪŋgwɪʃɪŋ] *adjective* **extinguishing agent** = one of several substances used to extinguish fires

extract 1 ['ekstrækt] *noun* part taken from a longer text; *the following paragraph is an extract from a flight manual* **2** [ɪk'strækt] *verb* **(a)** to obtain from a substance by chemical or mechanical action; *a dehumidifier extracts moisture from the atmosphere* **(b)** to take out or to obtain information from something; *extract the important information from a text*

extrapolate [ɪk'stræpəleɪt] *verb* to estimate by using known facts; to guess; *information given on a synoptic chart can be extrapolated, by the use of some simple guidelines*

extreme [ɪkˈstriːm] **1** *adjective* **(a)** most distant in any direction; the outermost or farthest; *the most extreme point on the map* **(b)** to the greatest or highest degree; very great; **extreme care must be taken** = the greatest care must be taken; **extreme difficulty** = great difficulty **2** *noun* either of the two things, values, situations, etc., situated at opposite ends of a range; *the extremes of boiling and freezing*; *the region experiences extremes of temperature*

eye [aɪ] *noun* organ (in the head) thanks to which you see; *see also* COORDINATION; VISION

Ff

F = FAHRENHEIT, FARAD

FAA *US* = FEDERAL AVIATION ADMINISTRATION

fabric ['fæbrɪk] *noun* material or cloth produced especially by knitting or weaving; *a breathing mask has a fabric carrying bag; pneumatic de-icer boots are made from vulcanized rubber fabric*

fabricate ['fæbrɪkeɪt] *verb* to make; to manufacture; *selected wing panels are fabricated entirely from magnesium alloys; the ease with which aluminium can be fabricated into any form is one of its most important qualities*

face [feɪs] **1** *noun* **(a)** the surface of an object; *the face of the earth; the exhaust cone prevents the hot gases from flowing across the rear face of the turbine disc;* **the north face of the mountain** = the vertical or near-vertical side facing north **(b)** *(of clock, instrument, etc.)* front part with dial, indicators, etc.; *the face of an instrument;* **the face of a clock** = the front part of the clock with numbers **(c)** front of the head - including the eyes, nose, mouth; **full face smoke mask** = protective mask for fighting fires which covers the whole face **2** *verb (of mountain side, building, surface, etc.)* to be turned towards a particular direction; to look in a particular direction; *hills and mountains which face the sun receive more intense radiation;* **the building faces north** = the building has its front towards the north

facilitate [fə'sɪlɪteɪt] *verb* to enable something to happen more easily or quickly; *a ramp is used to facilitate access to the wing; clearly marked exits facilitate rapid evacuation of passengers*

facility [fə'sɪlɪti] *noun* **(a)** ease in moving, acting, or doing something; aptitude or ability; **a facility in learning to fly** = a good natural ability for flying **(b)** installation or building which provides specific operating assistance; *DME (Distance Measuring Equipment) ground facility* **(c)** mode of operation which allows the user of equipment to do something; *the printer has a self-test facility* **(d)** facilities = something built to serve a particular function; equipment which can be used; a **clubhouse with good facilities** = a clubhouse with a number of features which can be used by members and guests, eg restaurant, bar, reading room, swimming pool; **medical facilities** = hospitals, clinics, etc.

facsimile [fæk'sɪmɪli or 'fæks] *noun* a fax, an exact copy of a document, drawing, etc., transmitted and received by a fax machine connected to a telephone link; *charts are transmitted by facsimile to meteorological offices; the civil aviation meteorological facsimile network is known as CAMFAX* (NOTE: it is also called **fax**)

fact [fækt] *noun* information presented as real; *temperature changes are an important fact in meteorology;* **in (point of) fact** = in reality; in truth

factor ['fæktə] *noun* **(a)** an important part of a result, a process, etc.; *visibility remains a very important factor in aviation;* **critical factor** = extremely important factor; **dominant factor** = most important factor; **safety factor** = something which plays an important part in safety **(b) by a factor of** = quantity by which a stated quantity is multiplied or divided, so as to indicate an increase or decrease in a measurement; **by a factor of ten** = ten times; *the rate is increased by a factor of 10*; **conversion factor** = formula or figure used for conversion of temperatures, distances, etc., from one system to another; *the conversion factor for converting UK gallons to litres is: x 4.546*; **load factor** = stress applied to a structure as a multiple of stress applied in 1g (acceleration due to earth's gravity) flight; *if a structure fails at 10,000 pounds load, an aircraft weighing 4,000 pounds will reach this load at a load factor of 2.5*

fade [feɪd] **1** *noun* **(a)** periodic reduction in the received strength of a radio transmission; *surface wave at night causes fade of the signal* **(b)** periodic reduction in braking power; *hard braking can cause fade and tyre burst through overheating* **2** *verb* **(a)** to lose brightness, loudness, or brilliance gradually; to become less strong; **the lights faded** = the lights became less and less bright, the lights dimmed; **the radio signal faded** = the radio signal became weaker and weaker

Fahrenheit ['færənhaɪt] *noun* scale of temperatures where the freezing and boiling points of water are 32° and 212° respectively; *compare* CELSIUS, CENTIGRADE (NOTE: used in the USA but now less common in the UK; usually written as an F after the degree sign: **32º F**)

fail [feɪl] *verb* **(a)** to stop working properly; **the brakes failed** = the brakes did not work; **the wing failed during a high-speed turn** = the wing broke during a high-speed turn **(b)** to receive an academic grade below the acceptable minimum in an examination or a course of study; **the trainee failed his navigation examination** = the trainee did not pass his navigation exam; **without fail** = certainly, definitely; *be here at 8 o'clock without fail*

fail safe system ['feɪl 'seɪf 'sɪstəm] *noun* system or device which has in-built safeguards against total failure; *the term fail safe means that the structure, though damaged, is capable of supporting a reasonable percentage of its design load*

failure ['feɪljə] *noun* **(a)** stopping; breakdown; *bearing failure*; *engine failure is sometimes accompanied by fire*; **power failure** = loss of engine power; loss of electrical power supply **(b)** not achieving the desired end or ends; *the failure of an experiment*; **failure to do something** = not having done something; *the steward's failure to remain at his station made the emergency situation worse* **(c)** failing a course, a test, or an examination; *his failure in the GFT (General Flying Test) meant that he didn't finish the course*

fair [feə] **1** *adjective* **(a)** *(weather)* free of clouds or storms; clear and sunny; **fair weather** = good weather **(b)** just; reasonable; free of favouritism or bias; **a fair exam** = an exam which tested students on what they had been taught, was of reasonable difficulty and duration and which did not trick the candidates; **it is fair to say that he should have done better** = it is reasonable to say that he should have done better **2** *verb* to join pieces so as to be smooth, even, or regular; *the aircraft's wing is faired into the fuselage*

fairing ['feərɪŋ] *noun* device to improve the flow of air over a surface; *there is a dorsal fairing at the base of the fin or vertical stabilizer*; **wheel fairings**, called **spats**, are fitted to light aircraft to reduce drag; *see also* SPAT; NACELLE

fairly ['feəli] *adverb* moderately; rather; **fairly high levels** = moderately high levels; **fairly simple** = moderately simple

fall [fɔːl] **1** *noun* **(a)** lessening in amount; drop; **fall in pressure** = drop in pressure **(b)** amount of rain or snow which comes down at any one time; *an overnight fall of snow* **(c)** *(season)* US autumn **2** *verb* **(a)** to become less in amount; to drop; to decrease; **atmospheric pressure is falling** = atmospheric pressure is decreasing **(b)** to be included within the range of something; *aircraft fall into a number of type categories*; *design methods fall into four groups*; *long-range high-frequency communications fall in the frequency bracket 2-25 MHz* **(c)** to drop or come down freely because of gravity; *light rain may fall occasionally* (d) to occur at a particular time; *New Year's Day falls on a Thursday this year* (NOTE: falling - fell - fallen)

false [fɒls] *adjective* not true; incorrect; *lightning may cause false readings from sensitive instruments*; false glide path information = incorrect glide path information

familiar [fə'mɪlɪə] *adjective* (a) often seen; common; *clouds are the most familiar visible meteorological feature* (b) known; *symbols and abbreviations which are strange at present become familiar after a time*; to be familiar with = to have some knowledge of something; *he is familiar with the procedure*

familiarize [fə'mɪlɪəraɪz] *verb* to familiarize yourself with = to get to know something well (NOTE: also written familiarise)

fan [fæn] *noun* circular device with rotating blades, powered by an engine or motor, for moving a gas such as air; *the compressor has large rotating fan blades and stator blades*

FAR *US* = FEDERAL AVIATION REGULATION

farad (F) ['færæd] *noun* SI unit of capacitance; *capacitance is measured in farads*

fasten ['fɑːsn] *verb* to secure or to close, as by fixing firmly in place; fasten your seat belt = put on and attach your seat belt; *if in-flight conditions require the captain to activate the fasten seat belt sign, all cabin service ceases and cabin crew take up their assigned seats and strap in*

fatigue [fə'tiːg] 1 *noun* (a) physical or mental tiredness resulting from exertion; *pilot fatigue was a contributing factor in the accident* (b) the weakening or failure of a material, such as metal, resulting from stress; *fan blades must be resistant to fatigue and thermal shock*; *titanium has good fatigue resistance*; fatigue crack = crack due to (material) fatigue

fault [fɒlt] *noun (electronics)* defect in a circuit or wiring caused by bad connections, etc.; *a fault in the automatic boost control unit was repaired*

faulty [fɒltɪ] *adjective* containing a fault or defect; imperfect; defective; *the faulty component was replaced*

fax [fæks] 1 *noun* (a) *(facsimile copy)* exact copy of a document, drawing, etc., transmitted and received by a fax machine connected to a telephone link (b) *(machine, system)* electronic apparatus linked to a telephone used to send and receive a fax; *charts are transmitted by fax to meteorological offices; see also* CAMFAX 2 *verb* to send a fax; *charts are faxed to meteorological offices*

feather ['feðə] *verb (in flight)* to feather a propeller = to turn the blades of a stopped propeller edge on to the airflow in order to reduce drag or wind resistance; *the feathered position not only reduces drag, but also minimizes engine rotation, thus preventing any additional damage to the engine*

feathering ['feðrɪŋ] *noun* the act of turning the blades of a stopped propeller edge on to the airflow in order to reduce drag; *feathering is accomplished by moving the pilot's control lever*; feathering gate = device on the propeller pitch control to prevent unwanted selection of the feathering position; feathering position = position of the propeller pitch control in which the blades are feathered

feature ['fiːtʃə] 1 *noun* (a) important, noticeable or distinctive aspect, quality, or characteristic; *sea breeze is a regular feature of coastal climates* (b) ground features = noticeable, important objects in the landscape, such as bridges, rivers, railway lines, etc., which are useful aids to navigation 2 *verb* to have as a particular characteristic; *many Rutan designs feature a canard wing*

Federal Aviation Administration (FAA) ['fedərl eɪvɪ'eɪʃn ædmɪnɪs'treɪʃn] *noun* (in the United States) the body responsible for the regulation of aviation; *the FAA issues licenses*

Federal Aviation Regulation

(FAR) ['fedərl eɪvi'eɪʃn reɡju'leɪʃn] *noun* (in the United States) regulation governing aviation; *FARs concern the following - Part 1, Definitions and Abbreviations; Part 61, Certification: Pilots and Flight Instructors; and Part 91, General Operating and Flight Rules*

feed [fiːd] *noun* supply of fuel, energy, etc. provided for use; *(of fuel system)* **gravity feed** = which uses the force of gravity to move the fuel from the tank to the carburettor

feedback ['fiːdbæk] *noun* **(a)** return of part of the output of a process or system to the input, especially when used to maintain performance or to control a system; *the LC ensures that a feedback signal of the monitored output frequency is sent back to the CSDU* **(b)** feedback mechanism

fiber ['faɪbə] *noun* US = FIBRE

fibre ['faɪbə] *noun* natural or synthetic filament like cotton or nylon; **carbon fibre** = (i) thin and very strong strand of pure carbon which can be combined with other materials to make them stronger; (ii) light and strong composite material containing strands of pure carbon (NOTE: the American English is **fiber**)

field [fiːld] *noun* **(a)** area of grass on farmland, in the countryside; *in the event of a power failure, it is important to select the most suitable field for a forced landing* **(b)** imaginary area; **magnetic field** = area of magnetic influence; *around any conductor carrying an electric current there will be a magnetic field*; **field of vision** = area in which something can be seen without moving the head or (in some cases) the eyes; visual field

fighter ['faɪtə] *noun* rather small, often single seat, type of aeroplane designed for warfare; *the F16 is an American-built fighter*

figure ['fɪɡə] *noun* **(a)** *(in a book)* diagram or drawing; *figure 1 shows a cross-section of an internal combustion engine* **(b)** number, especially in mathematical calculations; digit; **a head for figures** = good at figures, arithmetic, accounting, etc.; **a two-figure code** = a code with two numbers between 0 and 9 **(c)** *(geometry)* a form consisting of any combination of points or lines, such as a triangle

film [film] *noun* **(a)** thin skin or layer; *an electrical element made of gold film is sandwiched between the layers of glass* **(b)** thin covering or coating; *there is a film of oil between the piston and cylinder wall* **(c)** photographic film = celluloid material usually contained in a small metal cylindrical casing for use in cameras

filter ['fɪltə] **1** *noun* **(a)** material or device through which a liquid or a gas is passed in order to separate the fluid from solid matter or to remove unwanted substances; *fuel filter; oil filter*; **air filter** = device to filter solid particles out of the air in engine and ventilation systems; **filter element** *or* **filter cartridge** = removable paper or metal component in a filter housing which must be replaced periodically **(b)** electric, electronic, acoustic, or optical device used to reject signals, vibrations, or radiations of certain frequencies while passing others; *the tuner is a band pass filter which confines the bandwidth passed to the receiver to that required* **2** *verb* to pass a liquid or gas through a filter in order to remove unwanted substances; *fuel is filtered before entering the carburettor*

fin [fin] *noun* fixed vertical aerofoil at the rear of an aircraft; vertical stabilizer; *the fin provides directional stability about the vertical axis*

final ['faɪnl] **1** *noun* end part of a series or process **2** *adjective* which comes at the end; last; **final approach** = flight path in the direction of landing along extended runway centre line; **final assembly** = last in a series of stages of construction of an aircraft when all the pre-assembled parts are put together

fine [faɪn] *adjective* **(a)** of superior quality, skill, or appearance; **a fine day** = a day when the weather is good; **fine weather** = good weather **(b)** very small in size, thickness or weight; *cirrus cloud has a fine, hair-like appearance*; **fine powder** = powder consisting of very small particles; **fine spray** = spray consisting of very small drops of liquid **(c) fine wire** = very thin wire **(d)** referring to the pitch or blade angle setting of the propeller; *fine pitch enables full engine speed to be used on take-off and coarse pitch allows an economical engine speed to be used for cruising* (NOTE: the opposite for (c) is **thick**, and for (d) is **coarse**)

FIR = FLIGHT INFORMATION REGION

fire ['faɪə] **1** *noun* **(a)** area of burning; **an engine fire** = a fire in an engine; **fire deluge system** = system which extinguishes fire by spraying large quantities of water on it; *a lever actuates the fire deluge system*; **fire extinguisher** = portable device full of foam, water, powder, etc., for putting out fires; **fire triangle** = the illustration of the chemistry of fire as the three sides of a triangle representing fuel, oxygen and heat; *if fuel, oxygen or heat is removed from the fire triangle, combustion will cease* **(b)** St Elmo's Fire = luminous electrical discharge sometimes seen on aircraft during storms **2** *verb* to shoot with a (type of) gun; to launch (a flare, a rocket); *a parachute flare is a device which is fired to a height of 1,200 ft*

fireproof ['faɪəpruːf] *adjective* which is designed to resist the effect of fire; *a fireproof bulkhead is provided to separate the cool area of the engine from the hot area*

fit [fɪt] **1** *adjective* in good physical condition; healthy; *keep fit with diet and exercise* **2** *noun* the exactness with which surfaces are adjusted to each other in a machine; *there should be a loose fit between the cylinder and the piston, the difference being taken up by the piston rings* **3** *verb* **(a)** to be the correct size and shape for; *oxygen masks should fit the wearer properly* **(b)** to put on; to attach; *wheel fairings, called spats, are fitted to some light aircraft to reduce drag* (NOTE: **fitting - fitted**)

fitment ['fɪtmənt] *noun* act of attaching or fixing; *attachment points are supplied for the fitment of heavy equipment*

fitness ['fɪtnəs] *noun* state or condition of being physically fit, especially as the result of exercise and proper eating habits; *the age and physical fitness of some passengers can be a limiting factor in an evacuation*; **fitness to fly** = description of the physical or mental capabilities a person needs to fly an aircraft

FL = FLIGHT LEVEL

flag [flæg] *noun* **(a)** usually square or rectangular piece of cloth with a symbolic design or colour; *flags are flown from the signal mast* **(b)** small visual warning or indicating device on the face of an instrument; *there is a warning flag on the instrument if the VOR becomes unusable*

flame [fleɪm] *noun* the area of burning gases (usually yellow) seen when something is burning; *flames were seen coming from number 2 engine*; see also ARRESTER

flame out ['fleɪm 'aʊt] *verb (of combustion in a gas turbine engine)* to cease from some cause other than the shutting off of fuel; *air in the fuel line can cause an engine to flame out or stop*

flame-out ['fleɪmaʊt] *noun* ceasing of combustion in a gas turbine engine from some cause other than the shutting off of fuel (NOTE: also written **flameout**)

flammable ['flæməbl] *adjective* easily ignited and capable of burning fiercely and rapidly; easily set on fire (and therefore hazardous); *aviation gasoline is a flammable liquid* (NOTE: **flammable** and **inflammable** mean the same thing; the opposite is **non-flammable**)

flange [flænʒ] *noun* outside edge or rim of a part (such as beam, wheel, etc.); *the* *web connects the upper and lower flanges of a beam*

flap [flæp] *noun* movable control surface on the trailing edge of an aircraft wing, used primarily to increase lift and drag during final approach and landing; *flaps should be retracted immediately after landing to decrease lift and therefore increase brake effectiveness*; *see also* COWL

COMMENT: flaps are not usually used for takeoffs in light aircraft except when a short takeoff run is required. Flaps are not primary control surfaces of an aircraft

flare [fleə] *noun* **(a)** stage of the flight immediately before touchdown when the nose of the aircraft is raised into the landing attitude; *the approach, flare and landing can be carried out by automatic systems* **(b)** small rocket-like device with a bright light, for attracting attention; **parachute flare** = distress signal attached to a parachute to allow more time for the flare to be seen; *the parachute flare is a device which is fired to a height of 1,200 ft where a red flare and parachute detach*

flash [flæʃ] **1** *noun* **(a)** giving off light in sudden or periodic bursts; *lightning is accompanied by a brilliant flash*; *loss of vision may occur due to lightning flashes especially at night* **(b)** **flash point** = temperature at which fuel vapour or oil vapour will burst into flame **2** *verb* **(a)** to give off light in regular bursts; **warning lights flash** = warning lights go on and off rapidly **(b)** to appear or to happen suddenly; *the image flashed onto the screen*

flat [flæt] *adjective* **(a)** having a horizontal surface without a slope, tilt or curvature; *it has been shown that the flat chart misrepresents the globe-shaped earth*; **flat country** = country with no hills or mountains **(b)** *(of tyre)* having no air inside; *the flat tyre had to be changed because it had a puncture* **(c)** *(of battery)* electrically discharged or with no electrical charge left in it; *the engine wouldn't start because the battery was flat*

flatten (out) ['flætn] *verb* to make flat; *as altitude increases, the countryside appears to flatten out; the earth is spherical in shape but it is flattened at the poles*

flaw [flɔ:] *noun* imperfection in a material, often hidden, that may be an indication of future structural failure; **flaw detection** = process or system by which small weaknesses in metal structures are found

flew [flu:] *see* FLY

flexibility [fleksə'bılıti] *noun* **(a)** the amount or extent to which something can be bent or flexed; *wing structures must have flexibility in order to absorb sudden changes in loading* **(b)** the extent to which a system or device can change or respond to a variety of conditions or situations; adaptability; *the more reliable and quick fly-by-wire system allows a much greater degree of flexibility with aircraft stability* (NOTE: the opposite is **rigidity**)

flexible ['fleksəbl] *adjective* **(a)** not rigid, not stiff; **flexible pipes** = pipes made of soft material (such as rubber, plastic) **(b)** capable of responding to a variety of conditions or situations; adaptable; *AC electrical energy is more flexible and more efficient than DC* (NOTE: the opposite is **rigid**)

flight [flaıt] *noun* **(a)** motion of an object in or through the earth's atmosphere or through space; **theory of flight** = ideas and principles which contribute to our understanding of how things fly; **flight attendant** = member of the flight crew who looks after passengers, serves, food, etc.; **flight bag** = bag used by flight crew to carry manuals, documents, headset, etc.; **flight crew** = airline staff responsible for flying the aircraft; **flight data recorder (FDR)** = electronic device located in the tail section of an aircraft, that picks up and stores data about a flight (NOTE: it is often called the **black box** although it is not black); **flight deck** = place where the flight crew of an airliner sit while flying the aircraft; **flight gear** = (might include) flying suit, headset, case with charts, manual, etc.; **flight information region (FIR)** = airspace with defined limits which has an air traffic control information and alerting service; **flight level (FL)** = level of constant atmospheric pressure related to a reference datum of 1013.25 mb; *FL 250 = 25,000 ft*; **Flight Manual** = Pilots Operating Handbook **(b)** distance covered by a body, such as the distance covered by an aircraft through the atmosphere; *the flight from London to Paris took 55 minutes* **(c)** scheduled airline journey; *passengers for flight GF 008 to Amman should proceed to gate number 4*

float [fləʊt] **1** *noun* floating ball attached to a lever to regulate the level of a liquid in a tank, etc.; **float chamber** = part of a carburettor which houses a float; **float-operated switch** = shut-off valve operated by a float **2** *verb* to remain on the surface of a fluid without sinking; *because of the air-tight nature of the fuselage, most large aircraft will float for some time before sinking*

flow [fləʊ] *verb* **(a)** to move or run smoothly with continuity, as a fluid; *air flows over the wing surfaces and lift is produced* **(b)** to circulate; *liquid coolant flows around the engine* **2** *noun* (*of fluid*) continuous movement in a particular direction; *the flow of fuel from the fuel tanks to the engines*

flowmeter ['fləʊmiːtə] *noun* device for measuring the flow of a liquid or gas; *the oxygen flowmeter should blink once for each breath*

fluctuate ['flʌktʃueɪt] *verb* to vary or change irregularly; *the magnetic field will fluctuate at the supply frequency*

fluid ['fluːɪd] **1** *noun* substance whose molecules move freely past one another and that takes the shape of its container; a liquid or a gas; **de-icing fluid** = liquid for removing ice; **hydraulic fluid** = thin oil used in hydraulic braking systems, etc.

fluorescent ['flɔːresənt] *adjective* referring to the emission of electromagnetic radiation of visible light; *the fluorescent penetrant process of flaw detection uses a penetrant containing a fluorescent dye which fluoresces in ultra-violet light*

fly [flaɪ] *verb* to move through the air; to cause an aircraft to move through the air in a controlled manner; *an aeroplane may not fly over a city below such a height as would allow it to alight in the event of an engine failure*; *he's learning to fly*; **to fly in formation** = to fly as a number of aircraft in a group which maintains a particular pattern or arrangement in the air (NOTE: **flying - flew - flown**)

fly-by-wire ['flaɪbaɪ'waɪə] technology which interprets movements of the pilot's controls and, with the aid of computerized electronics, moves the control surfaces accordingly; *using fly-by-wire technology, the stalling angle cannot be exceeded regardless of stick input*; *the more reliable and quick fly-by-wire system allows a much greater degree of flexibility with aircraft stability*

COMMENT: fighters like the General Dynamics F16 and large transport aircraft such as the Boeing 777 and Airbus A320 have fly-by-wire systems

flying ['flaɪɪŋ] *noun* action of making an aircraft move through the air in a controlled manner; **hand flying** = flying (an aircraft) by moving the flight controls with the hands rather than by using the autopilot; *see also* TEST

foam [fəʊm] *noun* **(a)** a mass of bubbles of air or gas in a liquid film; *foam fire extinguishers*; *correct aircraft positioning will assist airport fire crews in the performance of their duties which would include covering the fuselage with foam, controlling the fire and providing escape routes* **(b)** any of various light, porous, semi-rigid or spongy materials used for thermal insulation or shock absorption; *polyurethane foam is used in packaging*

focal point ['fəʊkl 'pɔɪnt] *noun see* FOCUS

focus ['fəʊkəs] **1** *noun* point at which rays of light or other radiation converge; *the focus of a lens is also called the focal point*; **to come into focus** = to become clearer as through the viewfinder of a camera (NOTE: the plural is **foci** ['fəʊsaɪ]) **2** *verb* **(a)** to make things, such as light rays, converge on a central point; *a parabolic reflector focuses the transmission into a narrow beam* **(b)** to give an object or image a clear outline or detail by adjustment of an optical device; *focus the microscope in order to make the image easier to see* **(c)** to direct toward a particular point or purpose; *the crew focused all their attention on finding a solution to the problem*

fog [fɒg] *noun* condensed water vapour in cloud-like masses lying close to the ground and limiting visibility; *when visibility is less than 1,000 m owing to suspended water droplets in the atmosphere, the condition is known as fog*; **advection fog** = fog which forms when warmer moist air moves over a colder surface (land or sea); **evaporation fog** = steam fog; *evaporation fog is usually confined to water surfaces and adjacent areas of land*; **radiation fog** = fog caused by the cooling of the earth to below the dew point, combined with saturation and condensation and a light mixing wind; *radiation fog cannot form over the sea*; **steam fog** = evaporation fog - fog formed when cold air moves over relatively warm water

föhn [fɜn] *noun* warm dry wind that blows down the lee side of a mountain, particularly in the Alps (NOTE: also written **foehn**)

foot (ft) [fʊt] *noun* unit of length in the US and British Imperial Systems equal to 12 inches or 30.48 centimetres; **foot-pound** = the ability to lift a one pound weight a distance of one foot (NOTE: plural is **feet**; foot is usually written **ft** or **'** after figures: **10ft** or **10'**)

force [fɔːs] **1** *noun* **(a)** capacity to do work or cause physical change; *the force of an explosion* **(b)** power against resistance; *in small aerobatic aircraft, considerable force is needed on the control column when performing high-speed manoeuvres* **(c)** *(physics)* vector quantity that produces an acceleration of a body in the direction of its application; **centrifugal force** = force of acceleration away from the axis around which a body rotates; **centripetal force** = force of acceleration towards the axis around which a body rotates; **the force of gravity** = natural force of attraction which pulls bodies towards each other and which pulls objects on earth towards its centre (NOTE: we say **centrifugal force**, but **the force of gravity**) **2** *verb* **(a)** to use power against resistance; *because of distortion to the airframe, the pilot had to force the door open in order to exit the aircraft* **(b) to force someone to do something** = to use physical or psychological power to make someone do something they otherwise would not do; *the hijackers forced the crew to fly to Athens*

fore [fɔː] *adjective* located at or towards the front; **the fore and aft axis of the aircraft** = the longitudinal axis of the aircraft; **to come to the fore** = to take a leading part; *the jet engine came to the fore in the late forties*; *see also* AFT

forecast ['fɔːkɑːst] **1** *noun* prediction or foretelling, as of coming events or conditions; *weather forecast*; **forecast weather charts** = charts with information about the weather coming to a particular area; **terminal area forecast (TAF)** = weather forecast for the area around an airport **2** *verb* to estimate or calculate weather conditions by studying meteorological information; *rain is forecast for this afternoon* (NOTE: **forecasting - forecast** *or* **forecasted**)

form [fɔːm] **1** *noun* **(a)** document with blanks for the insertion of details or information; *insurance form; application form* **(b)** kind, type; *the ground automatic relief valve is a form of discharge valve*; *drizzle is the lightest form of precipitation* **(c)** shape of an object; *fluids take on the form of the container in which they are found*; **in the form of a triangle** = in the shape of a triangle **(d)** way in which a thing exists, acts, or shows itself; *water in the* *form of ice; fuel in the form of a spray* **2** *verb* **(a)** to come into being; *in certain conditions, ice forms on the leading edge of the wing; cumulous clouds only form in an unstable atmosphere* **(b)** to make a shape; *three points on the chart form a triangle* **(c)** to make up or constitute; *the classroom and accommodation building form the main part of the college*

formation [fɔːˈmeɪʃn] *noun* **(a)** coming into being or forming; **cloud formation** = natural production and development of clouds; **ice formation** = natural production and development of ice **(b)** *(of a number of aircraft)* **to fly in formation** = to fly in a group which maintains a particular pattern or arrangement in the air

former [ˈfɔːmə] **1** *adjective* having been in the past; **a former military pilot** = a pilot who used to be a military pilot **2** *noun* **(a)** the first of two things mentioned; *barometers are of two types - mercury and aneroid - the former being larger but more accurate than the latter* = the mercury barometer is larger and more accurate than the aneroid type **(b)** light secondary structure of the airframe which gives improved shape

QUOTE much has changed in the former Eastern European States, especially in terms of aviation operations and training
Civil Aviation Training

formula [ˈfɔːmjʊlə] *noun* mathematical rule expressed in symbols; *the formula for calculating speed is $D \div T = S$ (where D = distance, T = time and S = speed)* (NOTE: the plural is **formulas** or **formulae** [ˈfɔːmjʊliː])

forward [ˈfɔːwəd] **1** *adjective* at, near, or belonging to the front; *the forward section of the aircraft; forward and aft exits* **2** *adverb* US = FORWARDS (NOTE: the opposite is **aft**)

forwards [ˈfɔːwədz] *adverb* towards a position in front; *the throttles are moved forwards for take-off* (NOTE: American English is **forward**)

fouling [ˈfaʊlɪŋ] *noun* contamination of the spark plugs with oil or petrol so that they do not fire correctly; *the engine should be run at a positive idling speed to prevent spark plug fouling*

four-stroke [ˈfɔːˈstrəʊk] *see* STROKE

frame [freɪm] *noun* **(a)** structure that gives shape or support; *early aircraft fuselages were made of a frame covered by a fabric* **(b)** an open structure for holding, or bordering; *a door or window frame*

FREDA *see* MNEMONIC

freeze [friːz] *verb* to pass from the liquid to the solid state by loss of heat; *in certain conditions, rain droplets freeze rapidly on striking the aircraft* (NOTE: **freezing - froze - frozen**)

freight [freɪt] *noun* anything other than people transported by a vessel or vehicle, especially by a commercial carrier; cargo; *freight holds are usually located beneath the passenger cabins*

frequency [ˈfriːkwənsi] *noun* **(a)** the number of times or the rate at which something happens in a given period of time; *the frequency of flights to holiday destinations increases during the summer time* **(b)** the number of repetitions per unit time of a complete waveform, as of an electric current frequency; **frequency bracket** = range of frequencies; *VHF communications are allocated the frequency bracket 118-137 MHz*; **very high frequency omni-directional radio range (VOR)** = navigational aid (based on the ground) to help the pilot establish the bearings of the aircraft; *see also* BAND

COMMENT: the VOR projects 360 radials which can be followed to fly a particular path over the ground. VORs operate on VHF frequencies between 108.0 to 177.95 MHz

QUOTE a Baltimore man adjusted a baby alarm to improve its performance and found his youngster's squawks were being picked up by

incoming aircraft tuned to
the local NDB frequency

Pilot

frequent ['fri:kwənt] *adjective*
happening or appearing often; *frequent inspection*

friction ['frikʃn] *noun (physics)* a force that resists the relative motion or tendency to such motion of two bodies in contact; *energy is converted to heat through friction*

front [frʌnt] *noun* **(a)** *(of building)* the forward part or surface; *the entrance is at the front* **(b)** the area, location, or position directly before or ahead; **in front =** in a forward position relative to something else; *row 23 is in front of row 24* **(c)** *(meteorology)* the mixed area between air masses of different temperatures or densities; **cold front =** advancing mass of cold air, moving under and lifting warmer air; *a cold front brought rainy, windy conditions to the country*; **warm front =** advancing mass of warm air moving over a mass of cooler air; *see also* SURFACE

frontal [frʌntl] *adjective* **(a)** referring to the forward part or surface area of something; *the frontal area*; **frontal surface =** the boundary between two air masses **(b)** of or relating to a meteorological weather front; *a frontal storm*; **a frontal system =** a series of rain-bearing changes in the weather

frost [frɒst] *noun* deposit of very small ice crystals formed when water vapour condenses at a temperature below freezing; *frost had to be cleared from training aircraft which had been parked outside overnight*; *see also* HOAR

ft = FOOT

fuel ['fju:əl] *noun* substance (gas, oil, petrol, etc.) which is burnt to produce heat or power; *each wing tank holds 20 gallons of fuel*; *a fuel system includes tanks, fuel lines, fuel pumps, fuel filters and a carburettor or fuel injection system*; **fuel gauge =** instrument indicating fuel contents; **fuel injection =** spraying of fuel, under pressure, into the combustion chamber of an engine; **fuel injector =** device to introduce fuel under pressure into the combustion chamber of an engine; *a substance is added to the fuel to clean fuel injectors*; *see also* LINE; MIXTURE

fumes [fju:mz] *plural noun* smoke, gas or vapour given off by a substance, often unpleasant or harmful; *when the cabin is rapidly and completely filled by smoke and fumes, passengers will suffer from disorientation*

function ['fʌŋkʃn] **1** *noun* **(a)** specific occupation or role; *rota planning is one of the functions of the chief instructor* **(b)** purpose; *seals perform a very important function in a hydraulic system*; *the function of the flaps is to increase lift and drag* **2** *verb* **(a)** to act as; to serve the purpose of; *the escape slide also functions as a life raft* **(b)** to operate; to work; *the system functions well*

fundamental [fʌndə'mentl] *adjective* **(a)** of or relating to the foundation or base; *the fundamental laws of aerodynamics* **(b)** forming or serving as an essential component of a system or structure; central; *electricity is one of the fundamental types of energy that exist in nature*

fungal growth ['fʌŋgəl 'grauθ] *noun* type of organism which lives and multiplies in certain fuels; *fuel contains chemicals for the inhibition of fungal growth*

fuse [fju:z] *noun* safety device that protects an electric circuit from excessive current; *circuit breakers perform the same function as a fuse*

fuselage ['fju:zəlɑ:ʒ] *noun* central body of an aircraft, to which the wings and tail assembly are attached and which accommodates the crew, passengers, and cargo; *the fire started in the wing but soon spread to the fuselage*

Gg

g [dʒiː] **(a)** symbol representing the acceleration due to earth's gravity **(b)** = GRAM

G [dʒiː] = GIGA

gain [geɪn] **1** *noun* **(a)** increase; *there is a gain of heat by the earth due to solar radiation*; *a gain in altitude* = an increase in altitude **(b)** increase in signal power, voltage, or current; *the amplifier boosts the gain of the incoming signal* **(c)** benefit, advantage; *an additional gain from the horizontally opposed arrangement of cylinders is a smoother running engine* **2** *verb* **(a)** to increase; *he failed the test because the aircraft gained 100 ft in the 360° level turn* **(b)** *(a result in an examination)* to get or obtain; *he gained a pass in his meteorology exam*

gale [geɪl] *noun* very strong wind (force 8 on the Beaufort scale) usually blowing from a single direction; *gales are forecast for the area*

gallon ['gælən] *noun* **(a)** imperial gallon = unit of volume in the British Imperial System, used in liquid measure (and sometimes in dry measure), equal to 4.546 litres; *the system delivers fuel at the rate of 100 to 2,000 gallons per hour* **(b)** unit of volume in the US Customary System, used in liquid measure, equal to 3.785 litres

gamma rays ['gæmə 'reɪz] *noun* electromagnetic radiation given off by some radioactive substances; *gamma rays are given off when radioactive material breaks down*

gap [gæp] *noun* **(a)** space between objects or points; *an air gap igniter plug is similar in operation to a piston engine spark plug*; **air gap type spark plug** = spark plug with a gap between the electrodes across which the spark jumps **(b)** difference; *micro switches have a very small gap between make and break* **(c)** opening; *the pilot could see the airfield through a gap in the clouds*

gas [gæs] *noun* state of matter other than solid and liquid; *oxygen and nitrogen are gases*; **gas turbine engine** = engine with a turbine which is rotated by expanding hot gases; *see also* EXHAUST

gaseous ['gæsiəs] *adjective* relating to, or existing as a gas; *the atmosphere is the gaseous envelope surrounding the earth*

gasket ['gæskɪt] *noun* any of a wide variety of seals or packings used between matched machine parts or around pipe joints to prevent the escape of a gas or fluid; *seals, gaskets and packing make a seal by being squeezed between two surfaces*

gasoline ['gæsəliːn] *noun* US liquid (made from petroleum) used as a fuel in internal combustion engine; **aviation gasoline (AVGAS)** = fuel used in piston engined aircraft; *see also* PETROL

gate [geɪt] *noun* **(a)** device for controlling the passage of water or gas through a pipe; *the waste gate may be controlled manually by the pilot; during a descent from altitude, with low power set, the turbocharger waste gate is fully closed* **(b)** *(electronics)* a circuit with many inputs and one output that works only when a particular input is received; *a logic gate is almost the same as a switch* **(c)** device to prevent a lever from being moved to an incorrect setting; *it is necessary to move the rpm control lever through a feathering gate to the feathering position*

gauge [geɪdʒ] **1** *noun* **(a)** instrument for measuring or testing; *temperature gauge; pressure gauge*; **fuel gauge** = instrument to indicate fuel contents; **intake temperature gauge** = instrument to indicate the temperature of air entering an engine **(b)** unit of diameter or width; **heavy gauge wire** = thick wire **2** *verb* calculate approximately by using the senses; *in fog, it is difficult to gauge horizontal distances* (NOTE: **gauging - gauged**)

gear [gɪə] *noun* **(a)** toothed wheel that turns with another toothed part to transmit motion or change speed or direction; **reduction gear** = gears in an engine which allow the propeller to turn at a slower speed than the engine **(b) landing gear** = alternative name for the undercarriage; **main gear** = two main landing wheel assemblies; **nose gear** = landing wheel in the nose of the aircraft **(c) valve gear** = the mechanism for opening and closing valves **(d)** *(informal)* equipment and/or clothing; **flying gear or flight gear** = (might include) flying suit, headset, case with charts, manuals, etc.

COMMENT: the main landing gear are nearest the aircraft's center of gravity. Main landing gear are designed to withstand a greater landing shock than the nose wheel or tail wheel and consequently should make contact with the surface first when landing

gearbox [ˈgɪəbɒks] *noun* device to allow changes in the ratio of engine speed to final drive speed; *the auxiliary power unit*

(APU) is a small gas turbine engine which is connected to a gearbox

genera [ˈdʒenərə] *noun see* GENUS

general [ˈdʒenrəl] *adjective* which is concerned with or applicable to a whole group of people or things; **general description** = not a detailed description; **general flying test (GFT)** = test of aircraft handling skills for student pilots; **general principles** = main ideas; **general purpose switches** = all purpose switches; **general weather situation** = the overall weather picture without the detail; **as a general rule** = usually; **in general** = usually; **in general use** = used a lot

generate [ˈdʒenəreɪt] *verb* **(a)** to bring into being; *in an emergency, it may be necessary for crew to generate a little panic in passengers to motivate them to move* **(b)** to produce something such as heat or electricity as a result of a chemical or physical process; *the passage of air around the wing generates lift*

generation [dʒenəˈreɪʃn] *noun* **(a)** act or process of creating or making; **generation of ideas** = process of producing or getting ideas; **generation of electricity** = production of electricity **(b)** class of objects derived from an earlier class; **a new generation of computers** = computers which share a recent development in computer technology which separates them as a class from earlier computers

generator [ˈdʒenəreɪtə] *noun* power-operated device for making electricity; *starter generators are a combination of a generator and a starter housed in one unit*

genus [ˈdʒiːnəs] *noun (of cloud)* class, group, or family; *various types of cloud are grouped into ten basic cloud genera* (NOTE: the plural is **genera**)

geographic(al) [dʒiːəˈgræfɪk] *adjective* referring to geography; *a specific geographical area*; *the north geographic pole*

geography [dʒi'ɒgrəfi] *noun* **physical geography** = study of the earth's surface and its features

geometric [dʒi:ə'metrɪk] *adjective* referring to geometry; *a triangle is a geometric figure; geometric pitch (US) is the distance which a propeller should move forward in one revolution*

geometry [dʒi'ɒmətri] *noun* **(a)** study of the properties, measurement, and relationships of points, lines, angles, surfaces, and solids; *an understanding of geometry is essential to the student of navigation* **(b)** configuration; arrangement; *the geometry of the engine nacelle;* **variable geometry** = technology which allows the angle between wing and fuselage to be altered to give a more or less swept wing for better high-speed and low-speed flight characteristics

geostationary [dʒi:əʊ'steɪʃnri] *adjective* referring to an object, such as a satellite in space, which rotates round the earth at the same speed as the earth and is therefore stationary with reference to a point on the earth; *there are two main types of satellite that are used for collection and transmission of meteorological data, polar and geostationary*

geostrophic wind [dʒi:əʊ'strɒfɪk] *noun* wind which blows horizontally along the isobars, across the surface of the earth

GFT = GENERAL FLYING TEST

GHz = GIGAHERTZ

giga- (G) ['gɪgə] *prefix meaning* one thousand million

gigahertz (GHz) ['gɪgəhɜːts] *noun* frequency of 10^9 Hertz

given ['gɪvn] *adjective* **(a)** particular; specified; fixed; *at high altitudes, less fuel is consumed for a given airspeed than for the same airspeed at a lower altitude* **(b)** **given (that)** = taking into account; considering; *given the condition of the engine, it is surprising that it starts*

glare [gleə] *noun* strong blinding light; *glare can be caused by diffuse reflection of sunlight from the top of a layer of fog*

glide [glaɪd] *verb* to fly without power; *in the event of an engine failure, it is important to have enough altitude to be able to glide clear of houses, people, etc.*

glidepath *or* **glide path** ['glaɪd 'pɑːθ] *noun* path followed by the aircraft down the glideslope; **glidepath coverage** = vertical and horizontal dimensions of the glideslope radio beam

glider ['glaɪdə] *noun* fixed wing aeroplane, normally with no power plant propulsion; *nowadays, gliders are often made of composite materials*

glideslope *or* **glide slope** ['glaɪdsləʊp] *noun* part of the ILS (Instrument Landing System) which provides a radio beam at an angle of approximately 3° to the point of touchdown from the outer marker thus giving the pilot information about the height of the aircraft on final approach; *if, during the approach, the aircraft deviates beyond the normal ILS (Instrument Landing System) glide slope and/or localizer limits (and when below 500 ft above ground level). the flight crew are alerted by the respective deviation pointers changing colour from white to amber and flashing*

gliding ['glaɪdɪŋ] *noun* **(a)** flying in a glider; **gliding club** = association of members who fly gliders as a pastime **(b)** flying in a powered aircraft with the engine either switched off or not producing power (idling); *the best gliding speed for the aircraft is 75 knots*

COMMENT: on June 24th 1982, a British Airways 747 flying from Kuala Lumpur to Perth lost all power from all four engines for 13 minutes, yet landed safely in Jakarta: proof that even a large aircraft is capable of gliding

global ['gləʊbl] *adjective* worldwide; referring to something related to the whole earth; **global pressure patterns** = pressure patterns of the whole planet; **global**

positioning system (GPS) = satellite-based navigation system

globe [gləʊb] *noun* object shaped like a ball; sphere; *if the earth were a uniform globe, the average temperature would vary only with latitude*

GMT ['dʒiː 'em 'tiː] = GREENWICH MEAN TIME

go-around ['gəʊə'raʊnd] *noun (following a missed approach)* a climb into the circuit and manoeuvring into position for a new approach and landing; *because he was too high on the approach, the pilot executed a go-around*

govern ['gʌvn] *verb* to control or limit the speed, size or amount of something; *the size and number of valves required for a particular type of aircraft is governed by the amount of air necessary for pressurization and air conditioning; the type of undercarriage fitted to an aircraft is governed by the operating weight*

governor ['gʌvnə] *noun* device for controlling or limiting the speed size or amount of something; *overspeeding of the engine is prevented by a governor in the fuel system; see also* VALVE

gph ['dʒiː 'piː 'eɪtʃ] = GALLONS PER HOUR

GPS ['dʒiː 'piː 'es] = GLOBAL POSITIONING SYSTEM

GR *see* HAIL

grade [greɪd] *noun* **(a)** position in a scale of size or quality; *Kevlar 49 is the grade used in aircraft composites* **(b)** mark indicating a student's level of accomplishment; *students who were below a certain grade in the examinations were not allowed to continue the course*

gradient ['greɪdiənt] *noun (physics)* the rate at which a physical quantity, such as temperature or pressure, changes relative to change in a given variable, especially distance; *because there is a temperature gradient across each front it is possible to draw isotherms which reduce in value from warm to cold air; a pressure gradient occurs aloft from land to sea;* **wind gradient** = rate of increase of wind strength with unit increase in height above ground level; *after take-off, as the aircraft gains altitude, the ground speed may be affected by the wind gradient*

gradual ['grædʒuəl] *adjective* referring to something which progresses with continuous but unhurried certainty; *loss of cabin pressure may be gradual rather than sudden;* **gradual change** = change which takes place over a period of time

graduate ['grædjueɪt] *verb* **(a)** to be granted an academic degree or diploma; *she graduated from Oxford University with a first class honours degree* **(b)** to advance to a new level of skill, achievement, or activity; *after 50 hours of flying the single engine trainer, the student pilots graduate to flying the twin engine aircraft* **(c)** to divide into marked intervals, especially for use in measurement; *a thermometer has a scale graduated in degrees Celsius*

gram (g) [græm] *noun* unit of measurement of weight, equal to one thousandth of a kilogram

graph [grɑːf] *noun* diagram that shows a relationship between two sets of numbers as a series of points often joined by a line; *the graph shows the relationship between lift and drag at various airspeeds*

graphic ['græfɪk] **1** *adjective* **(a)** **graphic solution** = technique of using geometric constructions to solve problems; *one side of the calculator has a moveable slide which is used for the graphic solution of triangle of velocities problems* **(b)** described in vivid detail; *the eye witness provided a graphic description of the events leading to the accident* **2** *noun* picture used in a computer application; *the instructor's worksheets were greatly improved by the incorporation of graphics to aid comprehension of the subject matter*

graticule ['grætɪkjuːl] *noun* **(a)** series of fine lines in an optical instrument such as a telescope, used for measuring **(b)** the network of lines formed by the meridians and parallels of longitude and latitude of the earth on a flat sheet of paper; *a graticule of lines of latitude and longitude is imagined to cover the earth*

gravity ['grævɪti] *noun* **(a)** natural force of attraction which pulls bodies towards each other and which pulls objects on earth towards its centre; *in order for an aeroplane to fly, lift must overcome the force of gravity*; **acceleration due to earth's gravity (g)** = pulling force exerted on a body by the earth - the international standard value being 9.80665 m/s² **centre of gravity** = point of balance of an object; point through which the resultant force of gravity acts; *see also* FORCE **(b)** seriousness; *throughout the crisis caused by the engine failure, the passengers were unaware of the gravity of the situation*

great [greɪt] *adjective* **(a)** large in size, quantity, number, etc.; **great distances** = long distances; **a great deal of money** = a large sum of money; **great importance** = enormous importance **(b)** *(informal)* very good or enjoyable; exciting; **a great flight** = a very good flight

Greenwich Mean Time (GMT)

['grɪnɪtʃ 'miːn 'taɪm] *noun* local time on the Greenwich (UK) Meridian

> COMMENT: GMT is now called Coordinated Universal Time (UTC) and is also known as Zulu time. UTC is expressed in 24-hour format; for example, 7:00 P.M. is 1900 hours (say: nineteen hundred hours)

grid [grɪd] *noun* **(a)** pattern of equally spaced vertical and horizontal lines, sometimes used on a map; *grid lines facilitate the quick location of a point of reference* **(b)** metal cylinder in a cathode ray tube **(c)** pattern of equally spaced vertical and horizontal metal rods or bars; *lead-antimony alloy grid plates are components in a lead-acid battery*

ground [graʊnd] **1** *noun* solid surface of the earth; *hail being much denser and heavier than snow, falls at a much faster*

rate and can reach the ground even with *the 0° isotherm at 10,000 ft*; **low ground** = area of land which is not high (as opposed to mountains); *cold air flows downwards and accumulates over low ground*; **ground crew** = team of personnel who service and maintain the aircraft while it is on the ground; **ground elevation** = vertical distance, in feet, of the ground above sea level; **ground features** = bridges, rivers, railway lines, roads, etc. which are useful aids to navigation; **ground position** = the point on the surface of the earth immediately beneath the aircraft; **ground-running operation** = procedure of running the engine while the aircraft is stationary on the ground to check engine performance; **ground speed** = speed of the aircraft in relation to the ground; **ground temperature** = the temperature recorded by a thermometer placed at ground level; **ground visibility** = horizontal visibility near the surface of the earth **2** *verb* **(a)** to prohibit an aircraft or member of an aircrew from flying; *the pilot was grounded after failing a medical examination* **(b)** *US* to connect an electrical circuit to a position of zero potential; *while refuelling a light aircraft it is important to ground the airframe to prevent sparking caused by static electricity* (NOTE: **to earth** is preferred in British English)

group [gruːp] *noun* **(a)** number of individual items or people brought together because of similarities; **cloud group** = collection of different cloud types which have similarities, such as stratus clouds **(b)** collection of letters, numbers or symbols used in weather forecasting, etc.; **a four-digit group** = four single numbers found together; **a three-letter group** = three letters of the alphabet found together

growth [grəʊθ] *noun* increase in size, number, amount, etc.; *the growth of ice crystals*; *the growth of air travel*

guard [gɑːd] **1** *noun* **(a)** device to prevent injury or loss, etc.; *the thermocouple probes consist of two wires of dissimilar metal that are joined together inside a metal guard tube* **(b)** person who protects or keeps watch; *a security guard* **2** *verb* to protect from harm by watching over; **to guard against** = to take steps to ensure that something does not happen; *to guard against the risk of fire, passengers are requested not to smoke in the toilets*

guidance ['gaɪdəns] *noun* **(a)** helpful advice; *guidance is provided to assist people in filling in the form*; *the booklet contains guidance on the advisability of flying with a cold* **(b)** giving directions to an aircraft; **guidance system** = system which provides signals to the flight control system for steering the aircraft; *systems which are designed to carry out automatic landings under all visibility conditions must provide better guidance and control than a pilot*

guide [gaɪd] **1** *noun* something that directs or indicates; **intake guide vane** = device to direct the flow of air at the air-intake; **rough guide** = simple explanation to help a person to find his own way through more complex information **2** *verb* to direct; to indicate; *if there is smoke in the cabin, clear commands from the crew will help to guide passengers to the emergency exits*

gust [gʌst] **1** *noun* strong, sudden rush of wind; *on final approach, the pilot must be prepared to counteract the effect of gusts in order to maintain a smooth descent along the extended centreline of the runway*; *a gust of 30 feet per second*; **gust load** = increased load to the airframe caused by a sudden increase in wind strength **2** *verb* (*of wind*) to increase in strength suddenly; *wind is at 10 knots gusting to 20 knots*

gyro ['dʒaɪrəʊ] *noun* = GYROSCOPE

gyro- ['dʒaɪrəʊ] *prefix meaning* gyroscopic

gyrocompass ['dʒaɪrəʊ'kʌmpəs] *noun* gyroscopic compass; compass which uses gyroscopic directional stability rather than magnetism to indicate direction; *the gyrocompass should be checked against the magnetic compass and reset if necessary*

gyroscope ['dʒaɪrəskəʊp] *noun* device consisting of a spinning wheel, mounted on a base so that its axis can turn freely in one or more directions and thereby maintain its own direction even when the base is moved; *the traditional attitude indicator, heading indicator and turn-coordinator contain gyroscopes*; *see also* DIRECTIONAL (NOTE: often shortened to **gyro**)

COMMENT: a spinning gyro maintains its position even when an aircraft banks, climbs, or dives. Gyros drive the attitude indicator, direction indicator and turn coordinator to help pilots control an aircraft while flying in cloud or in poor visibility

gyroscopic [dʒaɪrə'skɒpɪk] *adjective* referring to a gyroscope; using the properties of a gyroscope; **gyroscopic compass** = GYROCOMPASS

gyroscopic precession ['dʒaɪrə'skɒpɪk prɪ'seʃn] *noun* characteristic of a gyroscope, that the force applied to a spinning gyroscope will act at a point 90° in the direction of rotation, not at the point where the force is applied; *forces of gyroscopic precession act on the direction indicator to keep it aligned vertically and horizontally*

Hh

hail [heɪl] *noun* precipitation as small pellets of ice; *precipitation is the falling of water, as rain, sleet, snow or hail onto the surface of the earth; although hail, and in particular, heavy hail is rare and of short duration, damage to an aircraft may be severe*

> COMMENT: in weather reports and forecasts, hail is indicated by the abbreviation 'GR'

hailstone ['heɪlstəʊn] *noun* small pellet of ice which falls from clouds; *a hailstone starts as a small ice particle in the upper portion of a cumulonimbus cloud*

hailstorm ['heɪlstɔːm] *noun* storm, where the precipitation is hail (not rain, not snow); *by flying through the hailstorm, damage was done to the leading edges*

hand-held ['hændheld] *adjective* which can be held in the hand; *nowadays, headsets are usually used in preference to hand-held microphones; see also* HOLD

handle ['hændl] **1** *noun* device for holding, or being operated, by the hand; *a door handle; a fire control handle* **2** *verb* **(a)** to touch with the hands; *cabin staff should not handle unwrapped food which is to be served to passengers* **(b)** to move or operate by hand; *the student pilot handled the aircraft well in the turbulent conditions* **(c)** to deal with; to manage; *flight crew must be able to handle any emergency when it occurs*

handling ['hændlɪŋ] *noun* **(a)** *(such as food)* touching with the hands **(b)** moving or operating by hand; **aircraft handling** = act of manoeuvring the aircraft in the desired manner; **adverse handling characteristic** = aspects of an aircraft handling which are poor; **baggage handling** = process by which passengers' baggage is loaded onto an aircraft, unloaded and moved to the airport terminal; *automation has speeded up baggage handling* **(c)** dealing with; managing; *handling difficult passengers is one of the responsibilities of cabin staff*

hands off ['hændz 'ɒf] *adjective & adverb (of working system)* where the operator does not control the operation which is automatic; *automatic flight control system capable of landing an aircraft hands off*

hangar ['hæŋə] *noun* large shelter for housing and maintaining aircraft; *light aircraft should be left with parking brakes off so that they can be moved quickly in the event of a fire in the hangar*

HASELL *see* MNEMONIC

haul [hɔːl] *noun see* LONG-HAUL; SHORT-HAUL

hazard ['hæzəd] *noun* a possible danger; *thunderclouds are of special interest to aircrew because of the hazards they may pose to aircraft in flight*

hazardous ['hæzədəs] *adjective* which can be risky or dangerous; *flying over mountainous terrain can be hazardous; structural icing is a hazardous phenomenon for rotary wing as well as fixed wing aircraft*

haze [heɪz] *noun* dust or smoke in the atmosphere; *haze can seriously reduce air-to-ground visibility*

head [hed] **1** *noun* **(a)** (i) top part of the body above the shoulders; (ii) person; **head count** = an easy way of counting large numbers of people **(b)** main end part or top of something; **cylinder head** = removable top part of a piston engine cylinder containing plugs, inlet and exhaust connections and valves; **pitot head** = externally mounted device which senses and sends airspeed information to the airspeed indicator in the cockpit; **head wind** = HEADWIND **(c)** leader, chief or director; **head of department** = most senior position in the department **2** *verb* to fly in a particular direction; **head north** = fly towards the north

heading [ˈhedɪŋ] *noun* direction in which the longitudinal axis of the aircraft is pointing, expressed in degrees from north; **heading bug** = a movable plastic marker on the horizontal situation indicator

> COMMENT: wind affects an aircraft in flight. Therefore heading does not always coincide with the aircraft's track. The pilot must head the aircraft slightly into the wind to correct for drift

heading indicator [ˈhedɪŋ ˈɪndɪkeɪtə] *noun* instrument which gives course or direction information such as horizontal situation indicator (HSI) or direction indicator (DI)

> COMMENT: the heading indicator is driven by a gyro and provides steady, exact indications of heading

head-on [ˈhedˈɒn] *adjective & adverb* a **head-on collision** = a collision between two things or vehicles coming from opposite directions; **to approach head-on** = to approach from opposite directions

headphones [ˈhedfəʊnz] *noun* small speakers with padding, worn over a person's ears (used for private listening, instead of loudspeakers); *headphones are used to monitor the signal*

headset [ˈhedset] *noun* headphones with microphone attached, used for RT (radiotelephony) communications; *headsets are usually used in preference to hand-held microphones*

headwind [ˈhedwɪnd] *noun* wind which is blowing in the opposite direction to the direction of movement or flight; **headwind component** = one of the three possible components of a wind, the other two being crosswind and tailwind (NOTE: it is also written **head wind**)

heap [hiːp] *noun* group of things piled or thrown one on top of another; **heap or cumulus cloud** = clouds which form only in an unstable atmosphere and, as the name suggests, often build vertically for great distances

heat [hiːt] *noun* warmth; being hot; *the heat generated by combustion is considerable*; **carburettor heat** = system to prevent icing of carburettors in small piston engine aircraft; *see also* LATENT HEAT **2** *verb* to make warm or warmer; *the air leaving the turbocharger is very warm and can be used to heat the cabin*

heater [ˈhiːtə] *noun* device for heating; *pitot heads contain heater elements to prevent icing*

heating [ˈhiːtɪŋ] *noun* warming; *the heating action of the sun*; *see also* KINETIC

heavy [ˈhevi] *adjective* having a lot of weight; **a heavy load** = a load of great weight; **heavy rain** = rain which is dense and distributes a lot of water over the surface of the earth in a relatively short time; *precipitation is classified as light, moderate or heavy*; **heavy gauge wire** = thick wire

heavy-duty [ˈheviˈdjuːti] *adjective* referring to something designed for hard wear or use; *a heavy-duty battery*; *longerons are heavy-duty steel members*

height [haɪt] *noun* vertical distance of a point, level or object measured from a particular point, eg sea level; *pressure decreases with increasing height*; **height**

of the aircraft = vertical distance, measured in feet, of the aircraft above the surface of the earth

held [held] *see* HOLD

helicopter ['hɛlikɒptə] *noun* aircraft with one or more rotors rotating around vertical axes which provide lift and control; *helicopter operations are carried out at the airport*

heliograph ['hiːliəgrɑːf] *noun* instrument with a mirror to send messages by reflecting the sun; *heliographs enable reflected sunlight to be directed to a ship or aircraft in periods of direct sunlight*

hemisphere ['hemɪsfɪə] *noun* half a sphere; **northern hemisphere** = top half of the earth; **southern hemisphere** = bottom half of the earth

Hertz [hɜːts] *noun* SI unit of frequency, defined as the number of cycles per second of time

HF = HIGH FREQUENCY

high [haɪ] **1** *adjective* **(a)** having great vertical distance; *a high mountain* **(b)** great; large; a lot; **a high degree of** = a lot of; *a high degree of concentration is required when learning instrument flying*; **high engine rpm** = fast engine speed; **high performance aircraft** = an aircraft capable of flying faster, higher or with more manoeuvrability than normal aircraft; **high pressure** = a lot of pressure; **a high price** = expensive; **high reliability** = good reliability; **high speed** = fast; **high temperature** = hot **(c)** **high frequency (HF)** = radio communications range of frequencies between 3-30 MHz; *see also* BAND **2** *noun* area of high atmospheric pressure; *there is a high over the British Isles*

hijack ['haɪdʒæk] *verb* to take over control of an aircraft by one or several unauthorized person or persons with the intention of forcing the crew to fly it to a different destination; *the airliner was hijacked on its way to Paris*

hijacking ['haɪdʒækɪŋ] *noun* taking over control of an aircraft by one or several unauthorized person or persons with the intention of forcing the crew to fly it to a different destination; *the crew must be alert at all times to the possibility of hijacking, bombs and stowaways*

hill [hɪl] *noun* easily seen, natural elevation, smaller than a mountain; *slopes on the side of a hill or mountain facing away from the sun receive less intense radiation*; *hill shading is produced by assuming that bright light is shining across the chart sheet so that shadows are cast by the high ground*

hinder ['hɪndə] *verb* to delay; to make progress difficult; *free flow of fuel may be hindered by a blockage in the fuel line*; *his illness hindered his progress on the course*

hinge [hɪndʒ] **1** *noun* device which allows a door, flap or lid to open and close on a stationary frame; *flying control hinges should be inspected before flight* **2** *verb* (of a surface such as a door) to move against a stationary frame; *access to the engine compartment is normally via hinged cowling panels*

hoar (frost) ['hɔː 'frɒst] *noun* frozen dew which forms on outside surfaces when the temperature falls below freezing point; *rapid descent from cold altitudes into warm moist air may produce hoar frost on the aircraft*

hold [həʊld] **1** *noun* area or compartment within the aircraft for carrying freight; *carry-on baggage is limited by regulations as to size and weight and items in excess of this should be stowed in the luggage hold* **2** *verb* **(a)** to keep and prevent from moving; *the function of the autopilot system is to hold the aircraft on a desired flight path by means of gyroscopes and/or accelerometers*; *if the operating pressure falls or fails, a mechanical lock holds the reverser in the forward thrust position* **(b)** to keep an aircraft in a particular position on the ground or in the air while waiting for further clearance from air traffic control; *it* *is normal practice for ATC to hold taxiing aircraft well clear of the glide path and localizer antenna when visibility is poor* **(c)** to have and keep in the hand; *hold the microphone in your right hand;* **hand-held** = (device) which can be held in the hand; *nowadays, headsets are usually used in preference to hand-held microphones* **(d)** *(qualification)* to have; **he holds an IMC rating** = he has an IMC rating (NOTE: holding - held)

holder ['həʊldə] *noun* **(a)** device for holding something; *a holder for a fire extinguisher* **(b)** person who has a title or qualification; **a licence holder** = a person who has a licence

holding ['həʊldɪŋ] *noun* **holding fuel** = extra fuel carried by an aircraft to allow for time spent in the hold waiting for air traffic control clearance; **holding pattern** = 'racetrack-shaped' flight pattern with two parallel sides and two turns, flown usually while an aircraft is waiting for clearance to land; **holding point** = a particular location, in the air or on the ground where aircraft spend time, waiting for further clearance from air traffic control

hollow ['hɒləʊ] *adjective* having a space within; not solid; *a hollow drive shaft* (NOTE: the opposite is solid)

home [həʊm] *noun* home airfield; *QUJ is the true bearing from aircraft to station or the true heading home to overhead the station in no wind conditions;* **home airfield** = the airfield which one returns to after a two-leg flight

homeward ['həʊmwəd] **1** *adjective* which is going towards home; *homeward journey* **2** *adverb* **homeward bound** = heading towards home

homewards ['həʊmwədz] *adverb* towards home; *they were heading homewards when the accident happened*

homing ['həʊmɪŋ] *noun* flight towards or away from a radio station while using direction finding equipment; *where an RBI is fitted, homing to an NDB can be*

made by initially turning the aircraft until the relative bearing is zero

homogeneous [həʊməʊ'dʒiːnɪəs] *adjective* of the same kind; *if the air over a large region were homogeneous, there would be no horizontal differences in surface temperature; the atmosphere is not homogeneous - pressure, temperature and humidity can all change with height*

horizon [hə'raɪzən] *noun* the line where the sky and the earth appear to join; **artificial horizon** = horizon line on the attitude indicator; **radio horizon** = line along which direct rays from radio frequency transmitter become tangential to the earth's surface; **visual horizon** = horizon which can be seen

horizontal [hɒrɪ'zɒntl] *adjective* parallel to the horizon; at right angles (90°) to the vertical; *the horizontal motion of air is known as wind*

horizontal axis [hɒrɪ'zɒntl 'æksɪs] *noun (X axis)* horizontal reference line of a graph; *the plot shows the effect of airspeed on lift with airspeed shown on the horizontal axis and lift on the vertical axis*

horizontal situation indicator (HSI) [hɒrɪ'zɒntl sɪtʃu'eɪʃn 'ɪndɪkeɪtə] *noun* cockpit instrument which gives the pilot information about the direction of the aircraft's flight path; *on the aircraft, the horizontal situation indicator is located on the instrument panel below the attitude indicator*

> COMMENT: the HSI combines the function of the heading indicator and a VOR/ILS display

horizontal stabilizer [hɒrɪ'zɒntl 'steɪbɪlaɪzə] *noun* tailplane; *the horizontal stabilizer provides stability about the lateral axis of the aircraft*

horn [hɔːn] *noun* **(a)** device for projecting sound; **warning horn** = device which emits a loud warning noise **(b)** **horn balance** = part of a control surface forward of the hinge line which reduces the force needed by the pilot to move the surface

horsepower (h.p. *or* **HP)** ['hɔːspaʊə] *noun* accepted unit for measuring the rate of doing work; *horsepower is defined as 33,000 foot-pounds of work done in one minute*

hose [həʊz] *noun* long, flexible pipe usually made of fabric, plastic or rubber for pumping gases or liquids; **refuelling hose** = flexible pipe used to pump fuel from the bowser to the aircraft

hot [hɒt] *adjective* very warm; with a high temperature; *hot weather*; **hot air** = air introduced to melt ice forming in the carburettor in a piston engine aircraft

hour ['aʊə] *noun* **(a)** period of time which lasts sixty (60) minutes; *it's a three-hour flight to Greece from London* **(b)** method of indicating time; *flight BA 321 landed at Heathrow at 10.30 hours*

house [haʊz] *verb* to contain; to accommodate; *the areas between the ribs in the wings are utilized to house fuel tanks*; *the wing tips house the navigation lights*

housing ['haʊzɪŋ] *noun* compartment; container; *the crankcase is the housing that encloses the various mechanical parts surrounding the crankshaft*; **engine housing** = engine compartment

hover ['hɒvə] **1** *verb* to remain stationary, relative to the earth, while in the air **2** *noun* remaining stationary, relative to the earth, while in the air; *during a hover, helicopter pilots must be able to co-ordinate movements of both hands and feet*

however [haʊ'evə] *adverb* but; yet; *the wind was gusty, however the landing was good*; *the incident was serious, however he escaped with only a warning*

HSI = HORIZONTAL SITUATION INDICATOR

humid ['hjuːmɪd] *adjective* referring to air containing a lot of water vapour; **humid weather** = weather which, although warm, feels damp and uncomfortable

humidity [hjuːˈmɪdɪti] *noun* measurement of how much water vapour is contained in the air; **the humidity is high** = there is a lot of moisture or water vapour in the air; **absolute humidity** = vapour concentration or mass of water in a given quantity of air; **relative humidity** = ratio between the amount of water vapour in the air and the amount which would be present if the air was saturated, at the same temperature and the same pressure (shown as a percentage)

hydraulic [haɪˈdrɔːlɪk] *adjective* referring to any system or device which uses fluids, such as oil, to transmit a force from one place to another using pipes; *a hydraulic pump*; **hydraulic tubing** = system of tubes or thin pipes connecting the main components of a hydraulic system; **hydraulic pressure** = the pressure exerted by hydraulic fluid; *see also* LABEL

hydro- ['haɪdrəʊ] *prefix meaning* water; *a hydro-mechanical governor*

hygrometer [haɪˈgrɒmɪtə] *noun* instrument used for measurement of humidity; *the most common type of hygrometer is the wet and dry bulb thermometer arrangement*

hypoxia [haɪˈpɒksiə] *noun* medical condition in which not enough oxygen is supplied to the body; *the symptoms of hypoxia are sometimes difficult to detect*

COMMENT: cabin pressurization or oxygen equipment is usually required for flying at altitudes at or above about 10,000 ft (3,048 m)

Ii

IAS = INDICATED AIRSPEED

ICAO [aɪˈkeɪəʊ] = INTERNATIONAL CIVIL AVIATION ORGANIZATION

ice [aɪs] *noun* frozen water; **clear ice** = ice which is glass-like rather than white; **dry ice** = solidified CO_2; **rime ice** = ice formed when individual droplets of water freeze rapidly on striking the aircraft surface

icing [ˈaɪsɪŋ] *noun* process by which part of the aircraft becomes covered in ice while in flight; *engine icing can be extremely hazardous to flight; airframe icing can be encountered in wave cloud; see also* ANTI-ICING; DE-ICING

ideal [aɪˈdiːl] *adjective* perfect; as good as can be expected or the best possible; **an ideal situation** = a very good situation; **ideal flying conditions** = very good flying conditions

ident [ˈaɪdent] *noun* = IDENTITY function on the transponder panel which helps a controller to identify the aircraft; *the ident is suppressed until the standby VOR is fully run-up and has passed its monitor checks*

identical [aɪˈdentɪkl] *adjective* exactly the same; **identical computers** = computers which are exactly the same

identification [aɪdentɪfɪˈkeɪʃn] *noun* process by which a person, aircraft, etc., is recognized; **identification of ground features** = means by which particular features on a chart, such as railway lines or bridges, are matched with the real feature on the ground; **identification beacons** = aeronautical beacons which give out a Morse signal which enables a pilot to establish his location in relation to the beacon; *civil and military aerodrome identification beacons can be distinguished by colour*

identifier [aɪˈdentɪfaɪə] *noun* grouped number/letter code by which a weather station or beacon can be recognized; *when a TAF requires amendment, the amended forecast is indicated by inserting AMD (amended) after TAF in the identifier and this new forecast covers the remaining validity period of the original TAF*

identify [aɪˈdentɪfaɪ] *verb* to recognize; *crew members can be identified by their uniforms; in conditions of poor visibility, it is sometimes difficult to identify ground features*

identity [aɪˈdentəti] *noun* name and details of a person, aircraft, etc.; *the air traffic controllers are trying to establish the identity of the aircraft; see also* IDENT

idle [ˈaɪdl] **1** *noun* state of an engine when it is running but not delivering power to move the vehicle or aircraft; **idle rpm** = speed at which a piston engine turns when it is not running fast enough to move the vehicle or aircraft i.e. on a light aircraft when the throttle is almost closed; **idle cut-off** = position on the mixture control of a light aircraft which allows the engine to be shut down without leaving a combustible fuel/air mixture in the engine **2** *verb (of an engine)* to turn over slowly without providing enough power to move the vehicle or aircraft; *after starting a piston engine from cold, it is good practice to allow it to idle for a short time before opening the throttle wide*

idling ['aɪdlɪŋ] *noun (of piston engine)* turning over slowly without providing enough power to move the vehicle or aircraft; **idling speed** = rpm (revolutions per minute) of the engine when it is idling; *after start-up, the engine accelerates up to idling speed; before the engine is stopped, it should normally be allowed to run for a short period at idling speed to ensure gradual cooling;* **positive idling speed** = specific idling speed selected; (as in piston engine aircraft) idling speed selected to avoid spark plug fouling which may occur at low rpm; *an adjustable stop on the throttle control ensures a positive idling speed*

IFR = INSTRUMENT FLIGHT RULES

ignite [ɪɡ'naɪt] *verb* to cause to burn; to burn; *the spark plug ignites the fuel/air mixture; the air/fuel mixture ignites*

igniter [ɪɡ'naɪtə] *noun* device for starting gas turbine engines; *an electric spark from the igniter plug starts combustion*

ignition [ɪɡ'nɪʃn] *noun* **(a)** the starting of burning of a substance; *satisfactory ignition depends on the quality of the fuel* **(b)** the moment, in an internal combustion engine, when a spark from the spark plug causes the fuel/air mixture to burn; *ignition should occur just before top-dead-centre* **(c)** *(system)* electrical system, usually powered by a battery or magneto, that provides the spark to ignite the fuel mixture in an internal-combustion engine; *ignition problems are a source of many engine failures* **(d)** switch that activates the ignition system; **the key is in the ignition** = the key is in its position in the ignition lock; **ignition key** = key used to switch on the ignition; **ignition lock** = key-operated switch for activating the ignition circuit of an aircraft or a vehicle

illuminate [ɪ'luːmɪneɪt] *verb* **(a)** to give light to an otherwise dark area; *a flare illuminates the ground below it* **(b)** to light a lamp; *when the aircraft is 5 knots above stalling speed, a warning lamp illuminates*

illumination [ɪluːmɪ'neɪʃn] *noun* light; *batteries provide about 20 minutes illumination for the lamp;* **daylight illumination** = the amount of light in normal daytime conditions

illustrate ['ɪləstreɪt] *verb* **(a)** to explain clearly, often by using pictures; to demonstrate; *contour charts illustrate the horizontal distribution of height above mean sea level* **(b)** to show as an example; *a number of aviation disasters have illustrated the importance of clear, correct use of language in R/T (Radiotelephony) communications*

illustration [ɪlə'streɪʃn] *noun* **(a)** picture which explains something; *the illustration on page 23 shows a cross section of a typical gas-turbine engine* **(b)** example; *the mechanics of the föhn wind provide a good illustration of the adiabatic process in action*

ILS ['aɪ 'el 'es] = INSTRUMENT LANDING SYSTEM

ILS glideslope ['aɪ 'el 'es 'glaɪdsləup] *noun* radio beam in ILS (instrument landing system) which gives vertical guidance; *the angle of the glide slope is usually about three degrees to the horizontal; see also* GLIDESLOPE

ILS locator beacon ['aɪ 'el 'es ləu'keɪtə 'biːkn] *noun* non-directional beacon used for final approach; *power output can be as little as 15 watts for an ILS locator beacon*

IM = INNER MARKER

image ['ɪmɪdʒ] *noun* reproduction of the form of an object or person; *although difficult to see, the photograph shows the image of the aircraft with part of the fin missing* (NOTE: it suggests that the image has no detail and that it is the shape which is important)

imaginary [ɪ'mædʒɪnri] *adjective* not real; *the equator is an imaginary line around the earth*

IMC = INSTRUMENT METEOROLOGICAL CONDITIONS

immediate [ɪ'miːdɪət] *adjective* **(a)** happening at once or instantly; *fire extinguishers should be ready for immediate use in the event of an emergency*; in the immediate future = in the very near future **(b)** nearby; close at hand; *the immediate area surrounding the earth is known as the atmosphere*

immerse [ɪ'mɜːs] *verb* to cover completely in liquid; to submerge; *fuel is pumped from the main tanks via fully immersed booster pumps mounted on the base of the fuel tank*

imminent ['ɪmɪnənt] *adjective* which will happen in a very short time; *the transmission made it clear that the aircraft was in imminent danger; a message from the flight deck informs cabin staff that take-off is imminent*

impact ['ɪmpækt] *noun* striking of one body against another; collision; **impact resistance** = the ability of a material to withstand impact without breaking or shattering; *Kevlar 49 has high impact resistance*; **on impact** = as soon as it hit something; *one of the tyres burst on impact (with the ground) super-cooled water droplets start to freeze on impact with an aircraft surface*

impair [ɪm'peə] *verb* to cause to become less effective; *constant exposure to very loud noise impairs the hearing; the pilot's vision may be temporarily impaired by lightning flashes; an incorrect grade of fuel impairs engine performance*

impairment [ɪm'peəmənt] *noun* lessening of effectiveness; *de-icing equipment is used to prevent impairment of the lifting surfaces through ice formation*

impart [ɪm'paːt] *verb* to give; to pass on; *a rotating propeller imparts rearward motion to a mass of air*

impedance [ɪm'piːdəns] *noun* total electrical resistance to current flow in an alternating current circuit; *impedance will vary with changes in frequency*

impede [ɪm'piːd] *verb* to obstruct progress; to hinder; *hills and mountains impede the horizontal flow of air*

impeller [ɪm'pelə] *noun* rotor used to force a fluid in a particular direction; *a turbocharger consists of a turbine wheel and an impeller fitted on the same shaft*

importance [ɪm'pɔːtəns] *noun* significance; strong effect or influence; *upper winds are of great importance in meteorology* (NOTE: the expressions: **of fundamental importance, of great importance, of prime importance, of utmost importance, of vital importance** all mean **very important**)

impose [ɪm'pəuz] *verb* **(a)** to force something upon a person or thing; *the trimmer is used to ease the loads imposed on the flying controls during flight* **(b)** to **impose a fine** = to require somebody to pay a sum of money as punishment; to **impose restrictions** = to place limitations on somebody's actions

improve [ɪm'pruːv] *verb* to make better; *turbochargers improve aircraft performance; the trainee's flying skills improved a lot in a short period of time*

improvement [ɪm'pruːvmənt] *noun* the act or instance of something becoming better; *an improvement in weather conditions enabled the flight to depart*

impulse ['ɪmpʌls] *noun* a force of short duration; *a magneto is designed to produce electrical impulses one after another at precise intervals, so that each separate impulse can be used to provide a spark at a spark plug*; **impulse magneto** = magneto with a mechanism to give a sudden rotation and thus produce a strong spark

inability [ɪnə'bɪlɪti] *noun* being unable to do something; *an inability of the engine to accelerate may be an indication of serious mechanical problems*

inactive [ɪn'æktɪv] *adjective* (*of system*) which is not switched on or is in a passive state; *at the time of the accident the autopilot was inactive*

inadvertent [ɪnəd'vɜːtənt] *adjective* not intended, not meant; accidental; *a safety mechanism prevents inadvertent retraction of the undercarriage while the aircraft is on the ground*

inbound ['ɪnbaʊnd] *adverb & adjective* towards destination; *the aircraft flies outbound from the beacon along the airway and inbound to the facility at the other end of the leg*; **inbound traffic** = aircraft flying towards an airfield

incapacity [ɪnkə'pæsɪti] *noun* inability to do what is needed; not having the necessary power to do something; **crew incapacity** = an injury, etc., to a crew member which prevents him or her from performing his or her normal duties; *accident research has shown that crew incapacity greatly increases the risk to passengers' safety*

inch (in) [ɪnʃ] *noun* British Imperial System unit of length, also used in the US, equal to 25.4 millimetres or 2.54 centimetres or 1/12 of a foot (NOTE: the plural is **inches;** usually written **in** or " with numbers - **5ft 6in** or **5' 6"** - say: five foot six inches)

incidence ['ɪnsɪdəns] *noun* **(a)** frequency of occurrence; *the incidence of structural failure has decreased with the introduction of modern construction materials and techniques* **(b)** **angle of incidence** = angle formed between the chord-line of the mainplane (wing) and the horizontal when the aircraft is in the rigging position

incident ['ɪnsɪdənt] *noun* event or happening which interrupts normal procedure; *a violent passenger had to be removed from the aircraft before departure and details of the incident were reported in the local newspapers*

QUOTE in 1995, a pilot flying above Las Vegas was struck by a laser beam and incapacitated for more than two hours. It was one of over fifty incidents involving

lasers and aircraft reported in the area that year

Pilot

inclination [ɪnklɪ'neɪʃn] *noun* slope or slant from the horizontal or vertical; **angle of inclination** = angle formed between a sloping or surface and a reference point or line which is either horizontal or vertical; *between any two meridians there is an angle of inclination one to the other which varies with latitude*

incline 1 [ɪn'klaɪn] *verb* to slope or slant from the horizontal or vertical; to tilt; *the runway inclines slightly upwards* **2** ['ɪnklaɪn] *noun* slope; slant; *there is a steep incline at the end of the runway*

include [ɪn'kluːd] *verb* to take in as a part; to count along with others; *solid particles in the atmosphere include dust, sand, volcanic ash and atmospheric pollution; a fuel system includes tanks, fuel lines, fuel pumps, fuel filters and a carburettor or fuel injection system* (NOTE: the opposite is **exclude)**

inclusive [ɪn'kluːsɪv] *adjective* taking in the extremes in addition to the part in between; **bearings 180° to 270° inclusive** = bearings 180° and 270° are part of the range of bearings mentioned (NOTE: the opposite is **exclusive)**

incoming ['ɪnkʌmɪŋ] *adjective* (radio waves, solar radiation, etc.) being received; *incoming transmissions; incoming signal; there is a fall of temperature until about one hour after dawn when incoming solar radiation balances outgoing terrestrial radiation* (NOTE: the opposite is **outgoing)**

incorporate [ɪn'kɔːpəreɪt] *verb* to include as part of something which already exists; *some types of outflow valve incorporate safety valves; warning lamps often incorporate a press-to-test facility*

QUOTE the instrument panel on the Mooney Encore has been re-engineered to incorporate improvements

Civil Aviation Training

incorrect [ɪnkə'rekt] *adjective* not correct; not right; *if the trim position is incorrect, a warning horn will sound when number three thrust lever is advanced*

increase 1 ['ɪnkriːs] *noun* **(a)** becoming greater or more; rise; *decreasing engine rpm results in an increase in the rate of descent* **2** [ɪn'kriːs] *verb* to become greater or more; to rise; *as you increase height, the countryside below you appears to flatten out* (NOTE: the opposite is **decrease** or **reduction** for (a), and **decrease** or **reduce** for (b))

increment ['ɪŋkrɪmənt] *noun* something added; *an aircraft encountering a 30 ft/sec gust would experience a load factor increment of 1.61*; *the minimum detection range of a pulse radar system is equal to half the pulse length plus a small increment*

incur [ɪn'kɜː] *verb* to acquire or to receive something (often unwanted); **to incur a financial loss** = to lose money (in a business or commercial sense); *fuel penalties can be incurred if fuel surplus to requirements is carried*; *in some aircraft, the datum shift is operated automatically to cater for any large trim changes incurred by operating undercarriage, flaps, etc.* (NOTE: **incurring - incurred**)

indefinite [ɪn'defənət] *adjective* without limits; **an indefinite period of time** = a period of time which, in reality, may have no end; **an indefinite contract** = a contract which is not limited to a particular period of time

independent [ɪndɪ'pendənt] *adjective* free from the influence or effects of other people or things; **independent system** = system which can operate by itself; *airspeed is independent of wind and is the same regardless whether the aircraft is flying upwind, downwind or at any angle to the wind*

index ['ɪndeks] *noun* alphabetical list of references to page numbers found at the end of a book or long document; **index letter** *or* **index number** = letter or number

which makes it easier to reference or look up information; *each observing meteorological station is shown on the chart as a small circle, identified by its own index number* (NOTE: the plural is **indexes** or **indices** ['ɪndɪsiːz])

indicate ['ɪndɪkeɪt] *verb* **(a)** to show; to point out; *a lamp on the instrument panel will indicate when the pump is operating*; *the needle indicated to zero* **(b)** to serve as a sign or symptom; *black smoke from the exhaust may indicate a rich mixture or worn piston rings*

indicated airspeed (IAS) ['ɪndɪkeɪtɪd 'eəspiːd] *noun* airspeed shown by the cockpit or flight-deck instrument; *the aircraft stalls at an indicated airspeed of 50 knots*

indication ['ɪndɪkeɪʃn] *noun* **(a)** showing; pointing out; *indication of altitude is given on the altimeter* **(b)** sign; symptom; *a drop in engine rpm is an indication of ice forming in the carburettor*; **audible indication** = sound such as a warning bleep; **visual indication** = something (such as a warning lamp, etc.) which is seen and which suggests a more serious cause; *distorted wing panels are often a visual indication of structural damage to the airframe*

indicator ['ɪndɪkeɪtə] *noun* something, such as an instrument, which shows information; **airspeed indicator (ASI)** = primary cockpit or flight deck instrument which gives the speed of the aircraft in relation to the surrounding air; **attitude indicator (AI)** = flight instrument which gives pitch and bank information; **direction indicator (DI)** = instrument which gives direction information; **heading indicator** = instrument which gives course or direction information such as horizontal situation indicator (HSI) or direction indicator (DI); **horizontal situation indicator (HSI)** = instrument which gives the pilot heading information about the direction of the aircraft's flight path; **warning indicator** = indicator which gives notice of a possible problem which may require some action; *see also* VASI; VISUAL

individual [ɪndɪ'vɪdʒul] **1** *adjective* existing as a separate thing; *the hydraulic braking system consists of a master cylinder with individual brake cylinders at each wheel; there is a maintenance manual for each individual engine* **2** *noun* separate human being considered as one rather than as a member of a larger group; *the instructor regards his trainees as a number of individuals rather than a group*

induce [ɪn'djuːs] *verb* to bring about; to cause to happen; *if a coil carrying a changing current is placed near another coil, the changing magnetic field cuts the other coil and induces a voltage in it; unequal deposits on moving parts can induce severe vibration especially on propellers and helicopter rotors*

induced drag [ɪn'djuːst 'dræg] *noun* part of total drag, created by lift; *there are two basic types of drag, induced drag and parasite drag*

COMMENT: induced drag is created when high-pressure air below a wing rotates around the tip to the low-pressure area above and increases as airspeed decreases and angle of attack increases. This is shown to good effect when Concorde lands in wet or humid weather conditions. Wing tip vortices generated by induced drag can be seen

inductance [ɪn'dʌktəns] *noun* measure of a conductor's ability to bring a voltage into itself when carrying a changing current, for example during short times when the circuit is switched on or off; *at low frequencies, the rate of collapse of the magnetic field will be slow and the inductance will be low*

induction [ɪn'dʌkʃn] *noun* **(a)** process by which the fuel/air mixture is drawn into the cylinders of an internal combustion engine; *the four strokes of the engine are induction, compression, combustion and exhaust* **(b)** production of electrical current in a conductor by a change of magnetic field; *a transformer is a static device that changes the amplitude or phase of an alternating voltage or current by electro-magnetic induction*

inductive [ɪn'dʌktɪv] *adjective* referring to the production of electrical current in a conductor by a change of magnetic field; *one side effect of low frequency in an inductive circuit is that excess heat may be produced*

inductor [ɪn'dʌktə] *noun* component in the ignition system that produces electrical current in itself by a change of magnetic field; *an inductor can be used to control some circuits by opposing changes in current flow*

inert [ɪ'nɜːt] *adjective* which does not react with other substance; **inert gas** = helium, neon, argon, krypton, xenon; *inert gases, dust, smoke, salt, volcanic ash, oxygen and nitrogen together constitute 99% of the atmosphere*

inertia [ɪ'nɜːʃə] *noun* the tendency of a body at rest to stay at rest or of a moving body to continue moving in a straight line unless acted on by an outside force; *inertia switches operate automatically when a certain g (acceleration due to earth's gravity) loading occurs*

inertial [ɪ'nɜːʃl] *adjective* referring to inertia; **inertial navigation system** = navigation system which calculates aircraft position by comparing measurements of acceleration with stored data, using gyros rather than radios

inferential [ɪnfə'renʃl] *adjective (of results, etc.)* obtained by deduction; *the inferential method of ice detection is used in flight trials for aircraft certification*

inflammable [ɪn'flæməbl] *adjective* easily set on fire; *petrol is an inflammable liquid;* **highly inflammable** = very easily set on fire (and therefore hazardous) (NOTE: **flammable** and **inflammable** mean the same thing; to avoid confusion, it is recommended to use **flammable**; the opposite is **non-flammable**)

inflate [ɪn'fleɪt] *verb* to blow air into something and thereby increase its size; *a sharp pull on the cord will discharge the gas bottle and inflate the life jacket* (NOTE: the opposite is **deflate**)

inflation [ɪn'fleɪʃn] *noun* **(a)** the act of blowing air into something - for example a balloon or a tyre - and so increasing its size; *tyre inflation pressures should be maintained within 4% limits* **(b)** continuing increase in the price of things and a decrease in the buying power of money; *annual inflation is 4%* (NOTE: the opposite is **deflation**)

in-flight ['ɪn'flaɪt] *adjective* which takes place during a flight; *in-flight emergency*; *in-flight oil loss*

influence ['ɪnfluəns] **1** *noun* power which affects people or things; *the Atlantic Ocean has a great influence on the climate of the British Isles* **2** *verb* to have an effect on; to change; *in an emergency, a crew member's power of command will influence the reaction of passengers*

inform [ɪn'fɔːm] *verb* to tell somebody something; *after a particularly heavy landing, the pilot should inform an engineer so that checks can be made to the aircraft structure*

information [ɪnfə'meɪʃn] *noun* collection of facts or data; *meteorological visibility gives information on the transparency of the atmosphere to a stationary ground observer* (NOTE: no plural)

infra- 'ɪnfrə] *prefix meaning* below or beneath

infrared *or* **infra-red** [ɪnfrə'red] *adjective* referring to the range of invisible radiation wavelengths from about 750 nanometres to 1 millimetre; *solar radiation is short wave and of high intensity while terrestrial radiation is infra-red*

infrequent [ɪn'friːkwənt] *adjective* not often; *in northern Europe, thunderstorms are infrequent in winter time*

ingest [ɪn'dʒest] *verb* to take in or to absorb (such as into a jet engine through the intake); *jet engines may be damaged by ingested chunks of ice*

ingestion [ɪn'dʒestʃn] *noun* taking in or absorbing (such as into a jet engine through the intake); *ingestion of birds may seriously damage the blades of turbo-fan engines*

inherent [ɪn'hɪərənt] *adjective* existing as a basic or fundamental characteristic; *a boiling point of 100°C is an inherent characteristic of water*

inhibit [ɪn'hɪbɪt] *verb* to prevent or to limit (the effect of) something; *cloud cover inhibits cooling of the earth's surface at night*

inhibition [ɪnhɪ'bɪʃn] *noun* prevention or limitation of (the effect of) something; *fuel contains chemicals for the inhibition of fungal growth*

inhibitor ɪn'hɪbɪtə] *noun* device or substance which prevents or limits (the effect of) something; **icing inhibitor** = substance added to fuel to prevent fuel system icing

initial [ɪ'nɪʃl] **1** *adjective* **(a)** relating to or occurring at the beginning; first; **initial climb** = period of climb immediately after take-off; **initial letter** = the first letter of a word; **initial stage** = first stage **2** *noun* the first letter of a word; **initials** = first letters of a name, such as JS for John Smith; *his name is John Smith, his initials are JS*

initiate [ɪ'nɪʃieɪt] *verb* to get something going by taking the first step; to start; *in a serious emergency, a member of the cabin crew may initiate an evacuation of the aircraft*

initiation [ɪ'nɪʃieɪʃn] *noun* getting something going by taking the first step; starting; *normally speaking, the captain is responsible for the initiation of emergency procedures*

initiative [ɪ'nɪʃətɪv] *noun* power or ability to begin or to follow through competently with a plan or task; *crew* *members must be able to act collectively and with initiative in unusual situations*

inject [ɪn'dʒekt] *verb* to force or drive a fluid into something; *an accelerator pump, operated by the movement of the throttle lever, injects fuel into the choke tube*

injection [ɪn'dʒekʃn] *noun* the forcing of fluid into something; *power output can be boosted to a value over 100% maximum power, by the injection of a water methanol mixture at the compressor inlet or at the combustion chamber inlet*; **fuel injection (system)** = system in which fuel is sprayed under pressure into the combustion chamber of an engine

injector [ɪn'dʒektə] *noun* device that will force or drive a fluid into something; *(in fuel injection system)* **fuel injector** = injector that sprays fuel into the combustion chamber of an engine

injury ['ɪndʒəri] *noun* damage or harm done to a person; *escape slides are designed to minimize the risk of injury to passengers when leaving the aircraft*

inland ʀ'ɪnlənd] *adjective & adverb* referring to the interior of a country or land mass; *sea fog can extend for considerable distances inland*

inlet ['ɪnlət] *noun* **(a)** opening which allows an intake of something; *turbine inlet*; *combustion chamber inlet*; *air enters the cabin through an inlet*; **inlet valve** = the valve in a piston engine which allows fuel to enter the cylinder **(b)** coastal feature such as at the mouth of a river

inner ['ɪnə] *adjective* positioned farther inside; **inner wing** = the part of the wing near the fuselage; **inner marker** = radio beacon placed between the middle marker and the end of the ILS runway; *see also* MARKER

inoperative [ɪn'ɒpərətɪv] *adjective* not functioning; *to prevent accidental retraction of the undercarriage, a safety switch is fitted in such a way to the oleo, that when it is compressed on the ground, the 'undercarriage up' selection is inoperative*

input ['ɪnpʊt] *noun* something, such as energy, electrical power or information, put into a system to achieve output or a result; *pumps require high input current*; *if the number of turns on the secondary winding is greater than the number of turns on the primary, the output voltage from the secondary will be greater than the input voltage to the primary*; **pilot control input** = movements on the flying controls made by the pilot

insert [ɪn'sɜːt] *verb* to put in(to); *to prevent tyre explosion due to overheating, fusible plugs are inserted into the wheel assemblies*; *insert your telephone number in the space provided on the form*; *insert the key in the lock and turn it*

insertion [ɪn'sɜːʃn] *noun* putting in(to); *there is a space on the form for the insertion of a postal address*; *when the contours for a particular pressure level have been drawn in, the chart is completed by insertion of spot temperatures and wind speed information*

insignificant [ɪnsɪg'nɪfɪkənt] *adjective* not important; of no consequence; *minor changes in wind speed or direction are insignificant*

inspect [ɪn'spekt] *verb* to look at something closely and to check for problems; to examine for defects; *propellers should be inspected prior to flight*

inspection [ɪn'spekʃn] *noun* careful check for problems; *before flight, the pilot should carry out a careful inspection of the aircraft*

instability ʀɪnstə'bɪlɪti] *noun* condition in which a body or mass moves easily, and with increasing speed, away from its original position; *atmospheric instability often results in strong vertical currents of air*; *the built-in instability of some modern fighter aircraft makes them highly manoeuvrable but difficult to control without fly-by-wire technology* (NOTE: the opposite of **instability** is **stability**; but the opposite of the adjective **stable** is **unstable**)

install [ɪn'stɔːl] *verb* to put in position, connect and make ready for use; *most carburettors are installed in a warm position to help against icing*; **installed battery** = a battery in position in the aircraft

installation ʀɪnstə'leɪʃn] *noun* (a) act of putting equipment or devices into position and connecting them for use; *the installation of the computer took three hours* (b) equipment or devices which are installed; *in some auxiliary-power-unit installations the air intake area is protected against ice formation by bleeding a supply of hot air from the compressor over the intake surfaces*; **airborne installation** = radio device in an aircraft which operates in conjunction with a ground installation; *the airborne installation comprises an antenna, receiver and indicator(s)*

instance ['ɪnstəns] *noun* example which is used to support or contest a statement; *failure to check fuel levels before take-off is an instance of bad airmanship*; **for instance** = for example

instant ['ɪnstənt] **1** *adjective* immediate; happening immediately; **an instant increase** = an immediate increase **2** *noun* very short period of time; *the pilot has to act in an instant to counteract the severe downdrafts of a microburst*

instinctive [ɪn'stɪŋktɪv] *adjective (of personal reaction)* which is natural, rather than thought-out; *in most modern light aircraft, use of the trim wheel is instinctive, i.e. forwards for nose down and backwards for nose up*

instruct [ɪn'strʌkt] *verb* to give information or knowledge, usually in a formal setting such as a lesson or briefing; *the safety officer instructs employees on the use of the breathing equipment*; *the training captain instructs trainee pilots in the simulator*

instruction [ɪn'strʌkʃn] *noun* (a) act of giving information or knowledge, usually in a formal setting such as a lesson or briefing; *trainees receive first-aid instruction* (b) information on how something should be operated or used; *you must follow the instructions*; **instruction manual** = book containing information on how something should be operated or used

instructor [ɪn'strʌktə] *noun* person who gives information or knowledge, usually in a formal setting such as a lesson or briefing; **chief flying instructor (CFI)** = senior rank of flying instructor; **flying instructor** = trained person, a pilot, who teaches people how to fly an aircraft; **ground instructor** = trained person who teaches support subjects, such as meteorology, in a classroom

instrument ['ɪnstrəmənt] *noun* device for recording, measuring or controlling, especially functioning as part of a control system; *airspeed is given on an instrument called the airspeed indicator*; **instrument error** = information provided by an instrument which has to be corrected to obtain a true value; *see also* **PRIMARY**

instrumentation [ɪnstrəmən'teɪʃn] *noun* set of specialized instruments on an aircraft; *instrumentation in some basic light aircraft is restricted to a few instruments only*; *some modern light aircraft have very sophisticated instrumentation*

instrument flight rules (IFR) ['ɪnstrəmənt 'flaɪt 'ruːlz] *noun* regulations which must be followed when weather conditions do not meet the minima for visual flight; *the flight from Manchester to Prestwick was conducted under instrument flight rules*

instrument flying ['ɪnstrəmənt 'flaɪɪŋ] *noun* flying using no references other than the flight instruments; *some conditions require instrument flying*; *when in cloud, instrument flying is required*

instrument landing system (ILS) ['ɪnstrəmənt 'lændɪŋ 'sɪstəm] *noun* aids for an instrument landing approach to an airfield consisting of a localizer, glideslope, marker beacons and approach lights; *the instrument landing system* *provides both horizontal and vertical guidance to aircraft approaching a runway*

> COMMENT: the ILS is the most used precision approach system in the world

instrument meteorological conditions (IMC) ['ɪnstrəmənt miːtiərə'lɒdʒɪkl kən'dɪʃnz] *noun* meteorological conditions of visibility and distance from cloud ceiling which are less than those for visual meteorological conditions; *the basic licence does not permit the pilot to fly in instrument meteorological conditions*

instrument rating (I/R) ['ɪnstrəmənt 'reɪtɪŋ] *noun* additional qualification added to a licence, such as PPL, allowing a pilot to fly in instrument meteorological conditions; *he gained his instrument rating in 1992*

> COMMENT: an instrument rating is required for operating in clouds or when the ceiling and visibility are less than those required for flight under visual flight rules (VFR)

insufficient [ɪnsə'fɪʃnt] *adjective* not enough; *insufficient height resulted in the pilot landing short of the runway*

insulate ['ɪnsjuleɪt] *verb* (a) to prevent the passing of heat, cold or sound into or out of an area (b) to prevent the passing of electricity to where it is not required, especially by using a non-conducting material; *bus bars are insulated from the main structure and are normally provided with some form of protective covering*

insulating ['ɪnsjuleɪtɪŋ] *adjective* (material) which insulates; **insulating tape** = special adhesive tape which is used to insulate electrical wires; *insulating tape was used to prevent the electrical wires from touching*

insulation [ɪnsju'leɪʃn] *noun* (a) act of or state of preventing the passing of heat, cold, sound or electricity from one area to another; *for continuous supersonic flight, fuel tank insulation is necessary to reduce*

the effect of kinetic heating (b) **insulation** *or* **insulation material** = material or substance used to insulate

insulator ['ɪnsjuleɪtə] *noun* substance which will insulate, especially which will not conduct electricity; *wood is a good insulator*

intake ['ɪnteɪk] *noun* opening through which a fluid is allowed into a container or tube; **air intake** = front part of a jet engine where air enters; **intake temperature gauge** = gauge indication temperature at the air intake side of an engine

integral [ɪntɪgrəl *or* ɪn'tegəl] *adjective* (necessary part) which completes the whole or which belongs to a whole; *meteorology is an integral part of a flying training course*; **integral fuel tanks** = tanks which are located within the structure of the aircraft

integrity [ɪn'tegrɪti] *noun* state of being complete and in good working condition; *the engine fire warning system is checked to test its integrity*; *the integrity of an aid used to conduct procedural approaches must be high*

intend [ɪn'tend] *verb* to have a particular plan, aim or purpose; *a battery is intended to supply only limited amounts of power*; **intended track** = desired course of flight

intense [ɪn'tens] *adjective* (a) extreme in amount; **intense heat** = very high heat; **intense wind** = very strong wind (b) **intense concentration** = very hard or deep concentration

intensity [ɪn'tensɪti] *noun* amount or strength of heat, light, radiation; *surface air temperatures depend mostly on the intensity and duration of solar radiation*

intention [ɪn'tenʃn] *noun* course of action one means or plans to follow; *it is not the intention of this chapter to give a detailed description of world weather*; *our intention is to provide safe, cost-effective flying*

inter- ['ɪntə] *prefix meaning* between

interact [ɪntə'ækt] *verb* to act on each other; *angle of attack and the profile of the wing section interact to produce lift; direct and reflected path signals can interact to cause bending of the localizer and/or generation of a false glidepath*

intercept [ɪntə'sept] *verb* to stop or interrupt the intended path of something; *when a radio transmission is made from a moving platform, there will be a shift in frequency between the transmitted and intercepted radio signals*

interconnect [ɪntəkə'nekt] *verb (of several devices)* to connect together; *the fire extinguishers for each engine are interconnected, so allowing two extinguishers to be used on either engine*

inter-crew ['ɪntə'kru:] *adjective* inter-crew communications = communications between members of the crew; *the lack of inter-crew communication contributed to the accident*

interfere [ɪntə'fɪə] *verb* to interfere with = to come between and thus create a problem; to get in the way of something; to stop something working properly; *an engine intake close to another surface, such as the fuselage tail section, must be separated from that surface so that the slower boundary layer air does not interfere with the regular intake flow*

interference [ɪntə'fɪərəns] *noun* prevention of reception of a clear radio signal; *certain equipment, such as generators and ignition systems, will cause unwanted radio frequency interference*; precipitation interference = interference caused by rain, snow or hail

interlock 1 ['ɪntəlɒk] *noun (security device)* series of switches and/or relays; *interlocks operate in a specific sequence to ensure satisfactory engagement of the autopilot* **2** [ɪntə'lɒk] *verb* to connect together parts of a mechanism, so that the movement or operation of individual parts affects each other; *the two parts interlock to create a solid structure*

intermediate [ɪntə'mi:diət] *adjective* **(a)** in a position between two others; intermediate approach = part of the approach from (the aircraft) arriving at the first navigational fix to the beginning of the final approach; *the outer marker provides height, distance and equipment function checks to an aircraft on intermediate and final approach*; *see also* APPROACH; FINAL **(b)** *(of studies, etc.)* (level) situated between beginners and advanced; he is at an intermediate stage in his studies = he is in the middle of his course of study; an intermediate level language student = second language learner who has reached a level between elementary and advanced level

intermittent [ɪntə'mɪtənt] *adjective* stopping and starting at intervals; *the cycle of induction, compression, combustion and exhaust in the piston engine is intermittent, whereas in the gas turbine, each process is continuous*

internal [ɪn'tɜ:nəl] *adjective* referring to the inside or interior of something; *internal damage; the piston engine is an example of an internal combustion engine* (NOTE: the opposite is **external**)

international [ɪntə'næʃnl] *adjective* between countries; international call = telephone call between people in two different countries

International Civil Aviation Organization (ICAO) [ɪntə'næʃnl sɪvɪl eɪvi'eɪʃn ɔ:gənaɪ'zeɪʃn] *noun* established in 1947 by governments that "agreed on certain principles and arrangements in order that international civil aviation may be developed in a safe and orderly manner..."; *air navigation obstructions in the United Kingdom are shown on ICAO aeronautical charts*

COMMENT: ICAO is based in Montreal (Canada)

international standard atmosphere (ISA) [ɪntə'næʃnl 'stændəd 'ætməsfɪə] *noun* internationally agreed unit of pressure (the atmosphere) used in the calibration of instruments and the measurement of aircraft performance; *for en route weather the datum chosen is international standard atmosphere at mean sea level*

interpolation [ɪntəpə'leɪʃn] *noun* the estimation of a middle value by reference to known values each side; *spot temperatures at positions other than those printed are obtained by interpolation*

interpret [ɪn'tɜːprət] *verb* to understand something presented in code or symbolic form; *aircrew must be able to interpret information printed on a contour chart*

interpretation [ɪntɜːprə'teɪʃn] *noun* understanding of something presented in code or symbolic form; *synoptic charts require interpretation in order to understand the information given*

interrogate [ɪn'terəgeɪt] *verb* to transmit SSR or ATC signals to activate a transponder; *secondary surveillance radar interrogates the aircraft equipment which responds with identification and height information*

interrogation [ɪntərə'geɪʃn] *noun* transmission of a SSR or ATC signal to activate a transponder; *a transponder replies to interrogation by passing a four-digit code*

interrogator [ɪn'terəgeɪtə] *noun* ground based surveillance radar beacon transmitter/receiver; *the questioner, better known as the interrogator, is fitted on the ground, while the responder, also known as the transponder, is an airborne installation*

interrupt [ɪntə'rʌpt] *verb* to break the continuity of something; *the conversation was interrupted by a telephone call; in the northern hemisphere, the westerly flow of air is interrupted by variations which occur in pressure patterns*

interruption [ɪntə'rʌpʃn] *noun* break in the continuity of something; *because of the summer holiday, there was an interruption in the flying training course*

intersect [ɪntə'sekt] *verb (of lines, etc.)* to cut across each other; *meridians intersect at the poles and cross the equator at right angles*

intersection [ɪntə'sekʃn] *noun* point at which two lines cross each other; *the aircraft came to a stop at the intersection between runways 09 and 16; the intersection of the drift line and the wind vector gives the drift point*

intertropical convergence zone (ITCZ) [ɪntə'trɒpɪkl kən'vɜːdʒəns 'zəʊn] *noun* boundary between the trade winds and tropical air masses from the northern and southern hemispheres; *the intertropical convergence zone is the zone in which the trade winds from the two hemispheres approach each other*

interval ['ɪntəvəl] *noun* **(a)** amount of space between places or points; *the intervals at which contours are drawn depends on the scale of the chart and this interval, known as the vertical interval, is noted on the chart* **(b)** period of time between two events; *a precise interval is essential to obtain correct ignition timing on all cylinders*

introduction [ɪntrə'dʌkʃn] *noun* **(a)** something written which comes at the beginning of a report, chapter, etc.; something spoken which comes at the beginning of a talk; *in his introduction, the chief executive praised the efforts of the workforce over the previous 12 months* **(b)** bringing into use; *the introduction of fly-by-wire technology has made the pilot's task easier*

inverse [ɪn'vɜːs] *adjective* reversed in order or effect; *there is an inverse relationship between altitude and temperature, i.e. temperature decreases as altitude increases*

inversion [ɪn'vɜːʃn] *noun* **(a)** atmospheric phenomenon where cold air is nearer the ground than warm air; *smog is smoke or pollution trapped on the surface by an inversion of temperature with little or no wind*; **inversion layer** = layer of the atmosphere in which the temperature increases as altitude increases **(b)** turning something upside down; *inversion of the aircraft in flight may result in fuel stoppage*

invert [ɪn'vɜːt] *verb* to turn upside down; *a glass tube is sealed at one end, filled with mercury and then inverted so that the open end is immersed in a bowl containing mercury*

investigate [ɪn'vestɪgeɪt] *verb* to examine or look into (something) in great detail; *if the starter engaged light stays on after starting, it means that power is still connected to the starter and, if it is still on after 30 seconds, the cause must be investigated*

investigation ʀinvestɪ'geɪʃn] *noun* detailed inquiry or close examination of a matter; **accident investigation** = process of discovering the cause of accidents; **Air Accident Investigation Branch** = part of the CAA of the United Kingdom responsible for establishing the cause of accidents

QUOTE accident investigation by the FAA and the German LBA revealed that the crashed aircraft had been completely repainted in an unauthorized paint shop
Pilot

investigator [ɪn'vestɪgeɪtə] *noun* person who investigates; *accident investigators found poor co-ordination between controllers*

invisible [ɪn'vɪzɪbl] *adjective* which cannot be seen; *oxygen is an invisible gas*

involve [ɪn'vɒlv] *verb* to include; *in large transport aircraft, because of the distance and numbers of people involved, effective and rapid communications are required between flight crew and cabin crew and between cabin crew and passengers; two aircraft were involved in an accident*

involved [ɪn'vɒlvd] *adjective* over-complex; difficult; *the procedure for replacing a lost passport is very involved*

inward ['ɪnwəd] **1** *adjective* directed to or moving towards the inside or interior; *to provide protection against smoke and other harmful gases, a flow of 100% oxygen is supplied at a positive pressure to avoid any inward leakage of poisonous gases at the mask* **2** *adverb US* inwards (NOTE: the opposite is **outward**)

inwards ['ɪnwədz] *adverb* towards the inside or the interior; *the door opens inwards* (NOTE: the opposite is **outwards**)

ion ['aɪən] *noun* atom or a group of atoms that has obtained an electric charge by gaining or losing one or more electrons; *negative ion; positive ion; ultra-violet light from the sun can cause electrons to become separated from their parent atoms of the gases in the atmosphere, the atoms left with resultant positive charges being known as ions*

ionization [aɪənaɪ'zeɪʃn] *noun* process of producing ions by heat or radiation; *the intensity of ionization depends on the strength of the ultra-violet radiation and the density of the air*

ionosphere [aɪ'ɒnəsfɪə] *noun* part of the atmosphere 50 km above the surface of the earth; *since the strength of the sun's radiation varies with latitude, the structure of the ionosphere varies over the surface of the earth*

ionospheric [aɪɒnə'sferɪk] *adjective* referring to the ionosphere; **ionospheric attenuation** = loss of signal strength to the ionosphere

I/R = INSTRUMENT RATING

irregular [ɪ'regjʊlə] *adjective* not regular; *pilots of long-haul flights are subject to an irregular sleep pattern*

irrespective (of) [ɪrɪ'spektɪv] *preposition* taking no account of; regardless of; *rescue flights continue their work irrespective of the weather conditions*

ISA ['aɪ 'es 'eɪ] = INTERNATIONAL STANDARD ATMOSPHERE

isobar ['aɪsəbɑː] *noun* line on a weather chart joining points of equal atmospheric pressure; *isobars are analogous to contour lines*

isobaric [aɪsə'bærɪk] *adjective* referring to isobar(s); (chart) which shows isobars; *the rules which apply to the measurement of wind velocities on isobaric charts apply equally to contour charts*

isolate ['aɪsəleɪt] *verb* to separate something from other things or someone from other people; *the low-pressure fuel cock isolates the airframe fuel system from the engine fuel system to enable maintenance and engine removals to be carried out*

isolated ['aɪsəleɪtɪd] *adjective* separate; **isolated rain showers** = well spaced out rain showers

isolation [aɪsə'leɪʃn] *noun* state of being separated from something or somebody; *isolation of the aircraft's passengers and crew from the reduced atmospheric*

pressure at altitude is achieved by pressurization of the cabin

isotach [aɪsə'tætʃ] *noun* line of equal wind speed on charts; *wind speed is normally given in the form of isotachs*

isotherm ['aɪsəθɜːm] *noun* line of equal temperature on charts; *ascent of stable air over high ground may result in a lowering of the 0°C isotherm*

issue ['ɪʃuː] **1** *noun* (*especially of magazine*) number; copy; *the article was in last month's issue of the magazine* **2** *verb* (**a**) to give out; *the captain issued the evacuate command* (**b**) to publish; *the magazine is issued monthly* (**c**) (*permits, licences, etc.*) to give out, to grant; *the Civil Aviation Authority issue licences*

ITCZ = INTERTROPICAL CONVERGENCE ZONE

item ['aɪtəm] *noun* single article or unit (in a collection, on a list, etc.); *before practising stalls, the pilot should secure all loose items in the cockpit*

Jj

J = JOULE

JAA = JOINT AVIATION AUTHORITIES

jack [dʒæk] *noun* powered device to move heavy components, such as control surfaces of large aircraft; **operating jack =** device which converts rotary motion into linear or reciprocating motion in order to move (heavy) control surfaces; **screw jack or screwjack =** lifting device working with rotary input; type of operating jack; *pitch trim is achieved by lowering or raising the tailplane leading edge with a screw jack powered by two hydraulic motors*

jacket ['dʒækɪt] *noun* **(a)** piece of clothing with long sleeves worn with trousers or skirt; short coat **(b) life jacket =** inflatable device (somewhat resembling a sleeveless jacket) to keep a person afloat in water **(c)** outer covering or casing; *liquid cooling of a piston engine is achieved by circulating a liquid around the cylinder barrels, through a passage formed by a jacket on the outside*

jam [dʒæm] *verb* to cause moving parts to become locked and unmovable; **a jammed door =** a door which has become fixed and unmovable; *the investigation revealed that the accident had been caused by the controls being jammed due to a spanner caught in the control cables*

JAR = JOINT AVIATION REQUIREMENTS

jeopardize ['dʒepədaɪz] *verb* to put in doubt or danger; *injury to a crew member will seriously jeopardize the successful*

evacuation of the aircraft (NOTE: also written **jeopardise**)

jet [dʒet] *noun* **(a)** strong fast stream of fluid forced out of an opening; *a jet of water from a pipe* **(b)** type of engine used to power modern aircraft which takes in air at the front, mixes it with fuel, burns the mixture and the resulting expansion of gases provides thrust; *the turbo jet engine was invented by Frank Whittle in 1941* **(c)** type of aircraft which has jet engines; *the de Havilland Comet was the first commercial turbo jet transport*; *see also* STREAM; TURBOJET

jet lag ['dʒet 'læg] temporary disturbance of body rhythms such as sleep and eating habits, caused by high-speed travel across several time zones; *when I fly to Canada, it always takes me a couple of days to recover from jet lag*

jet-propelled ['dʒetprə'peld] *adjective* **jet-propelled aircraft =** aircraft powered by jet engines

jettison ['dʒetɪzn] *verb* to throw off or release from a moving aircraft; *the undercarriage failed to retract and the captain had to jettison the fuel over the sea before landing the aircraft*

join [dʒɔɪn] *verb* **(a)** to connect; *join the two wires*; *with a pencil and ruler, join point A to point B* **(b)** to bring together to make one whole part; *wing panels are joined by rivets* **(c)** to become a member of a club, etc.; *he had to pay a membership fee to join the gliding club*

joint [dʒɔɪnt] **1** *noun* place at which two or more things are joined together;

fuselage frame rings are formed with only one joint **2** *adjective* combined, with two or more things linked together; shared by two or more people, etc.; *a joint effort*

Joint Aviation Authorities (JAA)

['dʒɔɪnt eɪvi'eɪʃn ɔː'θɒrətiz] *noun* body, consisting of European representatives, set up to control and regulate aspects of civil aviation in Europe; *the Joint Aviation Authorities is an arrangement between European countries which has developed since the 1970s*

COMMENT: as of May 1998, the member states of the JAA are:- Austria, Belgium, Cyprus, Czech Republic, Denmark, Finland, France, Germany, Greece, Hungary, Iceland, Ireland, Italy, Luxembourg, Malta, Monaco, Netherlands, Norway, Poland, Portugal, Slovakia, Slovenia, Spain, Sweden, Switzerland, Turkey and the United Kingdom

Joint Aviation Requirement (JAR)

['dʒɔɪnt eɪvi'eɪʃn rɪ'kwaɪəmənt] *noun* JAA requirement concerning design, manufacture, maintenance and operation of aircraft; *JARs of relevance to maintenance staff are JAR-145, JAR-OPS 1 and JAR-OPS 3*

joule (J) [dʒuːl] *noun* International System unit of electrical, mechanical, and thermal energy; *ignition units are measured in joules (1 joule = 1 watt per second)* (NOTE: usually written **J** with figures **25J**)

junction ['dʒʌŋkʃn] *noun* place where two things meet; *the junction of two wires*; **junction box** = electrical unit where a number of wires can be connected together

Kk

K = KELVIN

katabatic [kætə'bætɪk] *adjective* referring to a cold flow of air travelling down hillsides or mountainsides; *due to katabatic effects, cold air flows downwards and accumulates over low ground*; katabatic wind = wind which occurs when the air in contact with the slope of a hill is cooled to a temperature lower than that in the free atmosphere, causing it to sink; *compare* ANABATIC; *see also* NOCTURNAL

kelvin (K) ['kelvɪn] *noun* base SI unit of measurement of thermodynamic temperature (NOTE: temperatures are shown in kelvin without a degree sign: 20K. Note also that 0ºC is equal to 273.15K)

kerosene *or* **kerosine** ['kerəsi:n] *noun* thin fuel oil made from petroleum; *kerosene will only burn efficiently at, or close to, a ratio of 15:1*

Kevlar ['kevlə] *noun* light and very strong composite material; *Kevlar and carbon fibre account for a large percentage of a modern jet airliner's structure*

key [ki:] *noun* piece of metal used to open a lock; **ignition key** = key used to switch on the ignition

kg = KILOGRAM(S)

kHz = KILOHERTZ

kick-back ['kɪkbæk] *noun* (*on a propeller driven aircraft*) the tendency of the engine to suddenly reverse the rotation of the propeller momentarily when being started; *on most modern engines the spark is retarded to top-dead-centre, to ensure easier starting and prevent kick-back*

kilo- ['kɪləʊ] *prefix meaning* one thousand

kilo ['ki:ləʊ] *noun* = KILOGRAM; *this piece of luggage weighs 15 kilos*

kilogram (kg) ['kɪləgræm] *noun* measure of weight equal to one thousand grams; *this piece of luggage weighs 15 kg* (NOTE: written kg after figures)

kilohertz (KHz) ['kɪləhɜ:ts] *noun* unit of frequency measurement equal to one thousand Hertz

kilometre (km) *US* **kilometer** [kɪ'lɒmɪtə] *noun* measure of length equal to one thousand metres (0.621 miles) (NOTE: written km with figures: 150 km)

kinetic [kaɪ'netɪk] *adjective* referring to motion or something produced by motion; **kinetic energy** = energy of motion; *the efficiency of conversion of fuel energy to kinetic energy is termed the thermal efficiency*; **kinetic heating** = heating of aircraft skin by friction with the air as it moves through it

kit [kɪt] *noun* set of items used for a specific purpose; *a physician's kit containing surgical equipment would be available to a qualified doctor assisting crew with major medical problems*; **tool kit** = set of tools consisting of spanners, screwdrivers, pliers, etc.; **first-aid kit** = set of plasters, bandages, antiseptic cream, etc., for minor emergencies

knob [nɒb] *noun* (a) rounded handle; *door knob* (b) rounded control switch or dial; *when the control knob is moved from* *the central position, the ailerons are moved* **(c)** round button (such as on a receiver); *turn the knob to increase the volume*

knot (kt) [nɒt] **1** *noun* unit of speed equal to one nautical mile per hour (approximately 1.85 kilometres or 1.15 statute miles per hour); *wind speeds in aviation are usually given in knots*

COMMENT: American light aircraft manufactured prior to 1976 had airspeed indicators marked in statute miles per hour. Knot means 'nautical miles per hour'. It is therefore incorrect to say 'knots per hour'

knowledge [ˈnɒlɪdʒ] *noun* familiarity, awareness or understanding gained through experience or study; *a knowledge of the factors which affect surface temperatures will contribute a great deal to the understanding of meteorology*

kt = KNOT(S)

LI

label ['leɪbl] **1** *noun (something that identifies)* a small piece of paper or cloth attached to an article with details of its owner, contents, use, destination, etc.; *hydraulic tubing has a label with the word HYDRAULIC* **2** *verb* **(a)** to identify by using a label; *parts are labelled with the manufacturer's name* **(b)** to add identifying words and numbers to a diagram; *there is a standard way of labelling the navigation vector*

lack ['læk] *noun* absence of something; need of something; *the engine stopped because of a lack of fuel*

lag [læg] *noun* delay - time between the input and the output; *there is a time lag between the piston moving down and the mixture flowing into the cylinder*; *see also* JET LAG

Lambert's projection ['læmbəts prə'dʒekʃn] *see* PROJECTION

laminate ['læmɪneɪt] **1** *noun* sheet of man-made material made up of bonded layers; *direction of the fibres and types of cloth used in the laminate are all very important factors* **2** *verb* to make by using bonded layers of material; *laminated windscreens*

lamp [læmp] *noun* small light; **warning lamp** = small light (often red) which informs of a possible danger by lighting up; *the switch is connected to a warning lamp on the instrument panel which will illuminate if the oil pressure falls below an acceptable minimum*

land ['lænd] **1** *noun* solid ground (as opposed to the sea); *a large land mass*

such as Greenland **2** *verb* **(a)** to set an aircraft onto the ground or another surface such as ice or water, after a flight; **to force land the aircraft** = to land the aircraft when it can no longer be kept in the air for any particular reason **(b)** to arrive on the ground after a flight; *flight BA321 landed at London Heathrow at 1030 hours*; *see also* CRASH-LAND (NOTE: the opposite is take-off)

landing ['lændɪŋ] *noun* act of setting an aircraft onto the ground or another surface such as ice or water after flight; *take-off and landing are normally made into wind in order to reduce the length of the take-off and landing run*; *in order to achieve a safe landing in a cross wind, the correct techniques must be used*; **landing** = routine landing when the aircraft makes contact with the surface with more force than usual, thereby possibly causing damage to the undercarriage; *the pilot reported a heavy landing*; *see also* CRASH-LANDING

landmark ['lænmɑːk] *noun* something on the ground, such as a noticeable building, bridge, coastal feature, etc., which enables the pilot to know where he/she is; *railway lines are usually useful landmarks*

lapse rate ['læps 'reɪt] *noun* rate at which temperature changes according to altitude; **adiabatic lapse rate** = rate at which air temperature decreases as it rises above the earth's surface (as height increases, temperature decreases); **environmental lapse rate (ELR)** = rate at which the temperature of the air falls as one rises above the earth; *although there is an* average ELR of 1.98°C per 1,000 feet, in practice the ELR varies considerably with space and time

---

COMMENT: it has been found that when dry or unsaturated air rises, its rate of fall of temperature with height (i.e. lapse rate) is constant at 3°C per 1,000 feet. Similarly, descending air warms by compression at that rate. This dry adiabatic lapse rate is normally referred to as the DALR. Air rising and cooling often reaches its dew point temperature, becomes saturated and any further cooling results in condensation and the release of latent heat. Release of latent heat delays the cooling process and the lapse rate at low levels is reduced to 1.5°C per 1,000 feet. This temperature change is called the saturated adiabatic lapse rate and is normally referred to as the SALR

---

largely ['lɑ:dʒli] *adverb* mainly, mostly; *heat is transferred from the earth's surface upwards largely by convection*; *the southern hemisphere consists largely of oceans*

last [lɑ:st] **1** *adjective* coming or placed after all the others; **the last chapter =** (i) the final chapter in a book; (ii) the chapter before the one being read **2** *verb* **(a)** to continue for a period of time; *a gust is a sudden increase in wind speed above the average speed lasting only a few seconds* **(b)** to stay in good or usable condition; *a piston engine lasts longer if it is handled carefully and serviced regularly*

latent heat ['leɪtənt 'hi:t] *noun* heat taken in or given out when a solid changes into a liquid or vapour, or when a liquid changes into a vapour at a constant temperature and pressure; **latent heat of fusion =** quantity of heat required to convert ice, at its melting point, into liquid at the same temperature; **latent heat of vaporization =** quantity of heat required to convert liquid to vapour at the same temperature; **latent heat of sublimation =** quantity of heat required to convert ice to vapour at the same temperature

lateral ['lætrl] *adjective* referring to the side; *drift is the lateral movement of the aircraft caused by the wind*

lateral axis ['lætrl 'æksɪs] *noun* axis of the aircraft from wing tip to wing tip about which the aircraft pitches up and down; *see also* AXIS; PITCH

latitude ['lætɪtjuːd] *noun* angular distance north or south of the earth's equator, measured in degrees, minutes and seconds, along a meridian, (as on a map or chart, etc.); *parallels of latitude are imaginary circles on the surface of the earth, their planes being parallel to the plane of the equator; the centre of London is latitude 51°30'N, longitude 0°5'W; compare* LONGITUDE

latter ['lætə] **1** *adjective* referring to something coming at the end or finish; **the latter part of the take-off run =** part of the take-off run immediately before the aircraft leaves the ground **2** *noun* second of two things mentioned earlier; **of the Airbus A320 and A340, the latter is the larger aircraft =** the A340 is the larger of the two (NOTE: the first of two is called the **former**)

launch ['lɔːnʃ] **1** *noun* small boat often used to transport people from a larger boat or ship to the shore **2** *verb* **(a)** to slide or drop a boat into the water to make it ready for use; *while passengers are fitting life jackets, crew will open exits and launch the life-rafts* **(b)** to force something into motion; *to launch a rocket*

lavatory ['lævətri] *noun* = TOILET

law [lɔ:] *noun* **(a)** statement describing a relationship observed to be unchanging between things while certain conditions are met; basic principle of science or mathematics; *the law of gravity*; **Boyle's Law =** the volume of a given mass of gas, whose temperature is maintained constant, is inversely proportional to the gas pressure; *see also* BUYS BALLOT'S **(b)** set of agreed rules; *aviation law*

layer ['leɪə] *noun* **(a)** one horizontal part; *the lowest layer of the atmosphere is* *called the troposphere*; **layer cloud** = stratus cloud **(b)** thickness of something; *layers of fluid next to the surface over which it is flowing travels more slowly than layers further from the surface*

layout ['leɪaʊt] *noun* way in which things are arranged; **cockpit layout** = design of the cockpit and the particular placement of controls, instruments, etc.

LC = LOAD CONTROLLER

LCD = LIQUID CRYSTAL DISPLAY

lead [led] *noun* very heavy soft metallic element; **lead-free** = which does not contain lead; *low-lead or lead-free fuel is used in most modern piston engines*; **lead-acid battery** = system of lead plates and dilute sulphuric acid - used as a starter battery or traction battery (NOTE: chemical symbol is **Pb**; atomic number is **82**)

lead [liːd] **1** *noun* **(a)** electrical wire or narrow cable; *a lead connects the monitor to the computer* **(b) to take the lead** = to take control of a situation; *it is vital in any emergency situation that a crew member should take the lead* **2** *verb* **(a)** to guide or show the way by going first; *in an emergency situation the aircraft commander may lead his passengers to safety*; *in a smoke-filled cabin, floor lighting leads passengers to the emergency exit* **(b)** to cause; *in winter, the cold conditions often lead to frost and fog*; *contraction of metal parts and seals can lead to fluid leakage* (NOTE: **leading - led**)

leading edge ['liːdɪŋ 'edʒ] *noun* front part of the wing which meets the oncoming air first; *in icing conditions, ice may build up on the leading edges*

leak [liːk] **1** *noun* escape of liquid or gas from a sealed container; amount of liquid or gas that has escaped; **exhaust leak** = escape of exhaust gases; *any failure of the aircraft structure may cause a leak of pressurized air which might be very difficult to cure* **2** *verb* (of liquid or gas) to escape from a sealed container; *fuel may leak from a fuel tank if the drain plug is not seated correctly*

leakage ['liːkɪdʒ] *noun* escape of liquid or gas from a sealed container; *any internal or external leakage of fuel will cause a reduction in the operating period* (NOTE: leak is normally used for an individual instance while leakage is used more generally: **there is a fuel leak from the central tank; fuel leakage is a safety hazard**)

lean [liːn] *adjective* (a fuel/air mixture) in which the ratio of air to fuel is greater than usual; *moving the mixture control lever aft to the lean position reduces the amount of fuel mixing with the air*

lee [liː] *adjective & noun* (side) which is protected from the wind; *the air on the lee side is drier than that on the windward side*; *the flow of air over and to the lee of hills and mountains may cause particularly severe turbulence* (NOTE: the opposite is **windward side**)

leg [leg] *noun* part of a flight pattern that is between two stops, positions, or changes in direction; *an airfield traffic pattern is divided into take-off, crosswind leg, downwind leg, base leg and final approach*; **base leg** = part of the airfield traffic circuit flown at approximately 90° to the direction of landing, followed by the final approach; **crosswind leg** = part of the airfield traffic circuit flown at approximately 90° to the direction of take off and climb out, followed by the downwind leg; **downwind leg** = part of the airfield traffic circuit which runs parallel to, but in the opposite direction to, the approach to land which is made into wind

QUOTE their route was across the States to Canada, Greenland and the North Pole, into Norway, through Europe, back to Iceland, then two long legs across the Atlantic via South Greenland and back to Seattle

Pilot

legal ['liːgl] *adjective* lawful or within the law; *alcohol concentrations of 40 milligrammes per 100 millilitres, i.e. half the legal driving limit in the UK, are* *associated with substantial increases in errors committed by pilots*

legend ['ledʒənd] *noun* list explaining the symbols on a chart or a map; *a legend is usually to be found at the edge or on the reverse side of most topographical charts*

length [leŋθ] *noun* (a) measurement along something's greatest dimension; *the length of the aircraft; the runway length is 3 kilometres* (b) piece that is normally measured along its greatest dimension; *a length of pipe* (c) extent from beginning to end; *the length of a book* (d) extent or duration; the amount of time between particular points in time; the distance between two points; **the length of a briefing** = how much time the briefing takes; **the length of the working life of components** = how long the components last; **the length of a flight** = (i) the time it takes to complete a flight; (ii) the distance of the flight in nautical miles or kilometres; *the length of the flight meant that there was no time for a meal to be served to the passengers; the length of the flight is 100nm*

lengthen ['leŋθən] *verb* to make long or longer; *the mercury column shortens when cooled and, due to expansion, lengthens when heated* (NOTE: the opposite is **shorten**)

lengthy ['leŋθi] *adjective* (a) long; extensive; *he wrote a lengthy report* (b) long; which lasts for a long time; **lengthy meeting** = long meeting; **lengthy explanation** = long explanation (NOTE: often suggests a meeting or explanation which is longer than necessary and therefore uninteresting)

lengthwise ['leŋθwaɪz] *adjective* & *adverb* along the length of (something); *in a lengthwise direction*

lens [lenz] *noun* normally round, piece of glass with curved surfaces found in microscopes, telescopes, cameras, spectacles, etc.; **lens-shaped cloud** *or* **lenticular cloud** = cloud with slightly curved (outwards) upper and lower surfaces

lessen ['lesn] *verb* to make less; *reverse thrust is used to lessen the loads on brakes and tyres; clean filters lessen the possibility of blockage*

level ['levl] 1 *adjective* (a) **level with** = at the same height or position as something else; *in most light aircraft, the aeroplane will be in a climb if the engine cowling is level with the horizon* (b) having a flat, smooth surface; **a level runway** = a runway without bumps, etc. (c) on a horizontal plane; **a level runway** = a runway which does not slope (d) steady; referring to something with no sudden changes; **speak in a level voice** = do not raise and lower the sound of your voice; **the level tone of an engine** = the unchanging sound of an engine; **level head** = clear thinking; *it is essential that the crew keeps a level head in an emergency* 2 *noun* (a) position along a vertical axis; *ground level; reference level; the tropopause is the level at which the lapse rate ceases to be so important;* **flight level (FL)** = level of constant atmospheric pressure related to a reference datum of 1013.25 mb (FL 250 = 25,000 ft); **the fluid level in the reservoir** = the point where the surface of the fluid reaches up to; **high-level cloud** = high altitude cloud (b) position on a scale; *an advanced level of study* (c) relative amount, intensity, or concentration; *an unsafe level of contamination; a gas turbine engine has an extremely low vibration level; a reduced level of noise*

lever ['li:və] 1 *noun* (a) device with a rigid bar balanced on a fixed point and used to transmit force, as in raising a weight at one end by pushing down on the other; *push the lever fully up to activate the brake mechanism; push the button to release the lever* (b) handle used to adjust or operate a mechanism; *throttle lever; undercarriage selector lever; feathering is accomplished by moving the pilot's control lever* 2 *verb* to move as with a lever; *the door would not open so the emergency services had to lever it open with specialized equipment*

licence ['laɪsəns] *noun* document which is proof of official permission to do or to own something; **Airline Transport Pilot's Licence (ATPL)** = licence required for pilot-in-command or co-pilot of public transport aircraft (for two crew operations); **Commercial Pilot's licence (CPL)** = licence required for pilot-in-command of public transport aircraft certified for single pilot operations; **Private Pilot's Licence (PPL)** = basic licence for flying light aircraft; **licence holder** = (i) person who owns a licence; (ii) leather case, etc., in which to keep the licence document

COMMENT: each licence has its own specific requirements and privileges. In the UK, one of the fundamental differences between a Private Pilot's Licence and other types of licence is that the holder of a PPL is not allowed to fly for 'hire or reward', i.e. the pilot cannot receive payment for flying

license ['laɪsəns] **1** *noun* US = LICENCE **2** *verb* to give someone official permission to do or to own something; to issue a licence

lie [laɪ] *verb* **(a)** to be in a flat position, often horizontal; *seat rails are attached to the floor beams and lie level with the flooring* **(b)** to be situated; *great circles are represented by curves which lie on the polar side of the rhumb line* (NOTE: care should be taken with the verbs **to lie**, as defined here: **lie - lay - lain** and **to lie** meaning 'not to tell the truth': **lie - lied - lied** and **lay**, meaning 'to put down' as in **'lay the book on the table'**: **lay - laid - laid**)

life jacket ['laɪf 'dʒækɪt] *noun* inflatable device (sometimes resembling a sleeveless jacket) to keep a person afloat in water; *pull down the toggles to inflate the life jacket*

life raft ['laɪf 'rɑːft] *noun* small boat-like vessel for use on an emergency over water

life vest ['laɪf 'vest] *noun* life jacket; *you will find a life vest under your seat*

lift [lɪft] **1** *noun* **(a)** component of the total aerodynamic force acting on an aerofoil which causes an aeroplane to fly; *in level flight, a lift force equal to the weight must be produced; the pilot can achieve maximum lift by pulling hard back on the controls* **(b)** electrically operated machine for moving people or goods between the floors of a building (NOTE: American English is **elevator**) **2** *verb* to move to a higher position; *a foot-pound is the ability to lift a one pound weight a distance of one foot*

COMMENT: Bernoulli's principle states that if the speed of a fluid speed increases, its pressure decreases; if its speed decreases, its pressure increases. Wings are shaped so that the high-speed flow of air that passes over the curved upper surface results in a decrease in pressure. Lift is created because of the pressure differential between upper and lower surfaces of the wing. Lift is also created because the angle of attack allows the airflow to strike the underside of the wing. Daniel Bernoulli (1700-1782) was a Swiss scientist

light [laɪt] **1** *noun* **(a)** brightness produced by the sun, the moon, a lamp, etc. **(b)** electromagnetic radiation which can be sensed by the eyes; **artificial light** = light made by using electrical, gas, etc., power **(c)** source of light such as a lamp; *switch off the navigation lights*; **warning light** = warning lamp; small light (often red) which informs of a possible danger by lighting up; *at 5 knots above stalling speed, a warning light on the instrument panel will flash* **2** *adjective* **(a)** without much weight; not heavy; *aluminium is a light metal*; **light aircraft** = (generally speaking) small, single engine aircraft for private (not commercial) use **(b)** of little force or requiring little force; **a light wind** = a gentle wind; **light controls** = flying controls which do not need much pilot effort to move them **(c)** of little quantity; *light rain; light snow* **(d)** of thin consistency; **light oil** = oil which pours easily

lighting ['laɪtɪŋ] *noun* lights; system of lights; *cabin lighting is switched off for take-off and initial climb; emergency* *floor lighting guides passengers to the emergency exits*

lightning ['laɪtnɪŋ] *noun* powerful and sudden electrical discharge from a cloud; *lightning is the most visible indication of thunderstorm activity*; *see also* ACTIVITY

likely ['laɪkli] *adjective* probable; **rain is likely** = rain will probably fall; **icing is likely to occur in cumulonimbus clouds** = icing is often a problem if flying in cumulonimbus clouds

limit ['lɪmɪt] **1** *noun* point, line, etc.. past which something should not go; *there is a time limit of one hour for the examination*; *the minimum age limit for holding a PPL in the UK is 17*; **the upper limit of cloud** = the highest point at which there is cloud **2** *verb* to restrict or to prevent from going past a particular point; *the amount of cabin baggage is limited to one bag per passenger*

limitation [lɪmɪ'teɪʃn] *noun* act of limiting or the state of being limited; *limitation of the maximum engine rpm to a little above maximum engine cruise rpm prevents compressor stall at the higher rpm range*

line [laɪn] *noun* **(a)** thin continuous mark as made by a pencil, pen, etc. or printed; *draw a line from point A to point B* **(b)** real or imaginary mark placed in relation to points of reference; *an isobar is a line joining points of equal pressure*; *(of a runway)* **centre line or centreline** = painted or imaginary line running along the centre of the runway; **line of position or position line** = line along which an aircraft is known to be at a particular time usually by taking a VOR bearing; **line of sight** = clear path between sending and receiving antennas **(c) line feature** = useful navigational landmark such as a railway line, road or river **(d)** long row of people, etc.; *a line of people*; *a line of cumulous clouds* **(e)** row of written or printed words; *look at line 4 on page 26* **(f)** telephone connection to another telephone or system; *dial 9 to get an outside line* **(g)** electrical cable or wire; **power line** = thick cable, supported by pylons, which carries electricity for long distances from power stations to cities, etc.; **telephone line** = cable supported on pylons from one telephone exchange to another; *on final approach to an unfamiliar airfield, pilots of light aircraft should keep a sharp lookout for power lines and telephone lines* **(h) railway line** = railway track or train track; *a railway line is a useful landmark* **(i)** system of pipes; *a fuel line* **(j)** company which owns and manages a system of transportation routes; *a shipping line*; *an airline such as KLM or QANTAS*; *see also* CHORD LINE

line up ['laɪn 'ʌp] *verb* to move aircraft into position ready for departure; *line up with the nosewheel on the runway centre line*

linear ['lɪniə] *adjective* referring to a line; straight; *although air may appear to be still or calm it is, in fact, moving west to east in space, the linear velocity being zero at the poles and approximately 1,000 mph at the equator*; **linear scale** = horizontal or vertical straight-line, rather than circular, scale on an instrument

link [lɪŋk] **1** *noun* **(a)** *(between two devices)* connection; *light aircraft can be steered while taxiing via a direct link from rudder pedals to nosewheel*; **communication link** = telephone or radio connection - as between ground crew and flight deck while an aircraft is preparing for departure **(b)** *(between two situations)* relationship; *there is a link between alcohol abuse and pilot error resulting in accidents* **2** *verb* **(a)** to make a connection, to join; *the connecting rod links the piston to the crankshaft* **(b)** to establish a relationship between two situations; *they link alcohol abuse and pilot error*

linkage ['lɪŋkɪdʒ] *noun* system or series of mechanical connections such as rods, levers, springs, etc.; *throttle linkage*; *rudder linkage*; *the linkage from the control column to the control surfaces should allow full and free movement*

lip [lɪp] *noun* **intake lip** = the rim or edge of the air intake of a jet engine; *as sonic speed is approached, the efficiency of the intake begins to fall because of shock waves at the intake lip*

liquid ['lɪkwɪd] **1** *adjective* having a consistency like that of water; *liquid oxygen is stored in cylinders*; **liquid crystal display (LCD)** = liquid crystals that reflect light when a voltage is applied, used in many watch, calculator and digital displays; **liquid ice** = water **2** *noun* substance with a consistency like water; *water is a liquid - ice is a solid*; **a liquid fire** = oil or petrol fire

list [lɪst] **1** *noun* series of names, words, things to do, etc., arranged one after the other in a vertical column; **check list** = series of items to check before flight or in the event of an emergency, etc. **2** *verb* to write a series of names, words, etc. one after the other in a vertical column; *list the advantages of a stressed-skin construction*

liter ['liːtə] *noun US* = LITRE

lithium ['lɪθiəm] *noun* soft silvery metallic element, the lightest known metal (often used in batteries); *an alloy of aluminium and lithium* (NOTE: chemical symbol is **Li**; atomic number is **3**)

litmus ['lɪtməs] *noun* substance which turns red in acid, and blue in alkali; **litmus paper** = small piece of paper impregnated with litmus to test for acidity or alkalinity

litre (l) ['liːtə] *noun* measure of capacity; volume of one kilogram of water at 4°C (= 1,000cc or 1.76 pints) (NOTE: written **l** after a figure: **10l**; also written **liter** in American English)

live [laɪv] *adjective* carrying electricity; *live wire*

LMT = LOCAL MEAN TIME

load [ləʊd] **1** *noun* **(a)** weight or mass which is supported; *the load on the undercarriage decreases as lift increases and, when the aircraft rises into the air, the aircraft is supported by the wings*;

load bearing = which supports some weight; **bending load** = load causing bending of a structure; **breaking load** = load capable of being supported before a structure breaks; **compressive load** = load caused by forces acting in opposite directions towards each other; **shearing load** = load caused by sliding apart the layers of a structure; **tensile load** = load caused by forces acting in opposite directions away from each other; **torsion load** = load caused by twisting of a structure **(b)** force which a structure is subjected to when resisting externally applied forces; *the load on the control column is increased when the aircraft is flown out of trim* **(c)** something that is carried in the aircraft; *fuel load*; **load manifest or load sheet** = detailed list of cargo on a flight; **passenger load** = the number of passengers on board; *see also* ABNORMAL **(d) work load** = share of work done by a person, system or device **(e)** *(of electricity)* the power output of a generator or power plant; **load controller** = device which monitors the output of a generator **(f)** the resistance of a device or of a line to which electrical power is provided **2** *verb* **(a)** to put something into a container, often for the purpose of transportation; *the aircraft is loaded with fuel before take-off* **(b)** to transfer data from disk into a computer main memory; *she loaded the software onto the computer*

load factor ['ləʊd 'fæktə] *noun* stress applied to a structure as a multiple of stress applied in 1g (acceleration due to earth's gravity) flight; *the higher the angle of bank, the greater the load factor*

COMMENT: in straight and level, unaccelerated flight, the load factor is 1. When an aircraft turns or pulls up out of a dive, the load factor increases. An aircraft in a level turn at a bank angle of 60 degrees has a load factor of 2. In such a turn, the aircraft's structure must support twice the aircraft's weight

loading ['ləʊdɪŋ] *noun* **(a)** act or process of adding a load to an aircraft; **loading is in progress** = passengers, baggage, freight, etc., are being put on the aircraft **(b)** total aircraft weight or mass divided by wing area; *inertia switches operate automatically when a certain g (acceleration due to earth's gravity) loading occurs*; **centrifugal loading** = centrifugal force acting on something; *centrifugal loading moves the valve towards the closed position* **(c)** action of transferring data from disk to memory; *loading can be a long process*

lobe [ləub] *noun* one of two, four or more sub-beams that form a directional radar beam; *any system employing beam sharpening is vulnerable to side lobe generation at the transmitter*

local ['ləukl] *adjective* not broad or widespread; **local meteorological conditions** = weather conditions in the restricted area of a particular place; **local time** = the time in the country you are talking about; **local mean time (LMT)** = the time according to the mean sun

locality [ləu'kælıtı] *noun* small geographical area; *the highest point in a locality is marked by a dot with the elevation marked alongside*

localized ['ləukəlaızd] *adjective* restricted in area or influence; **a localized fire** = a fire which has not spread (NOTE: also written **localised**)

localizer [ləukə'laızə] *noun* component of the instrument landing system that provides horizontal i.e. course guidance to the runway; *if, during the approach, the aircraft deviates beyond the normal ILS glideslope and/or localizer limits, the flight crew are alerted* (NOTE: also written **localiser**)

locate [ləu'keıt] *verb* **(a)** to find the position of; *survival beacons transmit a signal which enables search aircraft/vessels to rapidly locate accident survivors still in the sea* **(b)** to position; *the digital flight data recorder is located in the tail section*

location [lə'keıʃn] *noun* **(a)** place where something can be found; *before take-off, cabin staff brief passengers on the location of emergency exits and life jackets* **(b)** finding where something is; *rapid location of survivors is important*

locator [lə'keıtə] *noun* non-directional beacon used as an aid to final approach; *terminal control areas require charts which show detail on a large scale - terminal VORs, locator beacons, ILS installations, holding patterns, arrival/departure and transit routes*

lock [lɒk] **1** *noun* device operated by a key for securing a door, etc.; **ignition lock** = key-operated switch for activating the ignition circuit of a car or aircraft **2** *verb* **(a)** to secure a door by turning a key in the lock; *lock the door before leaving the building* **(b)** to be in or to move into a secure position; **undercarriage down and locked** = confirmation that the undercarriage is secure in preparation for landing **(c)** to block; to prevent moving; *anti-skid braking systems units are designed to prevent the brakes locking the wheels during landing* **(d)** to lock on = to search for, find and follow a target with a thin radar beam

log [lɒg] **1** *noun* written record of a flight, flying hours, maintenance checks, etc., for an aircraft, engine or propeller; **navigation log** = written details of headings and times for a flight; *the flight crew route flight plan is a composite document which also serves as a navigation log* **2** *verb* to write an entry in a log book or on a log sheet; *he calculates headings to steer for each flight stage and logs them*

logic ['lɒdʒık] *noun* electronic circuits which obey mathematical laws; *circuit packs consist of basic decision-making elements, referred to as logic gates, each performing operations on their inputs and so determining the state of their outputs*

logical ['lɒdʒıkl] *adjective* referring to something which, because of previous experience or knowledge, is natural or expected; which makes sense; *pre-flight* *checks on light aircraft are made in a logical manner from one side of the aircraft to the other*

longeron ['lɒndʒrən] *noun* main structural part of an aircraft fuselage extending from nose to tail; *longerons are normally used in aircraft which require longitudinal strength for holds underneath the floor*

long-haul ['lɒŋ'hɔːl] *adjective* long distance; *crew flying long-haul routes have to adapt to time changes* (NOTE: the opposite is **short-haul**)

longitude ['lɒnʒɪtjuːd or 'lɒŋɡɪtjuːd] *noun* angular distance on the earth's surface, measured east or west from the prime meridian at Greenwich, UK, to the meridian passing through a position, expressed in degrees, minutes, and seconds; *the centre of London is latitude 51°30'N, longitude 0°5'W; compare* LATITUDE

longitudinal [lɒdʒɪ'tjuːdɪnl] *adjective* in a lengthwise direction

longitudinal axis [lɒdʒɪ'tjuːdɪnl 'æksɪs] *noun* axis of the aircraft which extends from the nose to the tail; *see also* AXIS; ROLL

long-range ['lɒŋ'reɪnʒ] *adjective* (a) covering a long distance; *long-range radar* (b) **long-range weather forecast** = covering a period more than 5 days ahead

lookout ['lʊkaʊt] *noun* (careful) watch; *keep a careful lookout for other aircraft*; **to be on the lookout for** = to watch carefully for (something)

loop [luːp] *noun* (a) **loop antenna** = circular shaped conductive coil which rotates to give a bearing to a ground station (b) flight manoeuvre in which the aircraft rotates, nose up, through 360° while holding its lateral position; **ground loop** = unwanted uncontrolled turn of an aircraft while moving on the ground especially when taking off or landing

lose [luːz] *verb* not to have something any longer; *(of aircraft)* **to lose altitude** = to descend from higher to lower altitude (NOTE: **losing - lost**)

loss [lɒs] *noun* no longer having something; *the pilot reported loss of engine power*; **loss of control** = no longer being able to control; **loss of life** = death (in an accident); **loss of a signal** = disappearance of a signal; *the term attenuation means the loss of strength of a radio signal*

loudspeaker [laʊd'spiːkə] *noun (part of a radio, etc.)* electromagnetic device that converts electrical signals into audible noise (NOTE: also called **speaker**)

lounge [laʊnʒ] *noun* **departure lounge** = room at an airport where passengers wait to board their aircraft; **VIP lounge** = special room at an airport for VIPs (very important persons)

louvre ['luːvə] *noun* thin, horizontal openings for air cooling; *cold air can be let into the cabin through adjustable louvres*

low [laʊ] **1** *adjective* (a) not high, not tall; *a low building*; **low cloud** = cloud relatively near the surface of the earth; **low ground** = area of land which is not high (as opposed to mountains) (b) not high; below normal; *an area of low pressure*; **low temperature** = temperature which shows that it is cold (c) quiet; not loud; **low sound** = sound that is not loud **2** *noun* area of low atmospheric pressure; **polar low** = area of low atmospheric pressure over Greenland, for example

lower ['laʊə] **1** *adjective* (a) referring to something that is at a low level or towards the bottom; *the lower layers of the atmosphere*; **the lower surface of the wing** = the underneath surface of the wing (b) referring to something which is below something else of the same sort; *air is cooler high up than at lower levels* (NOTE: the opposite is **upper**) **2** *verb* (a) to let down to a lower position; **lower the undercarriage** = move the undercarriage into position ready for landing; **lower the flaps** = set the flaps to a down position (b) reduce in amount or intensity; **to lower the temperature** = to reduce the temperature;

to lower the pressure = to decrease the pressure; **to lower the volume (of sound)** = to make (the radio, etc.) quieter or less loud

lubricate ['luːbrɪkeɪt] *verb* to oil or to grease moving parts in order to reduce friction; *oil passes through the hollow crankshaft to lubricate the big-end bearings; turbo chargers are lubricated by the engine oil system*

lubrication [luːbrɪ'keɪʃn] *noun* act or process of covering moving surfaces with oil or grease (to reduce friction); **lubrication system** = tank, pipes, pumps, filters, etc., which together supply oil to moving parts of the engine

luggage ['lʌgɪdʒ] *noun* baggage, i.e. cases and bags that someone takes when travelling; **hand luggage** = small bags that passengers can take with them into the cabin of an aircraft; *the amount of hand luggage is limited to one bag*; *see also* BAGGAGE

Mm

m = METRE

m = MINUTE

Mach ['mæk] *noun* ratio of the speed of an object to the speed of sound in the same atmospheric conditions; *Mach 2 equals twice the speed of sound*

> COMMENT: named after E. Mach, the Austrian physicist who died in 1916

machine [mə'ʃiːn] *noun* device with fixed and moving parts that takes mechanical energy and uses it to do useful work; *a drill is a machine for making holes in things; an electrical circuit is designed to carry energy to a particular device or machine which can then perform useful work*

magnesium [mæg'niːsiəm] *noun* light, silvery-white metallic element that burns with a brilliant white flame; **magnesium flare** = device for distress signalling at night; *to send off magnesium flares* (NOTE: chemical symbol: **Mg**; atomic number 12)

magnet ['mægnət] *noun* an object that produces a magnetic field, and attracts iron and steel; *magnetism in a magnet appears to be concentrated at two points called the poles*

magnetic [mæg'netɪk] *adjective* referring to a magnet or to something with a magnetic field; having the power of a magnet; *a freely suspended magnet - uninfluenced by outside forces - will align itself with the earth's magnetic lines of force which run from the north magnetic pole to the south magnetic pole*; **magnetic**

bearing = angle measured in a clockwise direction of a distant point, relative to magnetic north; **magnetic field** = area of magnetic influence; **magnetic north** = direction of the earth's magnetic pole, to which the north-seeking pole of a magnetic needle points if unaffected by nearby influences; **magnetic pole** = one of the two poles (north and south) which are the centres of the earth's magnetic field; **magnetic variation** or **magnetic declination** = differences in the earth's magnetic field in time and place; *to convert magnetic bearing into true bearing it is necessary to apply magnetic variation at the point at which the bearing was taken*; *see also* VARIATION

magnetism ['mægnətɪzm] *noun* force exerted by a magnetic field; *an electric current produces magnetism, and movement of a magnet can produce electricity*; **electro-magnetism** = magnetic field found around any conductor carrying current, the strength of which will depend on the amount of current flow

magnetize ['mægnətaɪz] *verb* to make magnetic; to convert an object or material into a magnet; *ferro-magnetic materials are easily magnetized*

magneto [mæg'niːtəʊ] *noun* device that produces electrical current for distribution to the spark plugs of piston aero-engines; **impulse magneto** = magneto with a mechanism to give a sudden rotation and thus produce a strong spark

> COMMENT: the crankshaft turns the magnetos, which provide the electrical energy to create a spark

from the spark plugs. This ensures that the spark plugs work even if the aircraft's battery and electrical system fail. Most aircraft have two magnetos per engine in case one fails

magnify ['mægnɪfaɪ] *verb* (**a**) to increase the size of, especially by using a lens, microscope, etc.; *it was only after the image was magnified that it was possible to see the flaw* (**b**) to increase the effect of something; *the stress level is magnified at times of high work load, for example, preparation for landing* (NOTE; **magnifying - magnified**)

magnitude ['mægnɪtjuːd] *noun* greatness in size or extent; *the magnitude of the pressure gradient force is inversely proportional to the distance apart of the isobars; when the surface wind speed reaches a certain magnitude the term gale is used*

maiden flight ['meɪdən 'flaɪt] *noun* first flight of a new aircraft; *the maiden flight of the A340 was in October 1991*

main [meɪn] *adjective* most important; principal; **main disadvantages** = principal negative points; **main gear** = undercarriage assemblies under each wing

mainplane ['meɪnpleɪn] *noun* aircraft wing (compared with the tailplane); *the region between the mainplane front and rear spars is commonly sealed off and used as tanks*

maintain [meɪn'teɪn] *verb* (**a**) to keep up or to carry on; to continue; **to maintain the present heading** = to continue on the same heading; **to maintain a constant selected engine speed** = not to change the engine speed (**b**) to keep in good mechanical or working order; *aero-engines must be maintained regularly to maximize engine life*

maintenance ['meɪntnəns] *noun* regular periodic inspection, overhaul, repair and replacement of parts of an aircraft and/or engine; *the gas turbine is a very simple engine with few moving parts when compared with a piston engine,*

giving it a high reliability factor with less maintenance; **maintenance crew** = ground staff whose responsibility it is to keep the aircraft serviceable; *the maintenance crew worked through the night to complete the work*; **maintenance manual** = manufacturer's instruction book of maintenance procedures

QUOTE poor maintenance training is expensive for the airline who notices the problem in late departures, longer than necessary maintenance periods and worst of all, crashes

Civil Aviation Training

major ['meɪdʒə] *adjective* (most) important; *there are two major cloud groups, stratus and cumulus*; **major airport** = large, important or international airport; **major problem** = serious problem (NOTE: the opposite is **minor**)

majority [mə'dʒɒrɪti] *noun* greater number or larger part - anything more than 50%; *the majority of passengers prefer to sit in a non-smoking area of the cabin* (NOTE: the opposite is **minority**)

malfunction [mæl'fʌŋkʃn] **1** *noun* failure to work or to function correctly; *the oil pressure and temperature of the CSDU can be monitored by the pilot and if a malfunction occurs, the pilot can then choose to disconnect the CSDU from the engine* **2** *verb* to function incorrectly or fail to function; *oscillating outputs from the alternators could cause sensitive equipment to malfunction*

mandatory ['mændætri] *adjective* compulsory, required or ordered by an official organization or authority; *fire detection systems in toilets are mandatory*

maneuver [mə'nuːvə] *noun US* = MANOEUVRE

maneuverability [mənuːvrə'bɪlɪti] *noun US* = MANOEUVRABILITY

maneuvering area [mə'nuːvrɪŋ 'eəriə] *noun US* = MANOEUVRING AREA

manifold ['mænɪfəʊld] *noun* system of pipes for a fluid from single input to multiple output or multiple input to single output; *inlet and exhaust manifolds of a piston engine*; **manifold pressure** = absolute pressure in the induction system of a piston engine measured in inches of mercury

manner ['mænə] *noun* way of doing something; *wind is said to be veering when it changes direction in a clockwise manner*; *pre-flight checks should be done in the correct manner*

manoeuvrability [mənuːvrə'bɪlɪti] *noun* ability and speed with which an aircraft can turn away from its previous path; *light training aircraft do not have great manoeuvrability but they are stable and therefore easier to fly* (NOTE: also written **maneuverability** in American English)

manoeuvre [mə'nuːvə] *noun* any deliberate or intended departure from the existing flight or ground path; **flight manoeuvre** = turns, loops, climbs, descents; **ground manoeuvre** = taxiing and turning onto runways and taxiways, etc. (NOTE: also written **maneuver** in American English)

manoeuvring area [mə'nuːvrɪŋ 'eəriə] *noun* part of the aerodrome for the take-off, landing and taxiing of aircraft (NOTE: also written **maneuvering** in American English)

manual ['mænjuəl] **1** *adjective* referring to the hands; done or worked by hand; *the electronic flight instrument system has two self-test facilities - automatic and manual*; **manual control** = hand-flying an aircraft equipped with an autopilot or automatic flight control system **2** *noun* reference book giving instructions on how to operate equipment, machinery, etc.; *maintenance manual; aircraft operating manual*

manually ['mænjuli] *adverb* by hand; *the system is switched on manually*

manufacture [mænju'fæktʃə] *verb* to make a product for sale using industrial machines; *the centrifugal compressor is usually more robust than the axial flow type and also easier to develop and manufacture*

map [mæp] **1** *noun* representation of the earth's surface on a flat surface such as a sheet of paper; *a map of Africa* **2** *verb* to make measurements and calculations of part of the earth's surface in order to produce a map; *a team of surveyors mapped the area*

margin ['mɑːdʒɪn] *noun* (a) blank space bordering the written or printed area on a page; *write notes in the margin of the book* (b) amount allowed in addition to what is needed; *safety margin*; *in certain configurations, it is possible for the buffet speed to be less than the required 7% margin ahead of the stall*

maritime ['mærɪtaɪm] *adjective* referring to the sea; **maritime wind** = wind blowing from the sea; *the Rocky Mountains of North America act as a barrier to the cool maritime winds from the Pacific Ocean*

mark [mɑːk] **1** *noun* (a) visible trace such as a dot or a line on a surface; *there are marks on tyres and wheel rims which are aligned and indicate the extent of tyre creep* (b) points (may be as a percentage) given for academic work; **pass-mark** = the mark which separates those who fail and those who pass an examination **2** *verb* (a) to make a visible line, dot, etc., on a surface; *mark the departure point on the chart* (b) to show or indicate; *the weather front marks the boundary between the two air masses* (c) to correct or check academic work done by a student; *the instructor marked the exam papers*

marked [mɑːkt] *adjective* very noticeable; clear and definite; **a marked increase** = a noticeable, therefore possibly large, increase; **a marked change in the weather** = a significant change in weather

marker ['mɑːkə] *noun* (a) something which acts as an indicator, such as of distance, position; **marker dye** = brightly coloured substance used by people adrift at sea to draw the attention of flight crews to their position **(b)** radio beacons, part of the ILS; **inner marker** = radio beacon placed between the middle marker and the end of the ILS runway; **middle marker** = ILS (marker) beacon on extended runway centre line, usually 3500 feet from the runway threshold; **outer marker** = ILS (marker) beacon, usually on centre line of approach at about 4.5 nm from the runway threshold

COMMENT: the outer marker (OM) is indicated on the instrument panel, by a blue light. The middle marker (MM) is indicated by an amber light and the inner marker (IM) by a white light

marshal ['mɑːʃəl] *verb* to direct aircraft into their parking positions on the apron by means of hand signals; *after taxiing, a marshaller marshals the aircraft to the disembarkation and unloading point*

marshaller ['mɑːʃlə] *noun* member of ground staff whose job is to direct aircraft into parking positions by means of hand signals

QUOTE when under a marshaller's control, reduce speed to a walking pace
Civil Aviation Authority, General Aviation Safety Sense Leaflet

marshalling signals ['mɑːʃlɪŋ 'sɪɡnəlz] *noun* hand signals used by a marshaller; *marshalling signals are used to direct aircraft on the ground*

mask [mɑːsk] **1** *noun* device to cover the face; **oxygen mask** = device to cover the nose and mouth which is connected to an oxygen supply; *anoxia at high altitudes can be overcome by breathing through an oxygen mask* **2** *verb* to hide; to cover up; *when practising instrument flying, the aircraft windows are masked to prevent the (student) pilot from seeing out of the aircraft*

mass [mæs] **1** *noun* **(a)** physical volume of a solid body; *mass is a basic property of matter and is called weight when it is in a field of gravity such as that of the earth*

(b) large body of something with no particular shape; *a land mass such as the continent of Africa*; **air mass** = very large mass of air in the atmosphere in which the temperature is almost constant, divided from another mass by a front; *air masses are divided into two types according to source region, and these are known as polar and tropical air masses*; **mass ascent** = slow ascent of a large body of air in regions of low pressure and of warm air rising over a cold air mass **2** *adjective* involving a large number of people, etc.; **mass exit** = the departure of everybody, or nearly everybody, from a room, etc.

mast [mɑːst] *noun* **(a)** vertical pole for a flag or antenna; *ice accretes on the leading edge of the detector mast*; **signals mast** = vertical pole on an airfield from which signal flags are flown **(b)** tube projecting from the underside of the aircraft from which liquid can drain well away from the airframe

master ['mɑːstə] **1** *adjective* main or principal; **master cylinder** = hydraulic cylinder from which pressure is transmitted to smaller slave cylinders; **master key** = key which can open a number of doors, etc.; **master switch** = most important of a number of switches operating a system **2** *verb* to overcome the difficulty of something; *it takes practice to master crosswind landings in light aircraft*

match [mætʃ] *verb* **(a)** to suit; to go well together; *the most important factor when matching a propeller to an engine is tip velocity* **(b)** to be equal to; *the polarization of the antenna must match that of the transmitter*

material [mə'tɪəriəl] *noun* substance out of which something can be made; *wood, fabric and paper are all free-burning materials*

matrix ['meɪtrɪks] *noun* grid-like arrangement of circuit elements; *oil coolers consist of a matrix, divided into sections by baffle plates*

matter ['mætə] *noun* **(a)** physical substance; *mass is a basic property of* *matter*; **foreign matter** = something unwanted which is found in a substance or a device (such as sand or water in fuel); *turbine blades can be damaged by foreign matter such as stones entering through the engine intake on take-off*; **solid matter** = solid substance **(b)** subject for discussion, concern or action; *safety is a matter of great importance* **(c)** trouble or difficulty; **what's the matter?** = what's the problem?; **it doesn't matter** = it isn't important, so don't worry

maximum ['mæksɪməm] **1** *adjective (of quantity, amount, etc.)* greatest possible; *the maximum daily temperature is 35°C*; *the maximum speed of the aircraft is 200 kt*; **maximum total weight authorized (MTWA)** = maximum authorized weight of aircraft fuel, payload, etc., given in the Certificate of Airworthiness (C of A) **2** *noun* the greatest possible quantity, amount, etc.; *there is a net gain of heat by the earth until terrestrial radiation balances solar radiation when the daily temperature is at its maximum*

mb = MILLIBAR(S)

mean [miːn] **1** *adjective* referring to something midway between two extremes; average; **mean daily temperature** = average daily temperature; **mean effective pressure (MEP)** = average pressure exerted on the piston during the power stroke; **mean wind** = average speed of a wind; *see also* MEAN SEA LEVEL; MEAN SUN **2** *noun* something having a position midway between two extremes; a medium or average; **arithmetic mean** = average value of a set of numbers **3** *verb* **(a)** to signify; to have as an explanation; *airspeed means the speed of the aircraft in relation to the air around it* **(b)** to intend to do something; *I meant to telephone the reservations desk this morning but I forgot* **(c)** to result in; *installing a new computer network means a lot of problems for everybody* (NOTE: meaning - meant)

means [miːnz] *noun* something which brings a result; a way of doing something; *a clear window fitted in the reservoir provides a means of checking hydraulic fluid level during servicing*; **by means of** = by using; *fuel is transferred from the tanks to the carburettor by means of pipes*; **there are various means for navigation** = there are various different methods used for the purposes of navigation (NOTE: the singular and plural are the same)

mean sea level (MSL) ['miːn 'siː 'levl] *noun* average level of the sea taking tidal variations into account; *below FL50 cloud heights are referred to a datum of mean sea level*

mean sun ['miːn 'sʌn] *noun* position of imaginary sun in solar day of exactly 24 hours, behind the real sun in February and in advance of the real sun in November; *local mean time (LMT) is the time according to the mean sun*

measure ['meʒə] **1** *noun* **(a)** indication or way of assessing; *the way he dealt with the in-flight emergency is a measure of his skill as a pilot* **(b)** reference for discovering the dimensions or amount of something; *the litre is a measure of capacity* **(c)** device used for measuring; **a 1-metre measure** = a long ruler (1 metre long) **(d)** action taken to get a result; *stricter safety measures were introduced* **(e)** amount of something; *to be a good pilot, you need a measure of self-confidence* **2** *verb* **(a)** to find the dimensions or amount of something; *to measure a distance*; *to measure an angle*; *to measure the speed of an aircraft*; *wind directions are measured from magnetic north* **(b)** to be of a certain size, length, quantity, etc.; *how much does the pipe measure?*

measurement ['meʒəmənt] *noun* **(a)** act of measuring; *measurement of relative humidity is done using an instrument called a hygrometer* **(b)** results of measuring; *the measurements of the room are: height = 4 metres, length = 10 metres, width = 4 metres*

mechanical [mɪ'kænɪkl] *adjective* referring to machines; *activation may be electrical or mechanical*; **mechanical**

advantage = increase in force gained by using levers, gears, etc.; **mechanical engineering** = the study of design, construction, and use of machinery or mechanical structures; *he gained a degree in mechanical engineering from university*; **mechanical linkage** = (in a light aircraft) system of rods, cables and levers which connect the control column in the cockpit to the control surfaces on the wings, tailplane and fin; **mechanical pump** = pump operated by the engine rather than by electrical power

mechanics [mɪ'kænɪks] *noun* **(a)** *(physics)* the study of the action of forces on matter or material systems **(b)** the way something works; *the mechanics of the föhn wind provide a good illustration of the adiabatic process*

mechanism ['mekənɪzm] *noun* **(a)** arrangement of connected parts in a machine or system; *the landing gear mechanism; the nose wheel steering mechanism* **(b)** physical process; *the mechanism by which thunderstorms develop*

medium ['miːdiəm] **1** *adjective* referring to something that has a position or represents a condition midway between extremes; *high, medium and low frequencies; medium level cloud* **2** *noun* substance through which something else is transmitted or carried; *tubes convey the cooling medium; the cooling medium for cooling oil can be ram-air or fuel*

mega- ['megə] *prefix meaning* large (NOTE: the opposite is **micro-**; note also that **mega-** is used in front of SI units to indicate one million: **megawatt** = one million watts)

megahertz (MHz) ['megəhɜːts] *noun* measure of frequency equal to one million cycles per second

melt ['melt] *verb* to become liquid by heating; *ice melts at temperatures above freezing*; **melting point** = temperature at which a solid turns to liquid; *magnesium has a melting point of 1204°F*

member ['membə] *noun* **(a)** main structural unit; *the skin in bonded to the internal members; a beam is a member which is designed to withstand loading applied at an angle to it, often perpendicular* **(b)** person who joins a club or organization; *he is a member of the gliding club* **(c)** person in a team or crew; *most large passenger aircraft are now operated by two crew members*

memorize ['meməraɪz] *verb* to fix in the memory; to learn by heart; *it is helpful if a student pilot can memorize certain items, such as downwind checks, early in his training* (NOTE: also written **memorise**)

memory ['memri] *noun* **(a)** mental ability of remembering and recalling past events or information; *he has a good memory* = he remembers things easily **(b)** part of a computer which is used for the fast recall of information; *the computer cannot run many programs at the same time because it doesn't have enough memory*

mental ['mentl] *adjective* referring to the mind or brain; *anoxia severely limits physical and mental performance*; **mental calculation** = calculation done in one's head, without using aids such as pen, paper or calculator

mention ['menʃn] *verb* to refer to something (briefly); *as mentioned in chapter 4; as I mentioned yesterday; no one mentioned the incident*

MEP = MEAN EFFECTIVE PRESSURE

Mercator's projection [mɜː'keɪtəz prə'dʒekʃn] *noun* map projection of the earth onto a cylinder so that all the parallels of latitude are the same length as the equator; *since meridians on this projection are represented by parallel straight lines, it is impossible to represent the poles on Mercator's projection*

COMMENT: named after the Latinized name of G. Kremer, the Flemish-born geographer who died in 1594

mercury ['mɜːkjuri] *noun* silver-coloured metallic element, liquid at room temperature, used in thermometers; *manifold pressure gauges are calibrated in inches of mercury* (NOTE: chemical element; symbol **Hg**; atomic number **80**)

meridian [mə'rɪdiən] *noun* imaginary great circle on the earth's surface passing through the north and south geographic poles; **prime meridian** = the meridian passing through Greenwich, UK, at longitude 0°, from which longitudes are calculated

mesh [meʃ] *noun* a net-like structure; **wire mesh** = metal sheeting made of criss-crossed wiring

message ['mesɪdʒ] *noun* short written, coded or verbal communication; *the crew can use the public address system to broadcast messages to the passengers*; *there's a message from Mr. Jones on your desk*

met = METEOROLOGY

metal ['metl] *noun* elements such as iron, gold, mercury, copper, aluminium

metallic [me'tælɪk] *adjective* referring to metal; like metal; **metallic materials** = aluminium, titanium, steel, etc.; *some fire extinguishers do not harm metallic, wooden, plastic or fabric materials*; **non-metallic materials** = wood, plastics, fabrics, etc., which are not made of metal

METAR ['miːtɑː] = AVIATION ROUTINE WEATHER REPORT

meteorological [miːtiərə'lodʒɪkl] *adjective* referring to meteorology; **meteorological chart** = chart with weather information on it; **meteorological conditions** = description of the weather in a given area; **meteorological forecast** = prediction of the weather to come; **meteorological visibility** = greatest horizontal distance at which objects can be seen and recognized by an observer on the ground with normal eyesight and under conditions of normal daylight illumination; *meteorological visibility is given in metres*

up to 5,000 metres, and thereafter in kilometres; *see also* MOTNE

meteorologist [miːtiə'rolədʒɪst] *noun* person who studies, reports and forecasts the weather; *the analysis of the surface chart is the procedure in which the meteorologist completes the chart by inserting the fronts and isobars in their correct positions*

meteorology [miːtiə'rolədʒi] *noun* science which studies weather and weather conditions; *terrestrial radiation plays an important part in meteorology*

meter ['miːtə] *noun* (a) US = METRE (b) device to measure current, rate of flow, vertical distance, speed, etc.; *a gas meter*; *see also* ALTIMETER; AMMETER; FLOWMETER; VOLTMETER

methanol ['meθənol] *noun* colourless, toxic, flammable liquid, CH_3, used as an antifreeze, a general solvent, and a fuel, also called methyl alcohol or wood alcohol; *power output can be restored, or can be boosted to a value over 100% maximum power, by the injection of a water/methanol mixture at the compressor inlet or at the combustion chamber inlet*

method ['meθəd] *noun* particular way of doing something, especially if it is well thought out and systematic; *the most common method of displaying radar information is on a cathode ray tube*

metre ['miːtə] *noun* international standard unit of length, approximately equivalent to 39.37 inches (NOTE: also written **meter** in American English)

MHz = MEGAHERTZ

micro- ['maɪkrəʊ] *prefix meaning* small (NOTE: the opposite is **mega-**; note also that the prefix **micro-** is used in front of SI units to indicate a one millionth part: **microsecond** = one millionth of a second)

microburst ['maɪkrəʊbɜːst] *noun* particularly strong wind-shear especially associated with thunderstorms; *the investigation revealed that the crew lost* *control of the aircraft as it flew through the microburst*

micro-switch ['maɪkrəʊswɪtʃ] *noun* miniature switch used to govern systems automatically; *operation of an aircraft may also be seriously affected by the freezing of moisture in controls, hinges and micro-switches* (NOTE: the plural is **micro-switches**)

mid- [mɪd] *prefix meaning* middle; **mid-summer** = middle of the summer

mid-air ['mɪd'eə] *adjective* **mid-air collision** = collision between aircraft in the air rather than on the ground

middle ['mɪdl] **1** *adjective* in the centre; **middle marker** = *see* MARKER **2** *noun* centre; *the seat in the middle of the row*

mile [maɪl] *noun* **statute mile** = non SI unit of length (= 1.609 kilometres); **nautical mile (nm)** = 1.852 kilometres; *see also* NAUTICAL

millibar (mb) ['mɪlibɑː] *noun* unit of atmospheric pressure equal to 1 thousandth of a bar; standard atmospheric pressure at sea level being 1013.25 millibars

millilitre (ml) ['mɪlilːtə] *noun* one thousandth of a litre (NOTE: also written **milliliter** in American English. Note also that it is usually written **ml** after figures: 35ml)

millimetre (mm) ['mɪlimiːtə] *noun* one thousandth of a metre (NOTE: also written **millimeter** in American English. Note also that it is usually written **mm** after figures: **35mm**)

min = MINIMUM

minima ['mɪnɪmə] *see* MINIMUM

minimal ['mɪnɪml] *adjective* small in amount, importance, etc.; *safety equipment carried on some light aircraft may be as minimal as a portable fire extinguisher; any attempt to increase range by applying power is of minimal benefit*

minimize ['mɪnɪmaɪz] *verb* to reduce or decrease to the smallest amount possible (NOTE: also written **minimise**)

minimum ['mɪnɪməm] **1** *adjective* smallest possible (amount, distance, etc.); *the minimum amount required; minimum weather requirements for a particular operation, such as runway visual range (RVR)* **2** *noun* smallest or least possible quantity or amount; *fires should be tackled with the minimum of delay; to keep the weight of the fuselage structure to a minimum, the difference between cabin pressures and the external atmospheric pressures should be kept to a minimum* (NOTE: the plural is **minima** or **minimums**)

minimum flying speed ['mɪnɪməm 'flaɪɪŋ 'spiːd] *noun* lowest true air speed (TAS) at which an aircraft can maintain height

minimum fuel ['mɪnɪməm 'fjuːəl] *noun* amount of fuel required to reach destination and land without delay

minor ['maɪnə] **1** *noun* person under the age of legal adulthood **2** *adjective* small in size or amount and therefore relatively unimportant; **minor repairs** = repairs which can be made quickly and with the minimum amount of equipment (NOTE: the opposite is **major**)

minus ['maɪnəs] **1** *preposition* reduced by; less; *6 minus 2 equals 4 (6 - 2 = 4)* **2** *noun* minus sign (-); *minus forty degrees Celsius (- 40° Celsius)*

minute 1 ['mɪnɪt] *noun* **(a)** time period of 60 seconds; *there are 60 minutes in one hour*; **wait a minute** = wait a while or a short period of time **(b)** unit of angular measurement equal to one sixtieth of a degree (written ' with figures); one degree equals 60 minutes; *20 degrees and 20 minutes east (20° 20'E)* **2** [maɪ'njuːt] *adjective* very small indeed; *metal fatigue begins as minute cracks, too small to be seen, at the point of maximum stress*

miscellaneous [mɪsə'leɪnɪəs] *adjective* various; mixed; not all the same; *the first aid box contains miscellaneous items for use in a medical emergency*

miss [mɪs] *verb* not to get or catch; *two passengers arrived so late that they missed the flight*

missed [mɪst] *adjective* **missed approach** = approach which does not result in landing, followed by a go-around; **missed approach procedure** = action and flight path to be followed after missed approach at a particular aerodrome

mist [mɪst] *noun* **(a)** visible water vapour, in the form of very fine droplets, in the atmosphere; *mist is thinner than fog*; **morning mist** = mist which usually disappears before midday, as the result of warming from the sun **(b)** liquid in spray form; *an air/oil mist* **2** *verb* **to mist up** = to become covered in tiny water droplets (of condensation) and therefore prevent clear vision through a surface; *the windscreen misted up*

mix [mɪks] *verb* to put together in order to form one mass; *it is a fact of nature that different air masses do not mix together; air is mixed with fuel which then flows into the cylinder through the inlet valve*

mixture ['mɪkstʃə] *noun* something which is the result of a number of things mixed together; **fuel/air mixture** = combination of fuel and air which is ignited in a piston engine to provide power; *(weak mixture)* **lean mixture** = a fuel/air mixture in which the ratio of air to fuel is greater than usual; **rich mixture** = fuel/air mixture in which the proportion of fuel is greater than normal; *see also* AUTOMATIC

mixture control ['mɪkstʃə 'kən'trəʊl] *noun* device for controlling the ratio of fuel to air entering an engine's carburettor or fuel injection system. The mixture control is a knob or lever marked in red usually to the right of the throttle lever; *in order to stop the engine, the mixture control should be moved fully aft*

COMMENT: aircraft engines operate at different altitudes and the pilot must adjust the mixture to produce the most efficient fuel/air mixture for the atmospheric density

mm = MILLIMETRE

MM = MIDDLE MARKER

mnemonic [nɪ'mɒnɪk] *noun* something (such as a word, sentence, little poem, etc.) to help the memory

COMMENT: some of the well known mnemonics: **ARROW** = Airworthiness Certificate, Registration Document, Radio Station Licence, Operating Handbook, Weight and Balance document - documents to be carried in (light) aircraft (U5); **BUMF checks** = Brakes, Undercarriage, Mixture, Fuel - downwind checks in a light, single engine aircraft with fixed pitch propeller; **FREDA** = Fuel, Radio, Engine, Direction indicator, Altimeter - airfield approach checks; **HASELL** = Height, Airframe, Security, Engine, Location, Lookout - pre-stall checks; **variation east, magnetic least: variation west, magnetic best** = mnemonic to help remember whether to add or subtract variation

mode [məʊd] *noun* **(a)** particular selected setting for the operation or functioning of equipment; *automatic mode; manual mode* **(b)** letter or number given to various pulse spacing of airborne transponders and ground interrogators; *mode A and mode C for altitude reporting, are used in air traffic control*

model ['mɒdl] *noun* simplified description of a system (often mathematical) to make calculation simpler; *the description of the weather patterns is a model only which, in reality, is modified greatly by a number of factors*

moderate 1 ['mɒdərət] *adjective* **(a)** referring to something well within limits, not extreme; **a moderate climate** = a climate which is not too hot, not too cold **(b)** the middle of three descriptions of intensity or amount i.e. light, moderate, severe; **moderate humidity** = humidity which is not light or severe; **light to moderate** = which varies between light and moderate; *light to moderate icing*; **moderate to severe** = which varies between moderate and severe; *moderate to* *severe turbulence* **2** ['mɒdəreit] *verb* to make less extreme; to become less extreme; *the south west wind moderates the climate of the UK; as the wind moderated, the aircraft was allowed to take off*

modern ['mɒdən] *adjective* referring to the present day; up to date; *modern engines are far more powerful than engines used in the past*

modification [mɒdɪfɪ'keɪʃn] *noun* alteration which improves; change in character or form which is normally an improvement; *there have been many modifications to the simple carburettor over the years; as a result of the crash, modifications were made to the rudder linkage*

modify ['mɒdɪfaɪ] *verb* to change or alter in order to improve; *the landing gear was modified to provide greater strength* (NOTE: **modifying - modified**)

modulate ['mɒdjuleɪt] *verb* to change the frequency, amplitude, phase, or other characteristic of an electromagnetic wave; *the ground station transmits a code in two short bursts, each of which is modulated with two tones*

modulation [mɒdju'leɪʃn] *noun* change of a property of an electromagnetic wave or signal, such as its amplitude, frequency, or phase; *pulse modulation is a series of quick, short bursts of energy which are radiated from an antenna which serves both the transmitter and the receiver*

module ['mɒdju:l] *noun* replaceable detachable unit; *all inputs are processed by the appropriate circuit modules of the generator control unit (GCU)*

moist [mɔɪst] *adjective* a little wet, damp or humid; *warm moist air from the Gulf of Mexico can extend into Canada*

moisture ['mɔɪstʃə] *noun* water or other liquid; *(in the air)* **moisture content** = amount of water in the atmosphere or as seen when it condenses onto cold surfaces; *when the air passing through the*

carburettor is reduced below 0°C (Celsius), any moisture in the air changes into ice

mold [məuld] *noun US* = MOULD

molecule ['mɒlɪkju:l] *noun* smallest particle into which an element or a compound can be divided without changing its chemical and physical properties; *the molecules of a gas move more quickly than the molecules of a liquid*

moment ['məumənt] *noun* **(a)** short period of time; *it only takes a moment to fill in the log book* **(b)** point in time; **at the moment** = at this particular time; *he's not in the office at the moment* **(c)** *(physics)* product of a quantity and its perpendicular distance from a reference point; *a load on the end of a beam creates a bending moment* **(d)** the tendency to cause rotation about a point or an axis; *the tailplane provides a pitching moment to keep the aircraft level*

momentum [mə'mentəm] *noun (physics)* measure of the motion of a body equal to the product of its mass and velocity; *in rain, the faster an aircraft travels the more water it meets and the greater the relative momentum of the water droplets*

monitor ['mɒnɪtə] **1** *noun* **(a)** screen for a computer; visual display unit (VDU) **(b)** **monitor system** = system for checking and warning; *electronic centralized aircraft monitor (ECAM)* **2** *verb* to check, on a continuing basis; *flowmeters are fitted which allow crew to monitor the flow of fuel to each engine*

monocoque ['mɒnəkɒk] *noun* three-dimensional body with all the strength in the skin and immediately underlying framework; *in monocoque construction there is no internal stiffening as the thickness of the skin gives the strength and stability*

monsoon [mɒn'su:n] *noun* wind from the south-west or south that brings heavy rainfall to southern Asia in the summer;

although the monsoon winds are thought of as being Asiatic phenomena, they do occur over Africa and parts of North America, especially the Gulf of Mexico; **monsoon season** = season of wind and heavy rainfall in tropical countries

Morse [mɔːs] *noun (still used for identifying some radio beacons)* code used for transmitting messages in which letters of the alphabet and numbers are represented by dots and dashes or short and long signals; *VOR (very high frequency omni-directional radio range) stations transmit a 2 or 3-letter aural Morse callsign on the reference signal at least every 30 seconds*

COMMENT: named after S. F. B. Morse, the American electrician who died in 1872

motion ['məʊʃn] *noun* movement; the act of changing position or place; **horizontal motion** = movement from side to side; **rotary motion** = circular movement; **vertical motion** = up and down movement

MOTNE (= METEOROLOGICAL OPERATIONAL TELECOMMUNICATIONS NETWORK EUROPE) network for the exchange of meteorological information needed by meteorological offices, VOLMET broadcasting stations, air traffic service units, operators and other aeronautical users

motor ['məʊtə] *noun* machine which provides power for moving a vehicle or device with moving parts; *an electric motor; a hydraulic motor* (NOTE: piston or jet power plants for aircraft are referred to as **engines** not motors)

mould [məʊld] **1** *noun* hollow shape for forming plastics, etc.; *moulds are used in the manufacture of plastic components* **2** *verb* to shape, often using a mould; *thermo-plastic material become soft when heated and can be moulded again and again* (NOTE: also written **mold** in American English)

mount [maʊnt] *verb* to fix to a support; *a propeller consists of a number of separate blades mounted in a hub*

mountain ['maʊntɪn] *noun* mass of rock rising (higher than a hill) above ground level; *they flew over mountains in the south of the country*; **Mountain Standard Time** = time zone of the west-central part of the USA and Canada, 7 hours behind GMT (Greenwich Mean Time)

mounted ['maʊntɪd] *adjective* fixed to a support; **rear-mounted** = mounted at the rear (of the aircraft); *some aircraft such as the Boeing 727 have rear-mounted engines*

mounting ['maʊntɪŋ] *noun* supporting component or attachment point; *Airbus aircraft have engine mountings under the wings*

movement ['muːvmənt] *noun* **(a)** change in place or position; *the upward movement of the piston compresses the fuel/air mixture*; **movement of the crankshaft** = the rotation of the crankshaft; **the downward movement of cool air** = the downward flow of cool air **(b) air traffic movements** = the number of aircraft taking off and landing; *an increase in air traffic movements*

MSL = MEAN SEA LEVEL

MTWA = MAXIMUM TOTAL WEIGHT AUTHORIZED

muff [mʌf] *noun (worn to protect against loud noise)* **acoustic ear muffs** = ear protectors or defenders

multi- ['mʌlti] *prefix meaning* multiple or many

multi-engine(d) ['mʌlti'endʒɪnd] *adjective* **multi-engine(d) aircraft** = aircraft with more than two engines

multiple ['mʌltɪpl] *adjective* many; *autoland system redundancy employs multiple systems operating in such a manner that a single failure within a system will have little effect on the* *aircraft's performance during the approach and landing operation*

multiplication [mʌltɪplɪˈkeɪʃn] *noun* mathematical operation to work out a specified number of times the value of a number; *the multiplication sign is x*

multiply [ˈmʌltɪplaɪ] *verb* to work out a specified number of times the value of a number; *to multiply 20 by 6 is to calculate what is 6 times 20 (6 x 20); 4 multiplied by 2 is 8 (4 x 2 = 8); to calculate fuel required, multiply the duration of the flight by the consumption of the engine at the required power*

multi-purpose [ˈmʌltiˈpɜːpəs] *adjective* suitable for many different uses; **multi-purpose tool** = tool which can be used in many different ways

mutual [ˈmjuːtjuəl] *adjective* directed and received in equal amount; **mutual inductance** = electro-magnetic field in one circuit caused by a quickly changing magnetic field in another circuit

Nn

N = NORTH

nacelle [næ'sel] *noun* streamlined housing for an engine; *the ram air intake is located in a wing leading edge or an engine nacelle fairing*

narrow ['nærəʊ] *adjective* small in width; not wide; *the narrow aisles of passenger aircraft makes it difficult to evacuate an aircraft quickly*; *narrow band of cloud*; *a narrow beam of electrons* (NOTE: the opposite is **wide** or **broad**)

NASA ['nɑːsə]   =   NATIONAL AERONAUTICS   AND   SPACE ADMINISTRATION

national ['næʃnl] *adjective* belonging to a country; *KLM is the national airline of the Netherlands*

National Aeronautics and Space Administration (NASA)
['næʃnl   eərə'nɔːtɪks   ənd   'speɪs ədmɪnɪ'streɪʃn] *noun* US organization for flight and space exploration

nature ['neɪtʃə] *noun* **(a)** the world: plants, animals and their environment in general; *electricity is one of the fundamental forces of nature* **(b)** sort, type; *action taken by the crew will depend on the nature of the emergency* **(c)** essential quality of something; *the convective nature of thunderstorms*; *magnesium is a fire hazard of unpredictable nature*

nautical ['nɔːtɪkl] *adjective* referring to the sea; *the terms pitch, roll and yaw are nautical in origin*

nautical mile (nm) ['nɔːtɪkl 'maɪl] *noun* 1.852 kilometres; *a nautical mile is the length of an arc on the earth's surface subtended by an angle of one minute at the centre of the earth*; *one knot is equal to one nautical mile per hour*; *compare* STATUTE MILE

NAVAID ['næveɪd] *noun* navigational aid

navigation [nævɪ'geɪʃn] *noun* theory and practice of planning, controlling and recording the direction of an aircraft; *the basis of air navigation is the triangle of velocities*; **navigation lights** = lights on an aircraft consisting of a red light on the left wing tip, a green light on the right wing tip and a white light on the tail

COMMENT: navigation lights must be used between sunset and sunrise

navigational [nævɪ'geɪʃnl] *adjective* referring to navigation; *the accuracy of modern navigational equipment is much greater than older systems*; **navigational aid (NAVAID)** = mechanical and electronic device to help a pilot navigate; *any type of navigational aid but particularly electronic aids, for example ADF (automatic direction finding) and NDBs (non-directional beacons)*; **navigational line** = line of position; line along which an aircraft is known to be at a particular time usually by taking a VOR bearing

NDB = NON-DIRECTIONAL BEACON

necessary ['nesəsri] *adjective* needed or essential; *a rich mixture is necessary at slow running*; **as necessary** = when needed; *warnings, cautions and advisory* *messages are displayed only when necessary*

necessity [nə'sɛsɪti] *noun* something essential; *student pilots should understand the necessity for treating thunderstorms with great respect*

needle ['niːdl] *noun* thin metal pointer in an instrument; *the needle indicated to zero*; **needle valve** = valve formed of a tapered needle projecting into a small opening in a tube, etc., usually connected to a float, which provides fine adjustment of fluid flow; *atmospheric pressure will allow the capsule to expand, causing the needle valve to move into the opening thus reducing the flow of fuel*

negative ['nɛɡətɪv] *adjective* (a) a value of less than 0; *in a reversing propeller, the propeller mechanism includes a removable ground fine pitch stop which enables the propeller to be set to a negative pitch* (b) referring to an electric charge of the same sign as that of an electron (indicated by the symbol -); **the negative terminal of a battery** = the terminal of a battery marked with the symbol - and normally coloured black (not red) (c) showing refusal; **a negative answer** = no (d) showing resistance or non-co-operation; *a negative attitude*

negligible ['nɛɡlɪdʒəbl] *adjective* small or unimportant to the extent that it is not worth considering; *atmospheric attenuation is negligible until the upper end of the UHF (ultra high frequency) band when it increases rapidly*; **negligible risk** = almost no risk

neoprene ['niəupriːn] *noun* type of synthetic rubber

net [nɛt] *adjective* after all necessary deductions; *(of engine)* **net dry weight** = basic weight of an engine without fluids (fuel, oil, etc.) and without accessories not essential for the engine to function

network ['nɛtwɜːk] *noun* (a) complex interconnected group or system; *a network of meteorological stations around the world exchange information* (b) system of

lines or channels which cross each other; *on a map, meridians of longitude and parallels of latitude form a network of lines called a graticule* (c) system of computers interconnected in order to share information

neutral ['njuːtrəl] *adjective & noun* (a) indicates an electrical charge which is neither positive nor negative (b) indicates the position of a switch or lever which leaves a system active but not engaged, for example an engine gear lever position in which the engine is disconnected from the driven parts (c) middle position of a control surface providing no aerodynamic effect other than that as part of the wing; *after a turn, the auto-control will return the ailerons to neutral as the aircraft returns to straight flight*

neutralize ['njuːtrəlaɪz] *verb (an acid)* to cancel the effect of; *spillage from a lead acid battery may be neutralized by washing with a diluted solution of sodium bicarbonate* (NOTE: also written **neutralise**)

nil [nɪl] *noun* nothing; zero; **nil drizzle** = no drizzle

nimbostratus [nɪmbəu'streɪtəs] *noun* cloud forming a low dense grey layer from which rain or drizzle often falls

nitrogen ['naɪtrədʒən] *noun* colourless, odourless gas which makes up four fifths of the earth's atmosphere; *some aircraft have high pressure air or nitrogen bottles provided in the undercarriage and flap circuits for emergency lowering* (NOTE: chemical symbol N; atomic number 7)

nm = NAUTICAL MILE

nocturnal [nɒk'tɜːnəl] *adjective* referring to the night; *because there is a requirement for a cold ground, a katabatic wind tends to be nocturnal, but if the slope is snow-covered, it can also occur during the day*

nominal ['nɒmɪnəl] *adjective* (a) not significant or not important; **a nominal increase** = a very small increase (b) named, specific; *as an installed battery becomes fully charged by the aircraft* generator, the battery voltage nears its nominal level and the charging current decreases

non- [nɒn] *prefix meaning* not or no

non-directional beacon (NDB) ['nɒndaɪ'rekʃnl 'biːkən] *noun* radio beacon transmitting a signal by which the pilot can determine his bearing; *ADF (automatic direction finding) position lines are obtained from ground installations called non-directional beacons (NDBs)*

non-essential ['nɒnɪ'sentʃl] *adjective* not necessary; *in order to ensure the shortest possible take-off run, all non-essential equipment was removed*

non-return valve ['nɒnrɪ'tɜːn 'vælv] *noun* valve which allows a fluid to pass in one direction only; *as the piston moves upwards in the cylinder, fluid is drawn in through a non-return valve*

non-smoking area ['nɒn'sməʊkɪŋ 'eəriə] *noun* area where smoking is not allowed; no smoking area; *see also* NO-SMOKING

normal ['nɔːməl] *adjective* referring to something which is usual and is to be expected; **normal room temperature =** temperature regarded as comfortable for usual daily activity; **under normal conditions =** when everything is as it usually is

north [nɔːθ] **1** *noun* compass point 360°; direction towards which the magnetic needle points on a compass; *fly towards the north; the wind is blowing from the north;* **north facing mountain side =** face of the mountain which looks towards the north **2** *adjective* **(a)** referring to areas or regions lying in the north; referring to the compass point 360°; *the north coast of France* **(b)** a **north wind =** a wind blowing from or coming from the north (NOTE: a wind is named after the direction it comes from) **(c)** northern part of a region or country; *North America* **3** *adverb (direction)* towards the north; *the aircraft was heading north; see also* COMPASS; MAGNETIC; TRUE

northbound ['nɔːθbaʊnd] *adjective* travelling towards the north; *a northbound flight*

north-east ['nɔːθ'iːst] **1** *noun* direction between north and east; *after take-off, the aircraft turned to the north-east* **2** *adjective* **(a)** situated in the north-east; *the north-east coast of England* **(b)** *(wind)* which blows from the north-east; north-easterly (wind); *a north-east wind* **3** *adverb* towards the north-east; *we are heading north-east*

north-easterly [nɔːθ'iːstəli] *adjective* **(a)** *(wind)* which blows from or which comes from the north-east; *a north-easterly wind was blowing* **(b)** *(direction)* which moves towards the north-east; *follow a north-easterly direction*

north-eastern [nɔːθ'iːstən] *adjective* referring to the north-east; situated in the north-east; *the north-eastern part of the United States*

northerly ['nɔːðəli] **1** *adjective* **(a)** situated towards the north; *the most northerly point of a country* **(b)** *(wind)* blowing from or coming from the north; **northerly airflow =** airflow coming from the north; *a northerly airflow from the polar regions* **(c)** *(direction)* moving towards the north; *we are flying in a northerly direction* **2** *noun* wind which blows from the north

northern ['nɔːðən] *adjective* referring to the north; situated in the north; *the northern hemisphere*

North Pole ['nɔːθ 'pəʊl] *noun* point which is furthest north on the earth; *from the UK the aircraft flew over the North Pole to Vancouver*

northward ['nɔːθwəd] **1** *adjective* going towards the north; **a northward flight =** flight heading towards the north **2** *adverb* *US* = NORTHWARDS

northwards ['nɔːθwədz] *adverb* towards the north; *one of the aircraft was flying northwards*

north-west ['nɔ:θ'west] **1** *noun* direction between north and west; *the aircraft turned towards the north-west* **2** *adjective* **(a)** situated in the north-west; *the north-west coast of England* **(b)** *(wind)* which blows from the north-west; north-westerly (wind); *a north-west wind* **3** *adverb* towards the north-west; *we are heading north-west*

north-westerly [nɔ:θ'westəli] *adjective* **(a)** *(wind)* which blows from or which comes from the north-west; *a north-westerly wind was blowing* **(b)** *(direction)* which moves towards or to the north-west; *follow a north-westerly direction*

north-western [nɔ:θ'westən] *adjective* referring to the north-west; situated in the north-west; *the north-western part of the United States*

nose [nəuz] *noun* extreme forward end of the aircraft; **nose gear** = nose wheel and supporting struts and linkages; **nose wheel** = undercarriage wheel at the front of the aircraft

no-smoking sign ['nəu'sməukɪŋ 'saɪn] *noun* (usually) lit-up sign warning passengers and crew that smoking is not allowed; *see also* NON-SMOKING

note [nəut] **1** *noun* **(a)** brief message on a piece of paper; *there's a note on your desk* **(b)** brief comments made on paper during a lecture; *make notes while you watch the video recording* **(c)** short comment or explanation in a text often at the end of a book or at the bottom of a page **(d)** piece of paper money; *a £10 note* (NOTE: American English is **bill**) **(e)** musical tone of definite pitch; *the note of the engine changes as rpm (revolutions per minute) is increased* **2** *verb* **(a)** to observe carefully, to take notice; *note that true north is always along a meridian* **(b)** to write down; *note the time of departure on the log sheet*; *note the time of any incident*

QUOTE immediately you become
unsure of your position, note
the time and, if you are in
touch with an ATC unit,

especially a radar unit, you
should request assistance
*Civil Aviation Authority, General Aviation
Safety Sense Leaflet*

notice ['nəutɪs] **1** *noun* **(a)** written or spoken announcement; **notice board** = usually wooden board in a corridor or classroom, etc., where information on paper can be displayed **(b)** formal warning or notification; **to give notice** = to inform an employee/employer in advance and in writing, of a termination to a period of employment; *as a result of the accident, the instructor was given three months' notice*; **the student pilot is grounded until further notice** = the student pilot cannot fly again until he is told by those in authority that he can continue **2** *verb* to observe; *while doing the pre-flight checks, Captain Smith noticed that there was a leak of hydraulic fluid from one of the brake cylinders*

noticeable ['nəutɪsəbl] *adjective* which catches the attention: which is easily noticed; **a noticeable increase** = an increase which is important enough to be observed; *there was a noticeable improvement in the trainee's recent exam results*

notification ['nəutɪfɪkeɪʃn] *noun* act or instance of informing someone; *notification of the new procedures will follow in a few days*; *he received notification that he had been accepted for the job*

notify ['nəutɪfaɪ] *verb* to inform; *students were notified of their exam results by post*; *the authorities must be notified of all in-flight incidents*

nozzle ['nɒzl] *noun* **(a)** projecting part with an opening at the end of a pipe, for regulating and directing a flow of fluid; *the nozzle of a portable fire extinguisher should be pointed at the base of the fire* **(b)** **propelling nozzle** = the extreme rear part of the jet engine where the jet exhaust enters the atmosphere

nucleus ['nju:kliəs] *noun* central part around which other parts are grouped; *an* *atom consists of a nucleus with orbiting electrons; condensation occurs on very small particles suspended in the air which are known as condensation nuclei* (NOTE: plural is **nuclei** ['njuːkliiː])

null [nʌl] *noun* an instrument reading of zero; **the null position** = the zero position; *nulls are used for direction sensing because they are better defined than the maxima*

numerical [njuː'merɪkl] *adjective* referring to numbers or digits; **a numerical value** = a number; *see also* ORDER

numerous ['njuːmərəs] *adjective* very many; a lot (of); *large transport aircraft have numerous quite large clearly marked exits to facilitate rapid evacuation of passengers*; *numerous refinements to the simple actuator will be found in use*

nut [nʌt] *noun* metal ring which screws on a bolt to hold it tight; *turn the nut anticlockwise to loosen it*

Oo

obey [ə'beɪ] *verb* **(a)** to carry out or comply with a command; *pilots must obey landing instructions* **(b)** to follow a physical law; *winds obey Buys Ballot's Law*

object 1 ['ɒbdʒekt] *noun* **(a)** thing you can touch, see, etc., and which has a particular form and dimensions; *any given object will collect more ice when travelling at high speed than at low speed* **(b)** intention or aim; *the object of the briefing is to inform all aircrew of the new procedures* **2** [əb'dʒekt] *verb* to raise or voice opposition; *staff objected to the introduction of longer working hours*

oblong ['ɒblɒŋ] **1** *adjective* rectangular; *an oblong piece of aluminium* **2** *noun* a rectangle

obscure [əb'skjuə] **1** *adjective* not clearly understood; **the explanation was obscure** = the explanation was difficult to understand because it wasn't clear **2** *verb* to make difficult to see; *deposits of ice crystals on the windscreen will obscure vision*

obscured [əb'skjuəd] *adjective* **sky obscured** = meteorological term to mean that fog, etc., prevents sight of the sky

observation [ɒbzə'veɪʃn] *noun* careful watching; *the type of cloud is established by observation and comparison with cloud photographs*

observe [əb'zɜːv] *verb* to watch carefully; *local wave action can be observed from a height of 200 feet; wing deflection can be observed from the passenger cabin*

observer [əb'zɜːvə] *noun* person working in a meteorological station who assesses weather conditions by visual means; *meteorological visibility is the greatest horizontal distance at which objects can be seen and recognized by an observer on the ground with normal eyesight and under conditions of normal daylight illumination*

obstacle ['ɒbstəkl] *noun* something which blocks a path or prevents progress; *low frequency transmissions can penetrate obstacles such as mountains; knowing the heights of obstacles en route, it must be ensured that in the event of an emergency, the flight may be continued in safety*

obstruct [əb'strʌkt] *verb* to block a path or to prevent the progress of something; *bags and luggage must not obstruct the aisles; a safety valve is normally provided, in case the water separator assembly becomes obstructed by ice*

obstruction [əb'strʌkʃn] *noun* **(a)** the act or process of obstructing; *the glidepath antenna cannot be placed close to the centre line of the runway because it would cause an obstruction* **(b)** something which blocks a path or prevents progress; *before start-up, the air intakes and jet pipes must be inspected, to ensure that they are free from any debris or obstruction*

QUOTE taxiways and aerodrome obstructions may be hidden by snow, so ask if you are not certain
Civil Aviation Authority, General Aviation Safety Sense Leaflet

obtain [ɒbˈteɪn] *verb* to acquire, to get; *telephone the meteorological office in order to obtain the latest weather forecast*; *the probes are positioned in the gas stream in order to obtain an accurate temperature reading*

obvious [ˈɒbvɪəs] *adjective* clear and easily seen or understood; *it is obvious that high ground will disturb the smooth horizontal flow of air*

occasion [əˈkeɪʒn] *noun* time at which an event or happening occurs; *in recent months the aircraft suffered two engine failures, on the first occasion the aircraft force landed safely*; *the maiden flight of an aircraft is a great occasion*; **on occasions** = sometimes

occasional [əˈkeɪʒnl] *adjective* happening from time to time; **occasional rain** = periodic rain; **occasional turbulence** = turbulence happening from time to time

occluded front [əˈkluːdɪd ˈfrʌnt] *noun* weather front created when air is forced upward from the earth's surface, as when a cold front overtakes and undercuts a warm front; *jet streams are very rare near occluded fronts because of the much smaller temperature gradient across the fronts*

occlusion [əˈkluːʒn] *noun* the forcing of air upward from the earth's surface, as when a cold front overtakes and undercuts a warm front; the formation of an occluded front; *if the air ahead of the warm front is less cold than the air behind the cold front, the cold front will undercut the less cold air and form a cold occlusion*

occupant [ˈɒkʊpənt] *noun* person who has a seat in an aircraft; **occupants** = the crew plus passengers; *in-flight emergency procedures are designed to successfully combat airborne emergencies which threaten the safety of the aircraft and its occupants*

occupy [ˈɒkjʊpaɪ] *verb* (a) to have a position; to be in a place; *the passenger is occupying the wrong seat* (b) to busy

oneself; *once an evacuation process is under way the crew will be fully occupied carrying out emergency drills*

occur [əˈkɜː] *verb* to happen; *heavy rains occur during the monsoon season*; *tropical revolving storms generally occur from June to October*; *an accident occurred on June 12th*

occurrence [əˈkʌrəns] *noun* happening, event; *there were a number of occurrences of hijacking in the eighties*; *the occurrence of the equatorial jet stream is due to a temperature gradient with colder air to the south*

ocean [ˈəʊʃn] *noun* (a) the body of salt water which covers the earth (NOTE: chiefly American usage: British English prefers the word **sea**) (b) any of the major sea areas of the world; *the Atlantic Ocean*

COMMENT: the five oceans are: the Atlantic, the Pacific, the Indian, the Arctic and the Antarctic (or Southern)

oceanic [əʊʃiˈænɪk] *adjective* referring to the oceans; *the trade winds maintain their direction over the oceanic areas, especially the Pacific, more than over land areas*; **an oceanic crossing** = a flight across sea or ocean

octa [ˈɒktə] *noun* = OKTA

odd [ɒd] *adjective* (a) strange, peculiar; *an odd situation*; *the fact that moist air is lighter than an equivalent volume of dry air seems odd to many people* (b) *(mathematics)* **odd number** = number which cannot be exactly divided by two, such as 1, 3, 5, 7, etc.; *a (battery) cell contains an odd number of plates*; **odd tenth** = odd decimal, such as .1, .3, etc.; *frequency allocation of localizers in the VHF band is 108-112 MHz at odd tenths e.g. 108.1 and 109.3, the even decimals being allocated to VOR facilities* (c) indicating a number a little greater than the approximate number given; **it is 60-odd miles to our destination** = it is a little more than 60 miles to our destination

offer [ˈɒfə] **1** *noun* (a) presentation for acceptance or rejection; *he made an offer* of $85,000 for the aircraft **2** verb **(a)** to show readiness to do something; *he offered to carry her suitcase* **(b)** to present for acceptance or rejection; *the company offered him a job and he accepted it* **(c)** to provide; *the battery offers a short term power capability*

official [ə'fiʃl] **1** *adjective* referring to an authority, such as the government or a recognized organization; **an official weather report** = a weather report produced by a meteorological station **2** *noun* person employed by a government authority or a corporation; *an official of the civil aviation department will be visiting today*

offshore [ɒf'ʃɔː] *adjective* **(a)** at a distance from the shore **(b) offshore wind** = wind which blows from the coast towards the sea (NOTE: the opposite is **onshore wind**)

ohm [əum] *noun* unit of measurement of electrical resistance; *see also* AMPERE

COMMENT: Ohm's Law states: the current in a circuit is directly proportional to the voltage causing it and inversely proportional to the resistance of the circuit

okta *or* **octa** ['ɒktə] *noun* unit of visible sky equal to one eighth of total area visible to the horizon; *the amount of cloud cover is given in oktas*

COMMENT: to measure cloud cover, the sky is divided into imaginary sections, each covering one eighth of the total. A cloudless sky is 'zero oktas', and a sky which is completely covered with clouds is 'eight oktas' or 'eight eighths'

oil [ɔɪl] *noun* thick mineral liquid used as a fuel or to make mechanical parts move smoothly; **engine oil** = oil used especially to lubricate engines

oleo ['əuliəu] *noun* telescopic strut in undercarriage which absorbs impact loads on landing; *a safety switch is fitted in such a way to the oleo, that when the oleo is compressed on the ground, the*

'undercarriage up' selection cannot be operated (NOTE: short for **oleo-pneumatic**)

OM = OUTER MARKER

omit [ə'mɪt] *verb* to leave out, not to include; *high charts show only information relevant to high altitude flights and many beacons/aids which are provided for low operations are omitted to keep the chart clear* (NOTE: **omitting - omitted**)

one-in-sixty rule ['wʌnɪn'sɪksti 'ruːl] *noun* in navigation, every 1° of track error, and every 60 nm (nautical mile) flown, results in the aircraft being 1 nm off track

onshore [ɒn'ʃɔː] *adjective* towards the coast; **onshore wind** = wind which blows from the sea towards the coast (NOTE: the opposite is **offshore wind**)

opacity [ə'pæsɪti] *noun* the state of not allowing light to pass through; *sometimes, it is possible to estimate the depth and opacity of the layer of mist or fog from the ground observations*

opaque [ə'peɪk] *adjective* which does not allow light to penetrate or pass through; *rime ice is an opaque, white, granular ice which forms on leading edges*; *see also* OPACITY

opening ['əupnɪŋ] *noun* **(a)** space which acts as a passage through which something or somebody can go; *an inlet valve opening* **(b)** formal start of operation; *the opening of the new flying school* **(c)** vacancy for a job; *there's an opening for a new chief ground instructor*

operate ['ɒpəreɪt] *verb* **(a)** to control the working of; to make something work; *the control column operates the ailerons and elevators*; *the flaps are operated by a switch* **(b)** to use; to manage; *the airline operates a fleet of Boeing aircraft* **(c)** to perform or function; *jet transports operate at high altitudes* **(d)** to perform a surgical procedure, by cutting into the body; *the surgeon operated on the patient*

operating ['ɒpəreɪtɪŋ] *adjective* **Pilot's Operating Handbook** = book giving details of an aircraft with recommendations and instructions regarding its use

operation [ɒpə'reɪʃn] *noun* **(a)** making something work; *the operation of the ignition system in a light aircraft is quite simple* **(b)** **flight operations** = flying and using of aircraft; *(in a commercial context)* **long-haul operations** = flying long distance routes; **overload operations** = operation of aircraft in unusual situations when take-off weight exceeds the permitted maximum **(c)** effect; **to come into operation** = to come into effect; *the new procedures come into operation on 1st January* **(d)** surgical procedure; *the doctor performed an operation* **(e)** *(mathematics)* procedure such as addition or subtraction

QUOTE periodically check the carburettor heating system and controls for proper condition and operation
Civil Aviation Authority, General Aviation Safety Sense Leaflet

operational [ɒpə'reɪʃnl] *adjective* **(a)** which is working or functioning; *air traffic control facilities were not operational at the time of the accident*; **the operational life of the aircraft** = the expected working life of an aircraft **(b)** ready for use; referring to an aircraft in a suitable condition to fly; **an operational aircraft** = an aircraft which can be used for its assigned purpose

operative ['ɒpərətɪv] *adjective* which is functioning or working; *the system is now operative after the recent maintenance*

operator ['ɒpəreɪtə] *noun* person who operates or uses equipment; *a ring graticule around the edge of the cathode ray tube enables the operator to read the bearing directly*

oppose [ə'pəʊz] *verb* **(a)** *(of a force)* to work against; *in level flight, the force of lift opposes the force of gravity* **(b)** to reject or be in conflict with; to try to prevent; *the local people oppose the building of the new runway* **(c)** as

opposed to = in contrast with; *over sea as opposed to over land*

opposite ['ɒpəzɪt] **1** *adjective* **(a)** situated or placed directly across from something; facing; **opposite sides of a building** = the back and front of a building **(b)** *(of two things)* **going in opposite directions** = (i) moving away from each other; (ii) moving towards each other **(c)** completely different; reverse; *for every action there is an equal and opposite reaction* **2** *noun* something completely different; the reverse; *the opposite of a katabatic wind is an anabatic wind*; *the opposite of starboard is port*

opposition [ɒpə'zɪʃn] *noun* in **opposition** = against; *drag acts in opposition to thrust*; *the EMF (electromotive force) which is produced by all motors, is in opposition to supply voltage and is directly proportional to motor rpm (revolutions per minute)*

optimum ['ɒptɪmʌm] **1** *adjective* referring to the point at which the condition or amount of something is the best; *the optimum altitude for jet aircraft is higher than that for piston engine aircraft* **2** *noun* point at which the condition or amount of something is the best; *generally speaking, engine output is at its optimum at cruising speed*

option ['ɒpʃn] *noun* choice or alternative; *on a bad approach, the pilot of a powered aircraft always has the option of going around*; *he was given the option of two aircraft to buy*

orbit ['ɔːbɪt] **1** *noun* path of a planet, or of a satellite. as it moves around another celestial body; *a year is the time taken for the earth to complete one orbit round the sun*; **to put into orbit** *or* **to send into orbit** = to launch into space so that is revolves around a celestial body (such as the sun) **2** *verb* to revolve around; *the earth orbits the sun* (NOTE: **orbiting - orbited**)

order ['ɔːdə] **1** *noun* **(a)** an instruction given as a command by somebody in authority; *the captain gave the order to evacuate the aircraft* **(b)** sequence of occurrence; *the firing order of sparking plugs in a piston engine is 1, 3, 4, 2*; **alphabetical order** = arrangement with words beginning with letter A coming first, followed by those beginning with letter B, then C, etc.; **numerical order** = arrangement with lowest numbers (1, 2, 3, etc.) coming first and highest numbers (25, 26, 27, etc.) coming last **(c)** condition or state; *although the aircraft is old, it is in good working order*; **out of order** = not working; *the telephone is out or order* **(d)** **in the order of** = approximately; *VOR (very high frequency omni-directional radio range) beacons of 200 watts have a range in the order of 200 nm (nautical miles)* **in order to** = for the purpose of; so as to; *indicated airspeed must be corrected in order to obtain true airspeed* **2** *verb* **(a)** to give a command; *before impact, the captain will order the crew to secure themselves at their assigned emergency stations* **(b)** to put in a sequence; *order the items in importance from 1 to 10*

organization [ɔːgənaɪˈzeɪʃn] *noun* **(a)** association of people working together for the same cause; *the World Meteorological Organization (WMO); the International Civil Aviation Organization (ICAO)* **(b)** putting into a structured and systematic form; *the organization of training materials for the new self-access learning centre is under way* **(c)** planning; *Captain Scott is responsible for the organization of examinations* (NOTE: also written **organisation**)

organize [ˈɔːgənaɪz] *verb* **(a)** to arrange into a system; *organize your notes so that you can find things easily* **(b)** to plan; *the trip was well organized and everybody enjoyed themselves* (NOTE: also written **organise**)

orientate [ˈɔːriənteɪt] *verb* to locate in relation to the compass; *the first step in map reading is to orientate the chart by relating the direction of land features to their representation on the chart*

orientation [ɔːrienˈteɪʃn] *noun* position in relation to the compass; *the horizontal situation indicator (HSI) presents a*

selectable, dynamic colour display of flight progress and plan view orientation*

orifice [ˈɒrɪfɪs] *noun* opening, mouth or vent; *the liquid expands and builds up a pressure differential across an orifice which leads to the expansion chamber*

origin [ˈɒrɪdʒɪn] *noun* **(a)** source, place where something starts; *an air mass takes on the characteristics of its place of origin* **(b)** base from which a map projection is drawn; *the value of convergence used is correct at the parallel of origin*

original [əˈrɪdʒənl] *adjective* before all others; first; *the atmosphere is said to be stable if, when a parcel of air is displaced vertically, it tends to return to its original level*

originate [əˈrɪdʒəneɪt] *verb* to create; to come into being; *tropical revolving storms originate within 5-15° of the equator*; *aircraft fires after an emergency landing, often originate in the wing area*

orographic [ɒrəˈgræfɪk] *adjective* referring to mountains; **orographic cloud** = cloud formed by air being forced upward over mountainous areas; **orographic uplift** = the lifting of air masses in contact with mountain regions

orthomorphic [ɔːθəʊˈmɔːfɪk] *adjective* of the correct shape; *an orthomorphic chart is one which has meridians and parallels which intersect at right angles and, at any point on the chart, the scale must be the same in all directions*

orthomorphism [ɔːθəʊˈmɔːfɪzm] *noun* shape representation on a map; *orthomorphism means that bearings may be measured correctly at any point on a chart*

oscillate [ˈɒsɪleɪt] *verb* **(a)** to move regularly between extremes **(b)** *(of electrical current)* to increase or decrease regularly as to produce oscillations; *instability protection is incorporated to guard against oscillating outputs from the alternators*

oscillation [ɒsɪ'leɪʃn] *noun* **(a)** *(of pendulum, etc.)* regular movement between extremes; *ridge waves can be thought of as oscillations about the stable state of the undisturbed air flow with the range of hills providing the disturbance* **(b)** regular increase and decrease of electrical current; *the supply is subject to oscillation*

oscillator ['ɒsɪleɪtə] *noun* electronic circuit that produces a pulse or a signal at a particular frequency; *the local oscillator replicates the radio frequency of the frequency generator at the transmitter*

out [aʊt] *adverb* **out of** = away from; no longer in; **out-of-balance turn** = a turn in which the aircraft 'skids' upwards and outwards from the turn or 'slips' inwards and downward; *during an out-of-balance turn, the ball in the slip indicator will be deflected to the left or right*

outbound ['aʊtbaʊnd] *adjective & adverb* towards destination away from a VOR; *the aircraft flies outbound from the beacon along the airway and inbound to the facility at the other end of the leg*; **outbound traffic** = aircraft flying away from an airfield

outbreak ['aʊtbreɪk] *noun* sudden start; *showers are local outbreaks of precipitation from detached cumulus or cumulonimbus*; *hand operated fire extinguishers are provided to combat any outbreaks of fire in the flight crew compartment and passengers cabins*

outer ['aʊtə] *adjective* **(a)** external; *pneumatic de-icer boots are made from vulcanized rubber fabric with an outer covering of neoprene* **(b)** positioned away from the centre; *winds near anticyclones are normally light near the centre, but tend to be stronger towards the outer edges* **(c) outer marker** = ILS (marker) beacon, usually on centre line of approach at about 4.5 nm from the runway threshold; *the outer marker provides height, distance and equipment function checks to an aircraft on intermediate and final approach; see also* MARKER; **outer wing** = part of the wing nearest the tip

outflow ['aʊtfləʊ] *noun* flow in an outward direction; *the outflow valve is controlled by the cabin pressure controller*

outgoing [aʊt'gəʊɪŋ] *adjective* going out; *there is a fall of temperature until about one hour after dawn when incoming solar radiation balances outgoing terrestrial radiation* (NOTE: the opposite is **incoming**)

outlet ['aʊtlət] *noun* passage for exit or escape; *the air leaves the compressor outlet and passes through a matrix assembly of the secondary heat exchanger; when the controlling super-charger outlet pressure is reached, the capsule is compressed sufficiently to open its bleed valve*

outline ['aʊtlaɪn] **1** *noun* **(a)** line around the shape of something; *warning labels have a solid red outline* **(b)** shape; *at low level, features are most easily recognized from their outline in elevation; cumulus cloud has detached domes or towers which are generally dense and have sharp outlines* **2** *verb* to explain simply and briefly; *the changes in conditions; outlined in the next paragraph*

output ['aʊtpʊt] *noun* product of a process; *air density will affect the output of the engine; the function of the supercharger is to increase the power output; the power output of an engine depends on the weight of mixture which can be burnt in the cylinders in a given time*

outward ['aʊtwəd] **1** *adjective* moving away from the centre or starting point; *the piston draws fluid into the cylinders on the outward stroke and expels fluid into the system on the inward stroke* **2** *adverb* US = OUTWARDS; *see also* BOUND

outwards ['aʊtwədz] *adverb* away from the centre or starting point; towards the outside; *the door opens outwards* (NOTE: the American English is **outward**)

overall 1 [əʊvə'ɔːl] *adjective* including everything; *the total aerodynamic losses* *result in an overall turbine efficiency of 92%; although the student failed in one of the five exams, his overall result was a pass* **2** [əʊvə'ɔːl] *adverb* generally; *overall, the test flight was a success* **3** ['əʊvɔːl] *noun* one-piece item of protective clothing; *the engineer was wearing an overall to prevent his clothes from getting dirty*

overalls ['əʊvɔːlz] *noun* protective trousers with a bib and straps over the shoulders; *wear overalls to protect your clothes*

overcome [əʊvə'kʌm] *verb* to beat, to conquer, to win against; *the effects of anoxia at high altitudes can be overcome by breathing through a mask; drag must be overcome with thrust in order for an aircraft to increase speed*

overhaul **1** [əʊvə'hɔːl] *verb (especially machine, engine, etc.)* to take apart and examine carefully in order to repair and clean, etc.; *to overhaul the system will take a couple of days* **2** ['əʊvəhɔːl] *noun (especially of machine, engine, etc.)* act of taking apart in order to repair and clean; *other than the oil pump and the generator rotor, there are no other moving parts in the system to wear or which require periodic overhaul*

overhead [əʊvə'hed] **1** *adjective* **(a)** vertically above the point where a course is measured or timed; *the aircraft started from overhead A at 1000 hours on a heading of 230°T* **(b)** above the level of people's heads; *overhead baggage lockers must be secured immediately prior to take-off* **2** *adverb* above one's head; *he noticed an aeroplane flying overhead*

overheat [əʊvə'hiːt] *verb* to get too hot; *an acceleration/deceleration control is fitted to prevent the turbine assembly from overheating during acceleration, and to prevent flame-out during deceleration*

overlap [əʊvə'læp] **1** *noun* part of one thing covering something else; *valve overlap* = the period when both the exhaust and inlet valves are open together, with the exhaust valve closing and the inlet valve opening **2** *verb* to have an area or

range in common with something else; to cover part of something else; *the maps overlap each other at the edges by three centimetres* (NOTE: **overlapping - overlapped**)

overload **1** ['əʊvələʊd] *noun* an excessive amount of work or electricity; *resettable circuit protective devices should be designed so that when an overload or circuit fault exists, they will open the circuit* **2** [əʊvə'ləʊd] *verb* **(a)** to load a device or system, such as an electrical circuit, with too much work; to demand more than a system is capable of; *operating pressure is maintained in that part of the system which leads to the selector valves, and some method is used to prevent overloading the pumps* **(b)** to load too heavily; *the aircraft failed to gain height after take-off because it was overloaded*

override [əʊvə'raɪd] *verb* to take over control of the operation of an automatic device or system; *a circuit-protective device must not be of a type which can be overridden manually* (NOTE: **overriding - overrode - overridden**)

overshoot ['əʊvəʃuːt] *verb (of aircraft)* to fly past a target; *the pilot tried to land but the aircraft overshot the runway* (NOTE: **overshooting - overshot**)

overspeed ['əʊvəspiːd] **1** *verb* to go too fast; *a fault in the constant speed drive unit causes the generator to overspeed* **2** *noun* going too fast; *overspeed is usually a fault in the constant speed drive unit which causes the generator to overspeed*

overspeeding [əʊvə'spiːdɪŋ] *noun* act of going too fast; *overspeeding of the engine is prevented by a governor in the fuel system*

overstress [əʊvə'stres] *verb* to subject to too much force; *it takes less g force (force of acceleration due to the gravity of the earth) to overstress a heavy aircraft than a light one*

owing to ['əʊwɪŋ 'tu] *preposition* because of; *integral tanks are now favoured for aircraft owing to the very* *high utilization of space and saving of weight; owing to the aerodrome being unserviceable, the landing was made at another aerodrome some distance away*

oxidation [ɒksɪˈdeɪʃn] *noun (chemical reaction)* combination of a substance with oxygen (with loss of electrons); *when aluminium surfaces are exposed to the atmosphere, a thin invisible oxide skin forms immediately that protects the metal from further oxidation*

oxide [ˈɒksaɪd] *noun* compound of an element with oxygen; *when aluminium surfaces are exposed to the atmosphere, a thin invisible oxide skin forms immediately that protects the metal from further oxidation*

oxidize [ˈɒksɪdaɪz] *verb* to form an oxide by the reaction of oxygen with another chemical substance; *over a period of time, the metal is oxidized by contact with air*

oxygen [ˈɒksɪdʒən] *noun* colourless, odourless gas - essential to human life - constituting 21% by volume of the earth's atmosphere; *our bodies can get oxygen through the lungs; at very high altitudes the flying pilot must be on oxygen at all times, unless an aircraft dispensation has been obtained* (NOTE: chemical symbol O; atomic number 8)

ozone (O_3) [ˈəʊzəʊn] *noun* poisonous form of oxygen found naturally in the atmosphere which is toxic to humans at concentrations above 0.1 parts per million; *the maximum concentration of ozone is between 20 and 25 km above the earth's surface*

Pp

PA ['piː 'eɪ] = PUBLIC ADDRESS

Pacific Standard Time   [pə'sɪfɪk 'stændəd 'taɪm] *noun* time zone of the west coast area of the USA and Canada, 8 hours behind GMT (Greenwich Mean Time)

pack [pæk] *noun* **(a)** detachable system; *circuit packs consist of basic decision-making elements, referred to as logic gates, each performing combinational operations; a power pack system is one in which most of the major components, with the exception of the actuators and, in some systems, the pumps, are included in a self-contained unit* **(b)** small package containing a set number of items; *the survival pack includes heliographs, sea marker dyes, day/night distress flares and parachute flares*

pair [peə] *noun* two matched items, similar in appearance and function; *a brake control valve usually contains four elements, one pair for the brakes on each side of the aircraft, to provide duplicated control*

panel ['pænl] *noun* **(a)** flat, often rectangular piece of the skin of the aircraft; *access to the engine compartment is normally via hinged cowling panels*; **access panel** = part of the aircraft skin which can be easily removed for inspection of inner components; **wing panel** = rectangular aluminium section of the aircraft skin of a wing; *wing panels of light aircraft are normally riveted together* **(b)** board with switches, dials, control knobs, etc.; *the pilot is trained to scan an instrument panel*

panic   ['pænɪk]   *noun*   sudden, overpowering fear or terror; *in order to prevent mass panic amongst passengers in an emergency situation, crew may have to use force*

PAR = PRECISION APPROACH RADAR

parachute ['pærəʃuːt] *noun* device used to slow down free fall from an aircraft, consisting of a light piece of fabric attached by cords to a harness and worn or stored folded until used in descent; **parachute flare** = distress signal, suspended from a parachute to allow more time for the flare to be seen, which is fired to a height of 1200 ft

parallel ['pærəlel] **1** *adjective* **(a)** *(of lines, routes, roads, etc.)* side by side and having the same distance between them at every point; *as one aircraft flew round to attempt another landing, a Boeing 757 was taking off on the parallel runway; the runway is parallel to the main road* **(b)** *(of electrical circuits)* **in parallel** = arranged so as to join at common points at each end; *when batteries are connected in parallel, voltage remains constant but capacity increase* **2** *noun* line which is parallel to another; **parallels of latitude** = imaginary lines of constant latitude around the earth's surface

parameter [pə'ræmɪtə] *noun* set of measurable values such as temperature which define a system and determine its behaviour; *parameters required by the crew to set and monitor engine thrust are permanently displayed on the screen*

parasite drag ['pærəsaɪt 'dræg] *noun* component of total lift, caused by friction between the airflow and the structure of the aircraft; *parasite drag increases as speed increases*

parcel ['pɑːsl] *noun* small package; **parcel of air** = small body of air; *when a parcel of air is heated, its volume increases and its density decreases thus there is a fall in pressure*

park [pɑːk] *verb* to leave (a car, an aircraft) in a (special) place when no one is using it; *park beside the Cessna 150*

parking brake ['pɑːkɪŋ 'breɪk] *noun* brake that is set, often by hand, when the aircraft is stationary for a period of time; *make certain that the parking brake is on before doing engine run-up checks; light aircraft should be left with parking brakes off so that they can be moved quickly in the event of a fire in the hangar*

partial ['pɑːʃl] *adjective* in part, not fully; **partial closing of an undercarriage door** = not full closing of the doors; **partial filter blockage** = incomplete blockage of filter

particle ['pɑːtɪkl] *noun* very small piece or part; *solid particles in the atmosphere include sand, dust, volcanic ash and atmospheric pollution; hailstones start as ice particles in the upper part of a cumulonimbus cloud*

particular [pə'tɪkjʊlə] *adjective* special, given, distinct; not general; *a particular time; a particular speed; the size and number of valves required for a particular type of aircraft is governed by the amount of air necessary for pressurization and air conditioning*

pass [pɑːs] **1** *noun* **(a)** *(permit)* badge or document which allows one to enter a restricted or prohibited area; *a security pass* **(b)** *(in an exam)* successful result **2** *verb* **(a)** to move; *tropical storms dissipate as they pass from sea to land; the air leaves the compressor outlet and passes through a matrix assembly* **(b)** to pass information = to give information; pass

your message = instruction to a pilot to give information via radio to an air traffic control (ATC) facility **(c) to pass an exam** = to succeed in getting a result over a pre-arranged limit **(d) to pass a book to someone** = to pick up and give a book to someone nearby **(e) to pass another aircraft** = to move past another aircraft

passage ['pæsɪdʒ] *noun* **(a)** movement; *the passage of air over a turbine is used to power a small emergency generator; the passage of a trough is marked by a sharp veer in the wind* **(b)** channel through which something can pass; *liquid cooling is achieved by circulating a liquid around the cylinder barrels, through a passage formed by a jacket on the outside* **(c)** part of a book or speech, etc.; *a passage from a training manual*

passenger ['pæsɪndʒə] *noun* person who travels in an aircraft, car, train, etc., and has no part in the operation of it; *the Piper Archer has seating for a pilot and three passengers*

passive ['pæsɪv] *adjective* receiving an action but taking no action; *in primary radar systems, the target is passive*; **passive state** = referring to a system or device which may be switched on or 'live' but not reacting to any input (NOTE: the opposite is **active**)

patch [pætʃ] *noun* small area; *a patch of fog; a patch of cloud; patches of early morning fog made identification of ground features difficult*

path [pɑːθ] *noun* route or course along which something moves; *projection of the path of the aircraft over the ground is called its track*; **approach path** = course taken by the aircraft in preparation for landing; **flight path** = line, course or track along which an aircraft flies; *see also* GLIDEPATH

pattern ['pætn] *noun* form or method which shows particular, consistent characteristics; **pressure pattern** = changes in pressure areas which take place regularly, for example every year; **traffic**

pattern = shape traced out on the ground, of aircraft track in the aerodrome circuit

pavement ['peɪvmənt] *noun (at aerodromes)* prepared concrete or tarmac surfaces for ground manoeuvring of aircraft, including taxiways and runways; *the bearing strengths of pavements intended for aircraft of 5,700 kg MTWA (maximum total weight authorized) or less are reported as the maximum allowable weight and maximum allowable tyre pressure*; pavement classification number (PCN) = number expressing the bearing strength of a pavement for unrestricted operations

payload ['peɪləʊd] *noun* the money-earning load carried by the aircraft including the passengers, baggage and freight; *the shape of an aircraft is determined by the requirement to provide an aerodynamic lift force great enough to support the weight of the aircraft and payload whilst in flight*

PCN   =   PAVEMENT CLASSIFICATION NUMBER

peak [piːk] *noun* highest point; *the intensity of solar radiation reaches a peak around noon*; peak value = maximum value

pedal ['pedl] *noun* foot-operated lever; rudder pedal = foot operated lever which moves the rudder; *just before take-off, the pilot should make sure that his feet are correctly positioned on the rudder pedals*

penalty ['penəltɪ] *noun* (a) unwanted result of an action; *the penalty of using a circular polarization transmission may be some loss of definition* (b) punishment; fine; *fuel penalties can be incurred if fuel surplus to requirements is carried*

penetrant ['penətrənt] *noun* something which forces or gets entry into an area or substance; *penetrant dye inspection is a non-destructive test used mainly for the detection of defects open to the surface; penetrant oil can be used to loosen rusty bolts, etc.*

penetrate ['penɪtreɪt] *verb* to force a way into; *cool air from the Atlantic can sometimes penetrate far into Europe; occasionally, thunder cloud will penetrate through the tropopause*

penetration [penɪ'treɪʃn] *noun* the act of forcing a way into or through; *long-range radars suffer little weather interference and have good cloud penetration characteristics*

per [pɜː] *preposition* for each, for every; *feet per minute (fpm) gallons per hour (gph)*

per cent [pə 'sent] *noun* out of each hundred; fifty per cent (50%) = half or 1/2 or 50 out of 100; twenty-five per cent (25%) = one quarter or 1/4 or 25 out of 100

percentage [pə'sentɪdʒ] *noun* (a) fraction with 100 as the understood denominator; *volumetric efficiency is usually expressed as a percentage* (b) part of the total; *only a small percentage of passengers take in the pre-departure safety briefing*

perform [pə'fɔːm] *verb* to do; *circuit breakers perform the same function as a fuse; the pilot performed a loop to conclude his flying display*

performance [pə'fɔːməns] *noun* ability of a system, such as an aircraft or an engine, to function as required; *the performance of the turbojet engine is measured in thrust produced at the propelling nozzle or nozzles*; high performance = a system which provides better-than-usual output; *an engine with a high performance; some high-performance engines have coolant and oil system thermostats which aid warming-up*

period ['pɪərɪəd] *noun* length of time; *a 24 hour period; a period of 3 minutes*

periodic [pɪərɪ'ɒdɪk] *adjective* referring to a particular length of time; periodic maintenance = maintenance made at a particular time interval; *periodic calibration of ILS (instrument landing system) installations is recommended*

permanent ['pɜːmənənt] *adjective* lasting or remaining without change; **permanent deformation** = damage (to a structure) which must be repaired by replacing the damaged part; **permanent magnet** = metal component which always has a magnetic influence (NOTE: the opposite is **temporary**)

permissible [pə'mɪsəbl] *adjective* allowable; not prohibited; *great care must be taken to ensure that the aircraft operates within regulated or permissible weight limits*

permission [pə'mɪʃn] *noun* consent or authorization; *a passenger who is drunk can be refused permission to board the aircraft*

permit 1 ['pɜːmɪt] *noun* document or pass which is proof of official permission to do or have something; *you need a permit to enter the restricted area*; **Permit to Fly (UK)** = issued by the CAA (Civil Aviation Authority) for aircraft which do not qualify for a Certificate of Airworthiness (C of A) **2** [pə'mɪt] *verb* to allow; *when oxygen mask are pulled down to the usable position, valves are opened which permit oxygen to flow*; *information passed to the operations department will be sufficient to permit the flight to be planned*

perpendicular [pɜːpən'dɪkjʊlə] *adjective* at right angles or 90° (to a base or a line); *the vertical grid lines are perpendicular to the horizontal ones*; *the air is acted upon by a force perpendicular to the isobars in the direction of low pressure*

persist [pə'sɪst] *verb* **(a)** to continue to exist; to last; *snow cover tends to persist on north-facing slopes of mountains* **(b)** to continue without giving up; *he persisted with his request until it was granted*

persistence [pə'sɪstəns] **(a)** *noun* continuing to exist; lasting; *the persistence and movement of cols is governed by the movement of the adjacent pressure systems* **(b)** continuing without giving up; *he managed to overcome his difficulties through persistence and hard work*

personnel [pɜːsə'nel] *noun* staff; body of people involved in a common purpose (such as work); *smoke masks are available for use by personnel within the aircraft*

PFCU = POWER FLYING CONTROL UNIT

phase [feɪz] *noun* **(a)** stage or part; *an emergency situation may occur during any phase of the flight* **(b)** *(electricity)* relationship between voltage and current; *the CSDU (constant speed drive unit) drive shaft turns the permanent magnet generator and single phase AC (alternating current) is induced in the winding on the stator* **(c)** **phase angle** = difference between two periodic phenomena expressed as an angle; **phase difference** = measure of phase angle from any VOR (very high frequency omni-directional radio range) radial related to that on bearing 360°

phenomenon [fɪ'nɒmɪnən] *noun* occurrence or circumstance which can be perceived by the senses; *metal fatigue is not a modern phenomenon*; *of all meteorological phenomena, thunderstorms present the greatest hazard to aviation* (NOTE: plural is **phenomena**)

physical ['fɪzɪkl] *adjective* **(a)** referring to matter and energy or the sciences dealing with them, especially physics; *oxygen and nitrogen together constitute 99% of the atmosphere and obey the physical laws as any other gas* **(b)** referring to the human body; *in some aircraft operating for long periods at high altitudes, physical discomfort may arise from low relative humidity*; **physical fitness** = state of health of the body

PIC = PILOT IN COMMAND

piece [piːs] *noun* bit, portion, part; *the upper and lower skin panel of each wing can be made in one piece*; **piece of equipment** = an item of equipment; *early rescue depends on rapid location of survivors and the survival beacon is the most important piece of equipment in this regard* (NOTE: **piece** is often used to show one item of something which has no plural: **a piece of equipment; a piece of information**)

pilot ['paɪlət] **1** *noun* **(a)** person who operates an aircraft in flight; **pilot in command (PIC)** = the pilot who has responsibility for the operation and safety of the aircraft during flight time; **Pilot's Operating Handbook** = book giving details of an aircraft with recommendations and instructions regarding its use **(b)** part of a system or device which leads the whole **2** *verb* to operate; to guide; *to pilot an aircraft*

> COMMENT: a pilot holding a private, or commercial pilot's licence may log as pilot-in-command time, only the flight time during which that pilot is the only operator of the aircraft's flying controls

pin [pɪn] *noun* short, usually cylindrical metal rod; **locking pin** = short metal device to prevent a nut from turning; **a 3-pin plug** = an electrical supply plug with three electrodes: live, neutral and earth

pinpoint ['pɪnpɔɪnt] **1** *noun* visual observation of the precise position of an aircraft; *the pinpoint is a very positive means of establishing position, as long as the feature is properly identified* **2** *verb* to draw attention to; *to pinpoint a problem*

pipe [paɪp] *noun* hollow cylinder or tube to convey a fluid; *a delivery pipe; an exhaust pipe*

pipeline ['paɪplaɪn] *noun* long hollow cylinder or tube to convey a fluid (such as oil or natural gas); *the incompressibility of liquids enables force to be transmitted long distances through pipelines*

piston ['pɪstən] *noun* solid cylinder that fits into a larger cylinder and moves under fluid pressure, as in petrol and diesel engines or compresses fluids, as in pumps and compressors; **piston engine** = petrol or diesel engine in which pistons are moved by combustion of fuel, this reciprocating movement producing rotating movement; **piston ring** = one of the metal rings which seals the space between the piston and the

cylinder wall; *there should be a lose fit between the cylinder and the piston, the difference being taken up by the piston rings*

pitch [pɪtʃ] **1 (a)** *noun* nose up/down movement of the aircraft about its lateral axis; *if the control column is moved forward or aft, the pitch attitude of the aircraft changes* **(b)** distance a propeller would advance in one rotation if there was no slip; **effective pitch** = distance the aircraft moves forward in flight for one rotation (360°) of the propeller; **fine pitch setting and coarse pitch setting** = angular propeller-blade settings; *variable pitch propellers were originally produced with two blade-angle settings - fine pitch to enable full engine speed to be used on take off and coarse pitch to allow an economical engine speed to be used for cruising*; **pitch angle** = the angle between the blade element chord line and the plane of rotation of the propeller; **pitch lock** = a means of holding the fine pitch stop in a prescribed position (NOTE: some manufacturers use the term to describe a device which locks the blades at whatever angle they are at if there is a failure of the pitch change mechanism); **pitch trim** = trim of the aircraft in the lateral axis so that there are no forward/aft forces on the control stick or yoke **2** *verb* (*of aircraft*) to move about the lateral axis; *move the yoke fore and aft to pitch down and up*

pitot ['piːtəʊ] *noun* pitot *or* Pitot tube = open-ended tube used to measure the speed of flow of a fluid; *device to sense pitot pressure created by the movement of air over the aircraft*; **pitot head** = externally mounted device which senses and sends airspeed information to the airspeed indicator in the cockpit; **pitot-static system** = pressure system for airspeed indicator, altimeter and vertical speed indicator

> COMMENT: the airspeed indicator is connected to the pitot tube and the static ports. The altimeter and vertical speed indicator are connected to the static ports only. The Pitot tube is

named after the French scientist Henri Pitot (1695-1771)

pivot ['pɪvət] **1** *noun* short rod on which another part rotates **2** *verb* to turn on a point; *the rocker arm pivots on a bearing and opens the valve*

place [pleɪs] **1** *noun* **(a)** space; area; *Greenwich is a place on the 0° meridian* **(b)** position; *decimal place* **(c)** in place of = instead; to take place = to happen; *the explosion took place just before the aircraft landed* **2** *verb* to put; *place the chart on the seat next to you*; *rotate the grid to place the wind direction under true*

plain [pleɪn] *adjective* without pattern or marking or writing; a plain sheet of paper = a sheet of paper with nothing on it

plan [plæn] **1** *noun* **(a)** drawing or diagram of a place viewed from above; *the horizontal situation indicator presents a selectable dynamic colour display of flight progress and plan view orientation* **(b)** scheme or programme worked out in advance of putting something into operation; flight plan = specified information about the intended flight of an aircraft, communicated orally or in writing to an ATC (air traffic control) facility **2** *verb* to organize a scheme or programme; *Jeppesen charts are used to plan and fly a safe route to a destination*

plane [pleɪn] *noun* **(a)** imaginary surface containing all the straight lines that connect any two points on it; *the planes of parallels of latitude are parallel to the plane of the equator; the pitch angle is the angle between the blade element chord line and the plane of rotation of the propeller* **(b)** aeroplane (NOTE: because of possible confusion with meaning (a), **plane** as in meaning (b) is considered bad usage by some; the word **aircraft** is preferred in that case)

planning ['plænɪŋ] *noun* making plans; *instructor in flight planning*

plant [plɑːnt] *noun* usually heavy equipment or tools used for doing something; power plant *or* powerplant =

engine used to move a vehicle or aircraft; *see also* POWERPLANT

plate [pleɪt] *noun* smooth, flat rigid object with the same thickness all over; *the basic construction of a lead-acid cell consists of a positive electrode and negative electrode, each of which is made up of lead-antimony alloy grid plates*

play [pleɪ] **1** *noun* slightly loose fitting of engineering parts which allows free movement; *some play should be felt in the aileron actuator rod linkage* **2** *verb* to play a part = to be part of a whole which has an effect on something; *contrast and colour play a part in identifying coastlines*

plot [plɒt] **1** *noun* graph; diagram that shows a relationship between two sets of numbers as a series of points joined by a line; *a plot of applied stress and resulting strain* **2** *verb* to calculate and mark a line on a graph or chart, etc.; to plot a course = to calculate and draw the desired route of an aircraft on a chart

plug [plʌg] **1** *noun* **(a)** device for making an electrical connection; *alternating current ground power can be fitted to an aircraft via a six-pin ground power plug; see also* PIN **(b)** device for igniting fuel in an engine; *an electric spark from an igniter plug starts combustion; the fuel/air mixture is ignited by a spark plug* **(c)** device to prevent liquid flowing out of a container; *oil drain plug; see also* SPARK PLUG **2** *verb* **(a)** to plug a hole = to fill a hole so that fluid cannot escape **(b)** to plug something in = to make an electrical connection often by inserting the plug on an electrical device (such as a computer) into an electrical supply socket

plunger ['plʌnʒə] *noun* machine part, such as a piston, that operates with a thrusting or plunging movement; *a flow indicator valve comprises a body, a spring-loaded plunger connected to an actuator arm, and a micro-switch*

plus [plʌs] *preposition* increased; with the addition of; added to; *at the selected decision height plus 50 feet, an aural alert*

chime sounds; four plus four equals eight
(4 + 4 = 8)

pneumatic [nju:'mætɪk] *adjective*
referring to air under pressure or
compressed air; *high-pressure pneumatic*
systems are generally fitted on the older
types of piston-engine aircraft to operate
the landing gear, wing flaps, wheel brakes

pneumatically [nju:'mætɪkli] *adverb*
by using air under pressure or compressed
air; *clamshell doors are hydraulically or*
pneumatically opened

PNR = POINT OF NO RETURN

pod [pɒd] *noun* streamlined casing or
housing; *the engine bay or pod is usually*
cooled by atmospheric air

point [pɔɪnt] 1 *noun* **(a)** particular figure
on a scale; *the melting point of ice is 0°C*
(Celsius) **(b)** particular place; a point on a
map = a particular place on a map;
departure point = place on the map
representing the place from which a flight
begins; **entry point** = position on the
ground above which an aircraft entering a
control zone crosses the boundary; **exit**
point = position on the ground above
which an aircraft leaving a control zone
crosses the boundary; **holding point** =
place, often designated Alpha, Bravo,
Charlie, etc., where aircraft wait before
entering the runway, as instructed by ATC
(air traffic control); **touchdown point** =
place on the runway where the aircraft
undercarriage first touches the ground on
landing **(c)** the sharp end of something; *a*
pencil point 2 *verb* **(a)** to direct (towards);
point the aircraft towards the airfield **(b)**
to indicate direction, often with a finger;
point to the east **(c) point out** = to draw
attention to; *the instructor pointed out the*
dangers of not keeping a good lookout

pointer ['pɔɪntə] *noun* indicating device
such as a needle on an instrument; *the*
pointer centralizes to indicate that the
aircraft is aligned with the runway centre
line; **cross-pointer indicator** = display
with crossing horizontal and vertical bars
to indicate aircraft position in relation to
the glideslope; *see also* ALTIMETER

point of no return (PNR) ['pɔɪnt ɔv
nəʊ rɪ'tɜːn] *noun* place on the route where
the aircraft does not have enough fuel to
return to the starting place; *the point of no*
return is calculated before departure to
cover the chance that both the terminal
airfield and its alternate become
unavailable during flight

polar ['pəʊlə] *adjective* **(a)** referring to
the north or south pole; *polar air; a polar*
region; the greatest horizontal gradients
of mean temperatures of a layer are found
at the boundaries between cold polar and
warm tropical air masses **(b)** referring to
the pole(s) of an electrical device or of a
magnet; *bar magnets attract each other*
because of polar differences

polarity [pə'lærəti] *noun* direction of
flow of flux or current in an object; *during*
discharge, when the polarity of the supply
changes, the stored energy is returned to
the supply; **polarity test** = test to see which
terminal is positive and which is negative
(positive polarity terminals are usually
marked red, negative are black)

polarization [pəlɔraɪ'zeɪʃn] *noun* **(a)**
characteristic of light or radio or other
electromagnetic waves in which the waves
are aligned in one direction and show
different properties in different directions;
the antenna must have the same effective
length and the same polarization as the
transmitter **(b)** partial or complete polar
separation of positive and negative electric
charge (NOTE: also written **polarisation**)

polarize ['pəʊləraɪz] *verb* **(a)** *(of*
broadcast signal waveforms) to align in
one plane; *the frequency allocation for*
VOR (very high frequency
omni-directional radio range) is
108-117.975 MHz (megahertz) and
transmissions are horizontally polarized
(b) to separate positive and negative
electric charges (NOTE: also written **polarise**)

pole [pəʊl] *noun* **(a)** north or south point
of the earth's axis; *a meridian is a line*
joining pole to pole; **North Pole** = point
which is furthest north on the earth; **South**
Pole = point which is furthest south on the

earth **(b)** *(electricity)* terminal (such as of a battery, etc.); *negative pole; positive pole* **(c)** long, rounded piece of wood or metal; *a flag pole*

pollution [pə'luːʃn] *noun* presence of abnormally high concentrations of harmful substances in the environment; **air pollution** *or* **atmospheric pollution =** pollution of the air by gas, smoke, ash, etc.; *solid particles in the air include dust, sand, volcanic ash and atmospheric pollution*

poor [pɔː] *adjective* bad; *poor weather conditions; poor visibility; air is a poor conductor* (NOTE: the opposite is **good**)

poppet ['pɒpɪt] *noun* **poppet valve =** intake or exhaust valve of a piston engine, operated by springs and cams

porous ['pɔːrəs] *adjective* referring to substances which allow fluid to pass through them; *the de-icing fluid passes through a porous plastic sheet*

port [pɔːt] *noun* **(a)** periodically opened entrance; *inlet port; as a piston in the pump moves outwards into its cylinder, it covers the inlet port and forces fluid out of the top of the cylinder* **(b)** the left side of an aircraft when facing forwards (when inside an aircraft); *unless an aircraft is flying in the same or exactly opposite direction to the wind, it will experience either port or starboard drift* (NOTE: the opposite is **starboard**)

portable ['pɔːtəbl] *adjective* capable of being carried in the hands; *a portable fire extinguisher; the aneroid barometer is a more portable device than a mercury barometer*

portion ['pɔːʃn] *noun* part or section; *a hailstone starts as a small ice particle in the upper portion of a cumulous cloud*

position [pə'zɪʃn] **1** *noun* **(a)** place or location; where something is; *the Greenwich or prime meridian and the equator are the axes of the system called latitude and longitude which is used for expressing position on the earth*; **position line =** line along which the aircraft is known to be at a particular time by taking a VOR (very high frequency omni-directional radio range) bearing; **position report =** report over a known location as transmitted by an aircraft to an ATC (Air Traffic Control) station **(b)** setting of a control, etc.; *the neutral position* **(c)** *(of person)* in a sitting position = seated **2** *verb* to place (something) in a special location; *the magnetic compass is positioned away from magnetic sources*

positive ['pɒzətɪv] *adjective* **(a)** definite, without doubt; *the pinpoint is a very positive means of establishing aircraft position* **(b)** *(mathematics)* referring to a number greater than zero; *oil is ducted to the front of the pitch change piston and the blades move to a positive angle*; *see also* IDLING **(c)** *(electricity)* referring to the + symbol; **positive terminal =** terminal (of a battery) marked +

possibility [pɒsə'bɪlɪti] *noun* chance occurrence; *anti-braking systems are designed to prevent the wheels from locking during landing thus reducing the possibility of wheel skid*

possible ['pɒsəbl] *adjective* capable of happening; *if possible, control surfaces should be moved by hand; there will be a possible delay; fire in a toilet could present difficulties due to the confined space and possible smoke accumulation*

potential [pə'tenʃl] **1** *adjective* capable of being, but not yet in existence; *a designated fire zone is a region where a potential fire risk may exist*; **potential danger =** possible future danger **2** *noun* *(electricity)* voltage; *precipitation static develops due to friction between the aircraft surface and precipitation causing the aircraft to become charged to a high potential*

pound [paʊnd] *noun* unit of weight equal to 16 ounces or 453.592 grams

powder ['paʊdə] *noun* substance made of ground or otherwise finely dispersed solid particles; *dry chemical* *fire-extinguishers contain a non-toxic powder*

power ['pauə] *noun* energy or force; **electric power** = electricity used to drive machines or devices; *see also* CRUISING

power-assisted ['pauəə'sıstıd] *adjective* **power-assisted controls** = controls which require less human effort to move

powered ['pauəd] *adjective* driven (by a type of energy or motor); *system powered by electricity*

powerplant ['pauəplɑːnt] *noun* engine used to move a vehicle or aircraft; *additional strength is required for the powerplant attachment point*

QUOTE by replacing the Rotax engine with a four-stroke Jabiru powerplant, the aircraft designers claim the aircraft will be provided with more power and increased all-round performance
*Flight International 16-22 July 1997 (NOTE: also written **power plant**)*

power supply ['pauə sə'plaı] *noun* electrical circuit that provides certain direct current voltage and current levels from an alternating current source for use in other electrical circuits; *if the power supply from the amplifier to the gauge fails, the needle slowly falls to zero*

PPL = PRIVATE PILOT'S LICENCE

PR ['piː 'ɑː] = PUBLIC RELATIONS

practicable ['præktıkəbl] *adjective* capable of being put into practice or effect; *some military aircraft use braking parachutes but this is not practicable on civil aircraft*

practical ['præktıkl] *adjective* referring to practice or action rather than theory; *for practical purposes, any straight line drawn on a Lambert's conformal projection, represents a great circle*

practice [præktıs] **1** *noun* **(a)** habitual or customary behaviour; *it is common*

practice for pilots to take turns to sleep on long-haul flights **(b)** performance or operation; **in practice** = when actually done; in reality; *frequency modulation (FM) in theory has a limitless number of sidebands, but in practice only the first eight pairs are significant* **2** *verb US* = PRACTISE

QUOTE if the aircraft has been standing overnight or longer, check the drains for water. This should, of course, be normal practice
Civil Aviation Authority, General Aviation Safety Sense Leaflet

practise ['præktıs] *verb* to do something repeatedly in order to improve; *in order to improve flying skills, a trainee pilot must practise regularly* (NOTE: also written **practice** in American English)

pre- [priː] *prefix meaning* before

pre-arrange [priːə'reınʒ] *verb* to decide or to plan in advance; to predetermine; *selective calling uses the four-letter code pre-arranged with the controlling authorities*

precaution [prı'kɔːʃn] *noun* action taken to prevent or avoid a dangerous situation or failure; *personnel concerned with fuelling should take every precaution to prevent outbreaks of fire*

precede [prı'siːd] *verb* to take place or to come before something else; *a period of calm often precedes a storm; when the RVR (runway visual range) is greater than the maximum value which can be assessed, the group will be preceded by the letter indicator P followed by the highest value which can be assessed*

precedence ['presıdəns] *noun* place or position before in time or importance; **to take precedence over** = to have priority over; to be more important than; *emergency landings take precedence over all others*

preceding [prı'siːdıŋ] *adjective* taking place or coming before (something else); *as mentioned in the preceding*

paragraph = as written in the paragraph before the one being read

precipitation [presɪpɪ'teɪʃn] *noun* water falling as rain, drizzle, hail, sleet (snow and rain) and snow from the atmosphere onto the surface of the earth; *cloud droplets are small and light at first, but when the droplets grow and become heavier, they fall as precipitation; precipitation is classified as light, moderate or heavy according to its rate of fall*

precise [prɪ'saɪs] *adjective* exact or accurate; *a pinpoint is an indication of the precise position of the aircraft; a precise interval is essential to obtain correct ignition timing on all cylinders during engine running*

precision [prɪ'sɪʒn] *noun* exactness or accuracy; *precision flying is only achieved by constant practice*; **with precision** = with exactness; **precision approach radar (PAR)** = ground-based primary radar system to give vertical and lateral information about an aircraft's final approach path

pre-departure ['priːdɪ'paːtʃə] *adjective* which takes place before a departure; *only a few passengers absorb the pre-departure safety information*

predetermine [priːdɪ'tɜːmɪn] *verb* to decide and set or fix beforehand

predetermined [priːdɪ'tɜːmɪnd] *adjective* decided and set beforehand; *when the roll control knob is returned to the central position, the aircraft rolls out on to a predetermined heading*

predict [prɪ'dɪkt] *verb* to foretell or to say beforehand; to foresee; *rain is predicted within the next hour; dead reckoning position is the position of the aircraft as predicted by calculation*

predictable [prɪ'dɪktəbl] *adjective* **(a)** reliably regular and therefore foreseeable; *only the high frequency band has predictable, reliable sky wave propagation by day and by night* **(b)** capable of being foreseen, expected or anticipated; *the*

accident was predictable = it was possible to know that the accident would happen before it happened

prediction [prɪ'dɪkʃn] *noun* telling or seeing beforehand; *the map display combines current ground speed and lateral acceleration into a prediction of the path over the ground to be followed over the next 30, 60 and 90 seconds*

predominance [prɪ'dɒmɪnəns] *noun* greatest importance or influence; *the predominance of a cold northerly airstream during the winter months*

predominant [prɪ'dɒmɪnənt] *adjective* which is most important or influential; which is more powerful than others; *the ocean surface usually consists of a predominant swell three or four feet high and 500 to 1,000 feet between crests*

predominate [prɪ'dɒmɪneɪt] *verb* to have greater number or importance; to be more powerful than others; *a cold northerly airstream predominates during the winter months*

prefer [prɪ'fɜː] *verb* to like more; to favour; *of the two basic types of fuel pump, where lower pressures are required at the burners, the gear-type pump is preferred because of its lightness* (NOTE: **preferring - preferred**)

preferable ['prefrəbl] *adjective* better than, more desirable; *three position lines are preferable to two; if there is a choice between two courses of action, the safest is the most preferable*

preference ['prefrəns] *noun* **in preference to** = by choice; in place of others; rather than; *for certain applications, eg landing gear and flaps, hydraulic systems are used in preference to mechanical or electrical systems*

prefix ['priːfɪks] *noun* part of word added at the beginning of a word to alter the meaning; *pre- is a prefix meaning: before* (NOTE: the plural is **prefixes**)

COMMENT: the prefixes for cloud types are: **alto-** medium level cloud

(6,500 feet to 23,000 feet); **cirro-** high cloud (16,500 feet and above); **nimbo-** any height, but rain bearing as for example **nimbo-stratus** rain carrying, low-level cloud; **strato-** low cloud (up to 6,500 feet)

pre-flight ['priː'flaɪt] *adjective* which takes place before a flight; **pre-flight briefing** = short instructional talk before a flight; **pre-flight checks** = checks made on the aircraft structure and systems before taking off; *during pre-flight checks, control surfaces should be moved by hand to ascertain that they have full and free movement*

pre-ignition ['priːg'nɪʃn] *noun* the ignition of the fuel/air mixture in the combustion chamber, occurring before the spark; *pre-ignition is often caused by a hot spot in the combustion chamber which ignites the mixture*

preparation ['prepə'reɪʃn] *noun* state of readiness or act of making something ready for use beforehand; *normal aircraft preparation are actions and precautions taken by the cabin crew on every flight to ready the aircraft for any abnormal or emergency situation which may occur during any phase of the flight*

prepare [prɪ'peə] *verb* (a) to make ready beforehand for a particular purpose, as for an event or occasion; *the instructor prepared the students for the exams*; **prepare for take-off** = be ready for take-off (b) to make by putting various elements or ingredients together; *regional area forecasting centres use information about upper wind speeds and temperatures to prepare specific forecasts and significant weather charts*

prescribe [prɪ'skraɪb] *verb* to set down as a rule or a guide; **prescribed procedures** = set or fixed pattern of doing something; *a means of holding the fine pitch stop in a prescribed position is also called 'pitch lock'*

pre-select ['priːsɪ'lekt] *verb* to select or to choose in advance

pre-selected ['priːsɪ'lektɪd] *adjective* selected or chosen in advance; *the CSU (constant speed unit) maintains the pre-selected propeller speed*

presence ['prezəns] *noun* existence; *the presence of cloud by day decreases the value of the maximum temperatures; a fuel sample hazy or cloudy in appearance would indicate the presence of water*

present 1 ['prezənt] *adjective* (a) in place, existing; *fuel, oxygen and heat must all be present for fire to exist* (b) period in time between the past and the future; **at the present time** = at this time; now; **present day aircraft** = modern aircraft; **present weather** = the weather at the moment of speaking **2** [prɪ'zent] *verb* (a) to create or to make; *a fire in a toilet could present difficulties; learning to fly presents a challenge*; **to present an opportunity** = to create or to give an opportunity (b) to give a prize or award; *charter passengers on Concorde were presented with a certificate as a souvenir of their flight*

presently ['prezntli] *adverb* (a) soon; *I'll be there presently* (b) *US* now, at the present time; **he's presently in France** = at the present time, he is in France; **a number of methods are presently in use** = a number of methods are currently in use

presentation [prezən'teɪʃn] *noun* showing, display; *the most widely acceptable presentation of flight fuel data is in a tabular form*

preset ['priː'set] **1** *verb* to set in advance; *radios allow the user to preset a number of different frequencies* (NOTE: **presetting - preset**) **2** *adjective* set in advance; *a pressure switch is a switch which is activated when a preset pressure is attained*

press [pres] *verb* to push or exert pressure on; *press the button*; **press to test/talk (PTT) button**

pressure ['preʃə] *noun* (a) (*physics*) force applied uniformly over a surface, measured as force per unit of area; **atmospheric pressure** = normal pressure of the air on the surface of the earth; **barometric pressure** = atmospheric pressure indicated by a barometer; **cabin pressure** = artificial atmospheric pressure created inside the cabin; pressure of air inside the cabin which allows people to breathe normally at high altitudes; **fuel pressure** = pressure exerted by fuel as it is pumped from the tanks to the engine; **tyre pressure** = air pressure in a tyre; *maximum allowable tyre pressure*; **pressure altimeter** = conventional altimeter which operates using atmospheric pressure; **pressure bulkhead** = lateral partition inside the aircraft which separates pressurized from non-pressurized areas; **pressure gauge** = instrument for measuring pressure; **pressure switch** = switch which is activated when a preset pressure is attained; *on some engines a fuel differential pressure switch fitted to the fuel filter senses the pressure difference across the filter element*; *see also* ABSOLUTE PRESSURE

pressure altitude ['preʃə 'æltitjuːd] *noun* altitude indicated when the altimeter is set to 1013.2 millibars; *when using flight levels, the altimeter should be set to 1013.2 mb to give the pressure altitude*

> COMMENT: pressure altitude is used in determining density altitude, true altitude, and true airspeed

pressurization [preʃərai'zeiʃn] *noun* **cabin pressurization** = maintenance of an acceptable atmospheric pressure in an aircraft while flying at high altitude; *at 35,000 ft (feet) passengers can breathe freely because of cabin pressurization* (NOTE: also written **pressurisation**)

pressurize ['preʃəraiz] *verb* to increase the pressure of; *when air pressure is used to transfer fuel, it will be necessary to pressurize the fuel tanks* (NOTE: also written **pressurise**)

prevail [pri'veil] *verb* to be most common or frequent; *hot dry conditions prevail in the Middle East in summertime*; the **prevailing** wind is from the **south-west** = the wind blows from the

south west more often than from any other direction

prevent [pri'vent] *verb* to keep from happening; *heated air provides sufficient heat in the outer skin to melt ice already formed and prevent any further ice formation*

previous ['priːviəs] *adjective* coming before; earlier; **the previous chapter** = the chapter before the one being read or referred to; **previous reports** = earlier reports

primarily ['praimrəli] *adverb* most often, mainly; *dry chemical fire extinguishers are primarily used for electrical fires*

primary ['praimri] *adjective* **(a)** first or most important; **of primary importance** = of greatest importance; **primary coil** = induction coil; **primary radar** = radar system which uses reflected radio signals

primary flight instruments ['praiməri 'flait 'instrəmənts] *noun* six instruments displayed on the instrument panel immediately in front of the pilot: airspeed indicator, attitude indicator, altimeter, turn coordinator, heading indicator, and vertical speed indicator; *when practising instrument flying, the attitude indicator is the most important of the primary flight instruments*

prime [praim] **1** *adjective* first; **prime importance** = greatest importance; **prime meridian** = Greenwich meridian (000°); **prime number** = number, which, if there is to be no remainder, is only divisible by itself and 1, such as 13, 17, 19, 23, 29 **2** *verb* to pump fuel spray into piston engine inlet manifold to make starting from cold easier; *during the summer, after the first flight of the day, it is not normally necessary to prime the engine*

primer ['praimə] *noun* **(a)** protective substance which is applied to a metal or wood surface before painting; *interior metal finishing is done with dust shedding gloss-paint over a primer* **(b)** small hand-operated pump, operated from the cockpit, to spray fuel into the piston engine inlet manifold to make starting from cold easier

principal ['prɪnsəpl] *adjective* main; *four principal control modes can be selected on the EFIS (electronic flight instrument system) control panel*

principle ['prɪnsəpl] *noun* a basic truth or law; *fire extinguishing is based on the principle of removing one of the three components necessary for fire to exist - fuel, oxygen and heat* (NOTE: do not confuse with **principal**)

prior ['praɪə] *adjective* earlier; previous; before (in time); *prior approval*; *prior permission*; **prior to** = before; *prior to our departure*; **prior to take-off** = before take-off

QUOTE the pilot remembered hearing the stall warning immediately prior to impact
Pilot

priority [praɪ'ɒrəti] *noun* order of importance or urgency; **high priority** = important or urgent in the circumstances; **low priority** = not important or urgent in the circumstances

probability [prɒbə'bɪlɪti] *noun* likelihood, chance of occurrence; *the probability of aquaplaning increases as the depth of tyre tread decreases*

probable ['prɒbəbl] *adjective* likely, most possible; *pilot error was the probable cause of the accident*

probe [prəʊb] *noun* metal sensing device; *ice is allowed to accumulate on a probe which projects into the airstream*

procedure [prə'siːdʒə] *noun* (a) series of actions taken to achieve something; *an emergency procedure* (b) process by which aircraft are brought into position for an instrument approach and landing; **procedure turn** = turn(s) made at 3° per second to align the aircraft with the runway

procedural [prə'siːdʒərl] *adjective* referring to procedure; **procedural approach** = specific approach made often

after procedure turns as part of timed, accurately flown flight pattern to prepare for a landing at a particular aerodrome; *it is important that the integrity of an aid used to conduct procedural approaches is high*

process ['prəʊses] *noun* series of actions or changes which achieve a particular result; *adiabatic process*; *combustion process*; *cooling process*

produce [prə'djuːs] *verb* (a) to create; *low altostratus clouds often produce rain* (b) to make or to manufacture; *most light aircraft are produced in the United States* (c) to show; *the pilot must produce his licence to the authorities within two weeks*

product ['prɒdʌkt] *noun* (a) something created or made by human or natural methods; *carbon monoxide is a product of the combustion process* (b) (*mathematics*) number obtained by multiplying two other numbers together; *the amount of power produced in a purely resistive circuit is a product of voltage and current (P = VI watts)*

production [prə'dʌkʃn] *noun* (a) creation; *the movement of air over the aerofoil is necessary for the production of lift* (b) making or manufacture; *production of aircraft in the factory came to a stop in 1974* (c) showing; *an authorized person may require the production of a certificate of airworthiness*

profile ['prəʊfaɪl] *noun* (a) outline or shape of something (seen from a side view); *the de-icing panels are formed to the profiles of the wing and tail unit leading edges into which they are fitted* (b) short description; *the handbook gives a short profile of the different aircraft types*

prognostic [prɒg'nɒstɪk] *adjective* referring to foretelling or foreseeing events such as the weather; **prognostic or forecast chart** = chart which predicts the weather for a given area; *prognostic or forecast charts are prepared, by the central meteorological office of each region, normally for periods up to 24 hours ahead*

programme ['prəʊgræm] *noun* ordered events to take place or procedures to be followed; a schedule; *every part of the aircraft must be designed to carry the load imposed on it and in order to determine such loads a programme of stress analysis is always carried out* (NOTE: also written **program** in American English)

progress ['prəʊgres] *noun* movement towards an end or aim; *the progress of an aircraft in flight*; **in progress** = taking place; **embarkation is in progress** = passengers are boarding the aircraft

progression [prə'greʃn] *noun* continuous series or sequence; *the instruments are checked in logical progression from left to right*

progressive [prə'gresɪv] *adjective (of movement)* gradual; in stages; *throttle movements should be kept to a minimum and be smooth and progressive*

prohibit [prə'hɪbɪt] *verb* to disallow or forbid; *smoking is prohibited in toilets*

project 1 ['prɒdʒekt] *noun* large-scale plan or scheme; *a project to modernize the airport* **2** [prə'dʒekt] *verb* **(a)** to protrude or jut out; *ice is allowed to accumulate on a probe which projects into the airstream* **(b)** to produce an image on a screen with a film or slide projector; *the instructor projected a diagram of the fuel system onto the screen*

projection [prə'dʒekʃn] *noun* production of an image on a surface; **Lambert's projection** = map projection of the earth based around two standard parallels of latitude; **Mercator's projection** = map projection of the earth onto a cylinder so that all the parallels of latitude are the same length as the equator; *see also* CONIC

prolong [prə'lɒŋ] *verb* to increase the duration or time (often unnecessarily); *to prolong the life of an engine*; *prolonged idling at low rpm (revolutions per minute) could cause spark plug fouling*

promulgate ['prɒməlgeɪt] *verb* to make known through official means; *the range promulgated for NDBs (non-directional radio beacons) in the United Kingdom is based on a daytime protection ratio between wanted and unwanted signals*

prone [prəʊn] *noun* **prone to** = likely to do something; subject to; more than usually affected by; *wing leading edges and engine intakes and propellers are prone to icing*

pronounced [prə'naʊnst] *adjective* noticeable or marked; *turbulence caused by convection is more pronounced over paved surfaces than over forest or grassy terrain*

propagation [prɒpə'geɪʃn] *noun (of radio waves)* transmission; *the speed of propagation of radio waves is slower over land than sea*

propel [prə'pel] *verb* to cause to move; *fronts are propelled by the wind behind them*; **jet-propelled aircraft** = aircraft powered by jet engines; *see also* PROPULSION; PROPULSIVE

propeller [prə'pelə] *noun* rotating shaft with blades which, together with the engine, moves an aircraft through the air; **propeller blade** = one of the elements of a propeller which generate lift when the unit is turning; **propeller pitch** = distance a propeller would advance in one rotation if there was no slip; *see also* BLADE; ROTATE

propelling nozzle [prə'pelɪŋ 'nɒzl] *noun* the extreme rear part of the jet engine where the jet exhaust enters the atmosphere

properly ['prɒpəli] *adverb* correctly; *when the chart is properly orientated, it is easier to compare distance between landmarks*; *the pinpoint is a very positive means of establishing position, as long as the feature is properly identified*

property ['prɒpəti] *noun* **(a)** characteristic or quality; *mass is a basic property of matter*; *one of the properties of mercury is that it is liquid at room temperature* **(b)** something owned; possessions; **personal property** = things belonging to a person

proportion [prə'pɔːʃn] *noun* **(a)** part of the whole compared with another part; *only a small proportion of passengers absorb the pre-departure safety information* **(b)** in proportion to = directly related to; *the force required to move the control column is in proportion to the force being exerted by the control surface*

proportional [prə'pɔːʃnl] *adjective* **(a)** comparable **(b)** related (to); (directly) proportional = directly related to (something); *the wind blows along contours with low values on the left, and the speed is directly proportional to the contour gradient*; inversely proportional = as when one thing increases and another decreases accordingly; *temperature is inversely proportional to altitude*; *the magnitude of the pressure gradient force is inversely proportional to the distance apart from the isobars*

propulsion [prə'pʌlʃn] *noun* act or instance of pushing or driving forwards; jet propulsion = jet power which provides thrust for an aircraft; *the first known example of jet propulsion produced by man was when Hero, a Greek engineer, made a machine as a toy in the year 120 BC* (NOTE: the verb is to propel)

propulsive [prə'pʌlsɪv] *adjective* pushing or driving; *the propeller is a means of converting engine power into a propulsive force called thrust* (NOTE: the verb is to propel)

protect [prə'tekt] *verb* to keep from harm, injury or damage; *gloves are worn to protect the hands in the event of a fire*

protection [prə'tekʃn] *noun* act or instance of keeping something from harm, injury or damage; corrosion protection = action and/or measures taken to prevent corrosion such as rust; fire protection = action and/or measures taken to prevent fire

protective [prə'tektɪv] *adjective* referring to something which keeps something else from harm, injury or damage; *busbars are insulated from the*

main structure and are normally provided with some form of protective covering

protrude [prə'truːd] *verb* extend above a surface; *prominent mountains frequently protrude above low-lying cloud and mist*

protrusion [prə'truːʒn] *noun* extension above a surface; something which protrudes; *when it has been necessary to physically remove a layer of snow, all protrusions and vents should be examined for signs of damage*

prove [pruːv] *verb* **(a)** to show that something is true; *the pilot proved that he was not at fault* **(b)** to be found to be, to be discovered to be; to prove useful = to be discovered as useful by experience; dry chemical extinguishers are used primarily for electrical fires and have also proved effective on liquid fires = it was discovered that, although these extinguishers were designed for electrical fires, they were good at putting out liquid fires such as petrol fires (NOTE: proving - proved - has proved, US also has proven)

provide [prə'vaɪd] *verb* to supply; to give; *radio altimeters provide a continuous indication of height above the surface immediately below the aircraft up to a maximum of 5,000 feet*; *flight crews are frequently provided with a full meteorological briefing*; *each tank is provided with a shut off valve*; *when aquaplaning, a tyre is not capable of providing directional control or effective braking*; provided that = on condition that; if; *the flight will take off on schedule provided that the weather improves*; providing that = on condition that; if; *the flame will be continuous providing (that) there is a continuous supply of fuel*

provision [prə'vɪʒn] *noun* **(a)** providing something; supply; *the provision of fresh air is important for passengers' comfort*; *catering companies are responsible for the provision of food*; *there is a generator for the provision of emergency power*; *the oil tank has provision for filling and draining* **(b)** (*of an agreement or contract*) legal statement which provides for something such as particular circumstances

proximity [prɒk'sɪmɪti] *noun* nearness in space or time; *the two aircraft were in close proximity*; **ground proximity warning system (GPWS)** = system in aircraft which warns pilot, by means of an audible signal, that the aircraft is below a preset height

PTT = PRESS TO TEST/TALK

public ['pʌblɪk] **1** *noun* people in general **2** *adjective* referring to the people in general; **public address (PA) system** = microphone, amplifier and loudspeaker set up to allow one person to be heard by a group of people; *the captain made a public address (PA) system announcement asking passengers to remain seated*; **public relations (PR)** = maintaining good relations with the public, especially to put across a point of view or to publicize a product; *the arrangements for the VIPs are being handled by the public relations department*

publication [pʌblɪ'keɪʃn] *noun* **(a)** making something public; publishing; *the publication of the latest figures* **(b)** book, magazine, chart, etc., which has been published; *the book is a CAA (Civil Aviation Authority) publication*

COMMENT: CAA (Civil Aviation Authority) publications are referred to as CAPs and each has a reference number for identification: 'the procedure for obtaining a bearing is described in CAP 413'

publish ['pʌblɪʃ] *verb* to prepare and issue a book, magazine, chart, etc., and sell or distribute it to the public; *all known air navigation obstructions in the UK are published in the Air Pilot*

pull out ['pʊl 'aʊt] *verb* to pull out of *or* from a dive = to return the aircraft to level flight after a nose-down flight path

pulse [pʌls] *noun* single vibration of electric current; short period of a voltage level; **pulse modulation** = use of a series of short pulses which are modified by an input signal, to carry information

pump [pʌmp] **1** *noun* device with rotary or reciprocating action which is used to move fluids along pipes or for compressing fluids; **fuel pump** = device to move fuel along pipes from the tanks to the engine; *see also* BOOSTER 2 *verb* to move or compress a fluid by means of a pump; *fuel is pumped from the tanks to the carburettor*

COMMENT: most modern aircraft are fitted with hydraulic pumps driven from the engine. Other types of pumps, may be found, but these are usually used to power emergency systems. Pumps can be driven directly from the engine gearbox, by an electric motor, or by air

pure [pjʊə] *adjective* unmixed; **pure aluminium** = aluminium which has not been combined with any other metal to create an aluminium alloy; *inner tubes for tyres are made of pure rubber*; *magnesium does not possess sufficient strength in its pure state for structural uses*

purpose ['pɜːpəs] *noun* **(a)** function; *the purpose of the engine is to convert heat energy to mechanical energy* **(b)** use; *for practical purposes, any straight line drawn on a Lambert's chart represents a great circle*; **general purpose** = for all-round or general use

pushrod ['pʊʃrɒd] *noun (part of the valve mechanism)* steel or aluminium rod which moves the rocker arm; *the camshaft operates the pushrod*

pylon ['paɪlən] *noun* **(a)** structure on the wing of an aircraft to support an engine; *most modern jet passenger transport aircraft have pylon mounted engines* **(b)** tall metal structure built to support electricity or telephone cables; *electricity pylons are difficult to see from the air so pilots of light aircraft should be particularly careful to note their positions*

pyrotechnic [paɪrə'teknɪk] *adjective* of or relating to fireworks; **pyrotechnic lights** = lights created by rockets or flares

Qq

Q-code ['kjuː'kəʊd] international telegraph code which is now used in RTF operations; **QDM** = magnetic bearing to D/F (direction finding) station; **QFE** = atmospheric pressure at aerodrome level; **QNE** = 1013.25 mb (millibars) - altimeter setting for FL (flight level) reading; **QNH** = atmospheric pressure at mean sea level; **QTE** = true bearing from D/F (direction finding) station; **QUJ** = true track (zero wind) to reach a destination

QFI = QUALIFIED FLYING INSTRUCTOR

quadrant ['kwɒdrənt] *noun* **(a)** device shaped like a quarter of a circle; **gated quadrant** = quadrant with a device preventing a lever from being moved to an incorrect setting; *the throttles, usually known as power levers, operate in a gated quadrant*; **throttle quadrant** = arc-shaped device in which the throttle levers move **(b)** compass **quadrant** = quarter part of a circle centred on a navigational aid

---
COMMENT: **NE quadrant** = 000° - 089°; **SE quadrant** = 090° - 179°; **SW quadrant** = 180° - 269°; **NW quadrant** = 270° - 359°
---

quadrantal [kwɒ'dræntl] *adjective* referring to a quadrant or to a quarter of a circle; **quadrantal error** = radio signal error caused by the metal structure of the receiving aircraft; **quadrantal height** = flight levels in each of the compass quadrants designed to provide safe separation for aircraft heading towards each other

qualified ['kwɒlɪfaɪd] *adjective* (person) who has gained a certificate after having completed a specialized course of study; **a qualified flying instructor (QFI)** = a pilot with an instructor's rating

qualify ['kwɒlɪfaɪ] *verb* **(a)** to add reservations or modify an earlier statement to make it less absolute; *fire in the wing or high seas may cause the captain to qualify the evacuation command, informing cabin crew of these conditions and allowing them to adjust the evacuation plan accordingly* **(b)** to study for and obtain a diploma which allows to do a certain type of work; *he qualified as an engineer in 1996*

quality ['kwɒlɪti] *noun* amount of excellence of something; *satisfactory ignition depends on the quality of the fuel*

quantity ['kwɒntɪti] *noun* size, extent, weight, amount or number of something; *a small quantity of illegal drug was found in the passenger's bag*

quarter ['kwɔːtə] *noun* one fourth of something (1/4 or 25%); *the fuel tank is only a quarter full*

Rr

radar ['reɪdɑ:] *noun* method of detecting distant objects and establishing their position, velocity, or other characteristics by analysis of very high frequency radio waves reflected from their surfaces; **radar beam** = shaft of radar waves directed towards a distant point; **airborne weather radar (AWR)** = radar installation in an aircraft to give the flight crew information about the en route weather; **primary radar** = system which uses reflected radio signals; **secondary radar** = system in which the active target replies to the interrogation unit; **tertiary radar** = long-range navigation aids; *see also* PRECISION (NOTE: an acronym for **RA**dio **D**etection **A**nd **R**anging)

radial ['reɪdɪəl] **1** *adjective* referring to lines of radius having a common centre; **radial engine** = engine in which the pistons are arranged like the spokes of a wheel **2** *noun* line of radio bearing from a VOR (very high frequency omni-directional radio range) beacon; *to get to a facility you must track the reciprocal of the VOR radial*

radiate ['reɪdɪeɪt] *verb* to send out rays or waves; *the earth radiates low intensity infra-red waves*; *short bursts of energy are radiated from an antenna*

radiation [reɪdi'eɪʃn] *noun* act or process of sending out rays or waves; **solar radiation** = radiation from the sun; **terrestrial radiation** = radiation from the earth; **radiation fog** = fog caused by the cooling of the earth to below the dew point, combined with saturation, condensation and light mixing air currents

radiator ['reɪdɪeɪtə] *noun* liquid-to-air heat exchanger that transfers engine heat to the outside air; *anti-icing additives are used in radiator coolants*; *see also* COOLANT

radio ['reɪdɪəʊ] *noun* wireless transmission through space of electromagnetic waves in the approximate frequency range from 10 kHz (kilohertz) to 300,000 MHz (megahertz); **radio aid** = navigation aid utilizing radio waves; **radio altimeter** = device for measuring the height of the aircraft above the earth using reflected radio waves; **radio horizon** = line along which direct rays from radio frequency transmitter become tangential to the earth's surface; **radio waves** = electromagnetic radiation waves; *the atmosphere absorbs radio waves*

radiotelephony (R/T)
[reɪdɪəʊtə'lefəni] *noun* transmission of speech by radio; *correct use of R/T phraseology avoids ambiguity*

radius ['reɪdɪəs] *noun* **the radius of a circle** = line drawn from a point on the circumference of a circle to the centre point (NOTE: plural is **radii** ['reɪdɪaɪ])

raft [rɑ:ft] *noun* flat-bottomed inflatable rubber craft for floating on water; **life raft** = small boat-like vessel for use in an emergency; **slide raft** = an escape slide which, when detached from the aircraft, can be used as a life-raft

rain [reɪn] **1** *noun* precipitation or water which falls from clouds in small drops; *rain is falling heavily*; *rain and weather present fewer problems for area radar compared to the other types* **2** *verb* to fall as drops of water from clouds; *it is raining*; *I don't think it will rain*

rainstorm ['reɪnstɔːm] *noun* heavy rain accompanied by wind; *in heavy rainstorm, the windscreen wipers may not be able to cope*

raise [reɪz] *verb* (a) to lift; raise the landing gear = retract the undercarriage (b) to increase; *to raise the temperature; to raise the pressure* (c) to cause (problems); *fuel vaporization can raise problems when starting the engine* (NOTE: do not confuse with the verb **to rise**; grammatically, the verb **raise** takes an object whereas the verb **rise** does not: **temperature rises; the sun's rays raise the temperature of the surface**)

ram [ræm] *noun* increase in air pressure caused by the forward speed of the aircraft; *due to ram effect from aircraft forward speed, extra air is taken into the engine*; **ram air** = airflow created by the movement of the aircraft which is used to cool, ventilate or drive turbines; *oil cooling is often achieved by using ram air or fuel*

ramp [ræmp] *noun* (a) inclined track for loading and unloading; *the height of the cabin floor to the ground on large jet transports means that injuries can occur by exiting through the doors when steps or ramps are not available* (b) US area of tarmac, concrete, etc., outside a hangar, used for parking aircraft (NOTE: also called **apron** in British English)

range [reɪnʒ] **1** *noun* (a) amount or extent of variation; *range of frequencies; range of temperatures* (b) row or chain of mountains or hills; *the Rocky Mountain range; valley winds require at least a reasonable pressure gradient, preferably along a range of hills which will produce a wind at right angles to the hills* (c) maximum distance an aircraft can fly on a given amount of fuel; *cruise level is selected to give the greatest fuel economy, i.e. the greatest range for least fuel* (d) maximum effective distance of operation; *precision approach radar (PAR) is subject to weather interference and has a limited range* **2** *verb* **to range from ... to ...** = to

vary from ... to ,,,; *temperatures range from 0°C (Celsius) at night to 40°C (Celsius) at midday*

rapid ['ræpɪd] *adjective* fast, with great speed; **rapid changes** = fast changes; *hoar frost is a light crystalline deposit which can form on the aircraft as a result of rapid descent from cold altitudes into warm moist air*

rapidity [rə'pɪdɪti] *noun* great speed; *spontaneous combustion occurs with such rapidity that there is an audible explosion*

rapidly ['ræpɪdli] *adverb* with great speed, quickly; *rime ice is formed when individual droplets of water freeze rapidly on striking the aircraft surface*

rare [reə] *adjective* uncommon, not often occurring; *smog or smoke fog is now rare because of pollution controls*

RAS = RECTIFIED AIR SPEED; *(ICAO)* RADAR ADVISORY SERVICE

rate [reɪt] *noun* quantity measured in relation to another measured quantity; **rate of climb** = speed of ascent measured in feet per minute; **rate of descent** = speed of descent measured in feet per minute; *if the speed of descent is too low, alter the throttle setting accordingly*; **flow rate** = amount of movement of a fluid through a system in a given time, such as gallons per minute; **lapse rate** = rate at which temperature changes according to altitude or rate of decrease of atmospheric temperature with increase in altitude

rather ['rɑːðə] *adverb* (a) to a certain extent; somewhat; a bit; quite; **rather cold weather** = weather which is quite cold, but not very cold (b) **rather than** = instead of; preferably; *air tends to flow around hills rather than rise over them*

rating ['reɪtɪŋ] *noun* (a) authorization on a licence, and forming part of the licence, giving special conditions or privileges; **instrument rating (IR)** = additional qualification gained from a course of training for instrument flying; **night rating** = additional qualification gained from a course of training for night flying **(b)** classification according to a scale; **octane rating** = the ability of the fuel to resist detonation, i.e. the higher the number, the greater is the fuel's resistance to detonation

ratio ['reɪʃiəu] *noun* relationship between two quantities expressed as the quotient of one divided by the other; *the air/fuel ratio is 15:1*; *chart scale is the ratio of the chart distance to earth distance* (NOTE: the ratio of 7 to 4 is written 7:4 or 7/4)

ray [reɪ] *noun* thin or narrow beam of light or other radiant energy; *cathode ray*; *the earth is heated by the rays of the sun*; *see also* X-RAYS

RBI = RELATIVE BEARING INDICATOR

re- [riː] *prefix meaning* again; **reassemble** = to assemble again; **rewrite** = to write again (NOTE: not all verbs beginning with **re-** have the meaning 'again', such as: **remember**)

reach [riːtʃ] *verb* **(a)** to arrive at (a place); *the aircraft reached its destination on time* **(b)** to get to a certain level; *up-currents in thunderstorms can reach 3,000 feet per minute*; *temperatures can reach 49°C (Celsius) in summertime in the Gulf region* **(c)** to extend; *the tops of thunderstorm clouds can reach through the tropopause*

react [riˈækt] *verb* **(a)** to act in response to an action; *because the rotors and stators of a compressor are of aerofoil shape, the airflow reacts in a similar way to the airflow over a wing* **(b)** *(of person)* to do or to say something in response to words or to an event; *the cabin crew reacted swiftly when the fire broke out* **(c)** *(chemistry)* **to react with something** = to change chemical composition because of a substance; *the electrolyte in the cells of a lead-acid battery reacts chemically with the plates*

reactance [riˈæktəns] *noun (electricity)* a component of impedance in an AC (alternating current) circuit; *reactance is a form of resistance which varies as the frequency changes*

reaction [riˈækʃn] *noun* response to an action or stimulus; *for every action there is an equal and opposite reaction*; *passenger reaction may be slower than usual in an emergency situation*; *quick reactions are needed in an emergency*

readily ['redɪli] *adverb* **(a)** promptly, immediately; *fire extinguishers must be readily available for use*; *ice melts very readily at 0°C (Celsius)* **(b)** it can readily be seen = it can be easily understood; *it can readily be seen from the preceding paragraph that density and pressure are linked*

reading ['riːdɪŋ] *noun* **(a)** information indicated by an instrument or gauge; **altimeter reading** = altitude indicated by the altimeter; **barometer reading** = barometric pressure indicated by the barometer **(b)** **map reading** = act or instance of interpreting information on a map

readout ['riːdaut] *noun* display or presentation of data from calculations or storage; *the rotating beam cloud base recorder/indicator operates continuously, day and night and produces an automatic readout of cloud base height*

rear [rɪə] **1** *noun* aft part, the part furthest from the front; *the rear of the aircraft* **2** *adjective* at the back; referring to the back; *the rear part of the aircraft is called the aft section*

rearward ['rɪəwəd] *adjective* towards the aft or the rear; *the expanding gas travels in a rearward direction*

reason ['riːzən] *noun* basis or motive for an action; *a rough surface is more susceptible to fatigue cracking than a smooth one and for this reason highly stressed members are often polished*

reasonable ['riːznəbl] *adjective* **(a)** acceptable; fair (amount); **a reasonable sum of money** = a sum of money which is not too high or which is acceptable **(b)** within the boundaries of common sense; *it would be reasonable to expect that radio frequencies would travel through the air* *in straight lines as a direct wave, but they bend, or refract*

receive [rɪ'siːv] *verb* to get, to obtain; *the sides of the hills and mountains which face the sun receive more intense radiation than flat surfaces because of the angle of exposure to the sun*

receiver [rɪ'siːvə] *noun* device that receives incoming radio signals and converts them to sound or light; *the transponder in the aircraft consists of a transmitter and a receiver*

recent ['riːsənt] *adjective* referring to a time immediately before the present; *recent engine designs include variable angle stator blades*; *a more recent development, is the barograph which utilizes the electrical output of the digital display barometer*; **recent weather** = significant weather observed in the period since the previous observation, but not now

reception [rɪ'sepʃn] *noun* act or instance of receiving radio signals; *the antenna is highly directive in transmission and reception*

reciprocal [rɪ'sɪprəkl] **1** *adjective* **reciprocal heading** = opposite heading, 180° from a given heading; *the reciprocal heading of 090° is 270°* **2** *noun* exactly opposite direction; *a wave transmitted vertically returns to earth on its reciprocal*

reciprocating [rɪ'sɪprəkeɪtɪŋ] *adjective* **(a)** *(of machine)* moving backwards and forwards or up and down; *a cam is an oval-shaped wheel which, when rotating, converts circular motion into reciprocating motion*

recognition [rekəg'nɪʃn] *noun* identification; *hydraulic fluids are coloured for recognition purposes*

recognize ['rekəgnaɪz] *verb* to know to be something that has been seen, heard, etc. before; to identify; *it may be difficult to recognize a particular stretch of coastline simply by its appearance* (NOTE: also written **recognise**)

recommend [rekə'mend] *verb* to say that something is worthy, desirable or suitable; *dry chemical extinguishers are recommended for use on aircraft brake fires; aircraft should be operated to the manufacturers recommended limits*

record 1 ['rekəd] *noun* **(a)** written account of facts and information for future reference **(b)** electronically stored data **2** [rɪ'kɔːd] *verb* **(a)** to write (information or data); *measure track angles and distances and record them in a log* **(b)** to capture and store electronically; *details of wind speed, direction, visibility, cloud cover, etc., are recorded onto a cassette*

recorder [rɪ'kɔːdə] *noun* device for capturing sound onto cassette or magnetic tape; *cockpit voice recorder*; **flight data recorder (FDR)** = electronic device located in the tail section of an aircraft that picks up and store data about a flight - also called **black box**

recording [rɪ'kɔːdɪŋ] *noun* action of writing or of picking up and storing information; *an anemograph is an instrument which maintains a continuous recording of wind direction and speed on a graph*

recover [rɪ'kʌvə] *verb* **(a)** to return to an earlier, normal condition or attitude; **recover from a stall** = return the aircraft to straight and level flight **(b)** to rescue and remove from a particular area, often the sea; *emergency services recovered two bodies from the wreckage of the helicopter*

recovery [rɪ'kʌvri] *noun* **(a)** a return to an earlier, normal condition or attitude; **recovery from unusual attitudes** = flight exercise requiring the student pilot to return the aircraft to its previous, normal, that is, straight and level attitude, after it has been in an unusual attitude **(b)** the rescue and removal from a particular area; *the recovery of survivors from the sea was carried out by helicopters*

rectangle ['rektæŋgl] *noun* 4-sided plane figure with 4 right angles, and with opposite sides of equal length; *the colour* *identification of refuelling equipment for AVGAS is: blue rectangle, red decal with AVGAS 100LL in white letters*

rectangular [rek'tæŋgjʊlə] *adjective* referring to something with the shape of a rectangle; *a rectangular wing panel*

rectification [rektɪfɪ'keɪʃn] *noun* changing an alternating current into direct current; *part of the generator alternating current (AC) is passed through a rectification circuit*

rectified airspeed ['rektɪfaɪd 'eəspiːd] *noun* indicated airspeed (IAS) corrected for instrument error and pressure error; *when rectified airspeed (RAS) is corrected for density error the resultant is known as the true airspeed*

rectifier ['rektɪfaɪə] *noun* electronic circuit that converts an alternating current supply into a direct current supply; *the ignition unit receives an alternating current which is passed through a transformer and rectifier*

rectify ['rektɪfaɪ] *verb* (a) to change alternating current into direct current; *AC (alternating current) output is rectified and regulated externally and returned as direct current to the stator field winding* (b) to correct; **to rectify a mistake** = to put right a mistake

reduce [rɪ'djuːs] *verb* to decrease, to make less; **reduce altitude** = descend; **reduce temperature** = make cooler (NOTE: the opposite is **increase**)

reduction [rɪ'dʌkʃn] *noun* decrease; *reduction in temperature, pressure, speed*; **reduction gear** = gears in an engine which allow the propeller to turn at a slower speed than the engine; *an air starter motor has a turbine rotor that transmits power through a reduction gear and clutch to the starter output shaft which is connected to the engine*

redundancy [rɪ'dʌndənsi] *noun* duplication of component parts of a system to enable the system to function even if one component fails; *with system redundancy, a single failure within a system will have*

little effect on the aircraft's performance during the approach and landing operation

redundant [rɪ'dʌndənt] *adjective* referring to a system which provides extra component parts to enable the system to function even if one component fails; *redundant structure design is composed of a large number of members all of which share a certain load so that if one of the members is lost, the load carried by the member is divided between all the others in such a way that the total load-carrying ability is reduced only slightly*

re-enter [ri'entə] *verb* to enter again; *for engine checks the aircraft should be headed into wind to prevent hot exhaust gases re-entering the engine*

refer [rɪ'fɜː] *verb* (a) to describe; to give a name to; *the term wind is used to refer to the horizontal motion of air* (b) to direct to a source of help or information; **refer to chapter 10 for more details** = look at or read chapter 10 for more information (NOTE: **referring - referred**)

reference ['refrəns] *noun* something used as a basis for further calculation or investigation; **visual reference** = anything seen and used as a guide to something else; *use the large building as a visual reference for the turn onto final approach*; **reference book** = book, such as a dictionary, where you can look for information; **reference point** = fixed datum near the centre of the airfield landing area; **reference signal** = signal against which telemetry data signals are compared; **by reference to** = by looking at and comparing

refinement [rɪ'faɪnmənt] *noun* improvement; *an internal locking device is one of the numerous refinements to the simple actuator*

reflect [rɪ'flekt] *verb* to throw back something such as radio waves or light; *snow surfaces reflect up to 90% of radiation while rock, sand and concrete reflect only 10-20%*

reflection [rɪ'flekʃn] *noun* throwing back of something such as radio waves or light; *glare caused by reflection of sunlight from the top of a layer of fog or haze can seriously reduce the air-to-ground visibility*

reflective [rɪ'flektɪv] *adjective* referring to something which throws back something such as radio waves or light; *reflective power means that at low angles of elevation of the sun, water reflects a great amount of solar radiation thus slowing down the rise in sea surface temperatures*

reflector [rɪ'flektə] *noun* device which throws back something such as light; *the shape of a water droplet makes it a good reflector, so water in the atmosphere absorbs and scatters radio waves*

refract [rɪ'frækt] *verb* to cause a wave, such as light or sound, to change direction or turn as it passes from one medium into another of different density; *a sky wave starts life as a direct wave and, on reaching the ionosphere, the direct wave is refracted and returns to the earth's surface*

refraction [rɪ'frækʃn] *noun* change in direction or turning of a wave, such as light or sound, as it passes from one medium into another of different density; **atmospheric refraction** = change in direction of waves due to variations in temperature, pressure and humidity, particularly at lower altitudes; **coastal refraction** = change in direction of waves when a signal crosses a coastline from sea to land; **ionospheric refraction** = change in direction as the wave passes through an ionized layer

refrigerant [rɪ'frɪdʒərənt] *noun* substance to provide cooling either as the working substance of a refrigerator or by direct absorption of heat; *heated air from the main air supply system passes through the evaporator matrix and by induction releases heat into the liquid refrigerant*

refuel *or* **re-fuel** [ri'fju:əl] *verb* to fill with fuel again; *fire risk is always present when you defuel and refuel*

regain [rɪ'geɪn] *verb* to obtain again or to acquire again; *the omni-bearing selector/course deviation indicator is a demand instrument which indicates which way to turn to regain the required bearing*

regard [rɪ'gɑ:d] **1** *noun* a particular point or aspect; **in this regard** = concerning this or with reference to this; **with regard to** = concerning or with reference to; *with regard to the turbo-propeller engine, changes in propeller speed and pitch have to be taken into account* **2** *verb* to look upon or consider in a particular way; *thoughtful concern for others is regarded as an essential component of good airmanship*

regardless [rɪ'gɑ:dləs] *preposition* in spite of, despite, with no thought of; **with fly-by-wire technology, the aircraft's stalling angle of attack cannot be exceeded regardless of control stick input** = the stalling angle of attack cannot be exceeded, despite or no matter what the pilot does with the flying controls

region ['ri:dʒən] *noun* **(a)** area; *the troposphere is deepest in equatorial regions and shallowest near the poles*; **flight information region (FIR)** = airspace with defined limits which has an air traffic control information and alerting service **(b)** **in the region of** = about or approximately; *the burning temperature of the fuel being in the region of 2,000°C (Celsius)*

register ['redʒɪstə] **1** *noun* official list or record; *the student's name was not on the register* **2** *verb* **(a)** to record; to indicate on an instrument; *during ground running checks, if oil pressure does not register within a few seconds, the engine should be stopped and the cause investigated*; *electrically operated pressure gauges register main and emergency system pressure* **(b)** to enter details on an official list; *to register an aircraft*

registration [redʒɪs'treɪʃn] *noun* entry of civil aircraft into records of national certification authority with details of letter and number code displayed on aircraft; **certificate of registration** = document as proof of registration

regular ['regjʊlə] *adjective* **(a)** occurring at fixed time intervals; **regular inspections** = inspections taking place at equal intervals of time **(b)** ordinary or standard; *throughout the tropics and sub-tropics, the sea breeze is a regular feature*

regulate ['regjuleɪt] *verb* to control, to adjust to a specific requirement; *controllable cowl flaps regulate the amount of air flowing across the cylinders*

regulation [regju'leɪʃn] *noun* **(a)** act or instance of controlling or adjusting to a specific requirement; *regulation of cabin temperature is controlled by the manual setting of a mechanically controlled switch* **(b) regulations** = rules or laws; **safety regulations** = rules or laws which must be followed to make a place safe; *equipment and furnishings on modern jet transports must comply with safety regulations concerning fire resistance*

regulator ['regjuleɪtə] *noun* device used to control the flow of fluids or electric current; **voltage regulator** = device to control the level of voltage

reinforce [riːɪn'fɔːs] *verb* to make stronger or to strengthen; *typical skin materials used in aircraft are made from epoxy resins which are reinforced with glass, carbon or Kevlar fibres*

reinforced [riːɪn'fɔːst] *adjective* made stronger or strengthened; **reinforced plastics** = plastic materials used with glass fibres to repair some types of aircraft structure

reinforcement [riːɪn'fɔːsmənt] *noun* strengthening; *there is reinforcement around each opening in the pressure cabin, such as the cabin door, escape hatch and windows*

relate [rɪ'leɪt] *verb* **(a)** to make a connection or link; to associate; *orientating the chart relates the direction of land features to their representation on*

the chart and aids recognition **(b)** to **relate to** = to concern or to be about; *Kepler derived the laws which relate to the motion of planets in their orbits*

relation [rɪ'leɪʃn] *noun* **(a)** (natural or logical) association between things; *the relation between thrust and drag;* **this bears no relation to that** = this is not connected with that in any way **(b) in relation to** = with reference to; *the range at which objects can be recognized is affected by the direction of viewing in relation to the position of the sun or the moon; the VOR station on the ground does the calculation and, depending on where the aircraft is in relation to the VOR station, it will receive signals which define the bearing of the aircraft from the VOR*

relationship [rɪ'leɪʃnʃɪp] *noun* natural or logical association between things; *there is a close relationship between altitude and pressure*

relative ['relətɪv] *adjective* **(a) relative airflow** = airflow over an aerofoil often related to the chord line of the aerofoil; **relative bearing** = bearing of a radio station or object with reference to the aircraft's heading; **relative density** = the ratio of density of a liquid with reference to water, or gas with reference to air; **relative humidity** = ratio between the amount of water vapour in the air and the amount which would be present if the air was saturated, at the same temperature and the same pressure (shown as a percentage); **relative wind** *US* = RELATIVE AIRFLOW **(b) relative to** = compared to; with reference to; *ground-speed is the speed of the aircraft relative to the ground*

relay ['riːleɪ] *noun* device which responds to a small current or voltage change by activating switches or other devices in an electric circuit; *thermocouple detectors operate a sensitive relay or electronic circuit when a predetermined temperature is exceeded;* **pressure relay** = component which transmits fluid pressure to a direct reading pressure gauge, or to a pressure transmitter which electrically indicates pressure on an instrument on the hydraulic panel

release [rɪ'liːs] **1** *noun* freeing from something that holds; *air rising and cooling often reaches its dew point temperature, becomes saturated and any further cooling results in condensation and the consequent release of latent heat* **2** *verb* to free from something that holds; *push the button to release the lever*; **release the brakes** = let the brakes off; **release the pressure** = allow pressure to reduce

relevant ['relɔvɔnt] *adjective* pertinent, having a connection with the matter in hand; **relevant information** = useful information which is related to the matter in question; *high charts show only information relevant to high altitude flights and many beacons and aids which are provided for low operations are omitted to keep the chart clear*

reliability [rɪlaɪɔ'bɪlɪti] *noun* dependability, trustworthiness; *the gas turbine is a very simple engine with few moving parts giving it high reliability with less maintenance*

QUOTE where a State introduces drug testing, high standards of medical reliability must be maintained

INTER PILOT

reliable [rɪ'laɪɔbl] *adjective* dependable, trustworthy; *the gas turbine is a very simple and reliable engine*

relief [rɪ'liːf] *noun* **(a)** variations in elevation of the surface of the earth; *relief is usually represented on aeronautical charts by contours, gradient tints or hill shading* **(b)** lessening (of pressure); **relief valve** = valve which opens at maximum safe pressure and closes again upon return to normal operating conditions; *a reservoir contains a relief valve to prevent over-pressurization*

relieve [rɪ'liːv] *verb* to cause a lessening; to remove excess pressure or tension; *safety valves relieve excess cabin pressure*; *a trim tab on the elevator relieves the forward and aft forces on the control stick or yoke*

relight [rɪ'laɪt] *verb* to ignite again; *the ability of the engine to relight will vary according to the altitude and the forward speed of the aircraft*

rely [rɪ'laɪ] *verb* to be dependent on; *pressure carburettors do not rely on venturi suction to discharge fuel into the airstream*

remain [rɪ'meɪn] *verb* to stay, to continue to be; *during the evacuation, crew must remain at their assigned stations and redirect passengers*; *the fuel/air ratio does not remain constant, but, as the speed increases, the mixture gets richer*; *the audible fire warnings may be cancelled but the red warning light will remain on*

remainder [rɪ'meɪndɔ] *noun* **(a)** something left after excluding other parts; the rest; *the auxiliary power unit is usually found in the tail section, separated from the remainder of the fuselage by a firewall* **(b)** *(mathematics)* number left over when one number is divided by another

remote [rɪ'mɔut] *adjective* **(a)** far away, and not near anything else; *a remote area*; *when the destination is a remote island, the calculation of the point of no return (PNR) becomes essential* **(b)** operated or controlled from a distance; *remote cabin pressure controllers* **(c)** a remote chance = a small but unlikely possibility

removal [rɪ'muːvl] *noun* taking something away; moving something from the position occupied; *the repair to the aircraft required the removal of the engine*

remove [rɪ'muːv] *verb* to take (something) away; to move (something) from the position occupied; *filters are fitted in lines in a hydraulic system, in order to remove foreign particles from the fluid*; *the engine will have to be removed for repair*

render ['rendɔ] *verb* **(a)** to make; to cause to become; *the failure of any* component in the fire detection system will render the system inoperative; tropical air moving northwards is subjected to surface cooling and rendered increasingly stable in its lower layers **(b)** to give; **to render assistance** = to provide help; only when all possible assistance has been rendered inside the cabin will crew themselves evacuate

repair [rɪ'peə] **1** noun mending; returning something to good condition after damage; the repair to the nosewheel took three hours **2** verb to mend; to return (something) to good condition after damage; after the wheels-up landing, the flaps had to be repaired

QUOTE Mr Pike elected to await repairs instead of taking up the offer of alternative flights, and found himself the only passenger aboard the Jumbo as it flew back to Heathrow four hours late

Pilot

repeat [rɪ'piːt] verb **(a)** to do again; the first officer repeated the transmission; the trainee had to repeat his navigation examination **(b)** to occur again; metal fatigue is induced by repeated stress cycling **(c)** to say again; could you repeat that please? I didn't hear; the message was repeated a few minutes later

repel [rɪ'pel] verb to push away by a force; like poles (ie, north and north, or south and south) of a magnet repel each other (NOTE: repelling - repelled)

repellent. [rɪ'pelənt] noun substance used to resist the effect of something; rain repellent is sprayed onto the windscreen and spread by the wipers

replace [rɪ'pleɪs] verb to take the place or to fill the place of; to put in the place of; as warm air rises, cold air moves in to replace it; the term Greenwich Mean Time (GMT) is being replaced by the term Coordinated Universal Time (UTC)

replacement [rɪ'pleɪsmənt] noun replacing something (with something else); the replacement of moist air by dry air is

the only sure way of dispersing advection fog

reply [rɪ'plaɪ] **1** noun answer; response; secondary surveillance interrogation is made on 1030 MHz (megahertz) and the reply on 1090 MHz (megahertz) **2** verb to answer; to respond; he replied to the letter (NOTE: **replying - replied**)

report [rɪ'pɔːt] **1** noun **(a)** official account of an occurrence; incident report; weather report **(b)** position report = report over a known location from an aircraft to air traffic control (ATC) **2** verb to write or tell information (in an official manner); the observer measures this distance in a number of directions and reports the minimum value as the meteorological visibility; an accident must be reported

represent [reprɪ'zent] verb to indicate or to show, using signs or symbols; 1 cm (centimetre) on the chart represents 500,000 cm (centimetres) on the earth; on a Mercator projection, meridians are represented as parallel straight lines

representation [reprɪzen'teɪʃn] noun way of showing, using signs or symbols; the synoptic chart provides a representation of the weather over a large area at a particular time

representative [reprɪ'zentətɪv] **1** adjective **representative of** = typical; which is an example of what all others are like; surface air temperatures are taken in such a way as to be representative of the air temperature near the surface yet unaffected by the direct surface heating or cooling effects **2** noun airline representative = person who acts on behalf of an airline; person who works for an airline; passengers should assemble in the departure lounge where an airline representative will meet them

request [rɪ'kwest] **1** noun (polite) demand; something asked for; ATC (air traffic control) received a request from the pilot for departure clearance; on request = when asked; a personal flying log book must be retained for production

on request by an authorized person **2** *verb*
to ask for something; *the pilot requested
vectors to enable him to locate the airfield*

require [rɪ'kwaɪə] *verb* **(a)** to need;
*dynamic seals require lubrication to
remain effective* **(b)** to impose an
obligation, to compel by law; *transport
operations over water require the carriage
of life rafts, life jackets, survival beacons
and pyrotechnics*

requirement [rɪ'kwaɪəmənt] *noun* **(a)**
necessity; *planning for an in-flight
emergency is a standard requirement of
pre-departure preparation* **(b)** legal
requirement = an obligation by law **(c)**
requirements = something which is
demanded or required; *the airframe had to
be built to very specific requirements*

re-register [riː'redʒɪstə] *verb* to register
again; *the aircraft had to be re-registered
because of an administrative error*

rescue ['reskjuː] **1** *noun* freeing from
danger; *early rescue depends on the rapid
location of survivors* **2** *verb* to free from
danger; *passengers were rescued from the
burning aircraft*

reserve [rɪ'zɜːv] **1** *noun* something kept
back for possible future use; reserve fuel =
fuel used only in a situation when the
aircraft has to be in the air for a longer time
than expected, as because of a go-around or
diversion **2** *verb* to keep something, such
as a seat, for somebody; *seats 23A and 23B
are reserved for Mr and Mrs Smith*

reservoir ['rezəvwɑː] *noun* container for
holding a store of fluid; *a reservoir
provides both storage space for the system
fluid, and sufficient air space to allow for
any variations in the volume of the fluid
in the system*

reset [riː'set] *verb* to set again;
*instruments which need resetting in flight
must be accessible to the crew* (NOTE:
resetting - reset)

resettable [riː'setəbl] *adjective* which
can be reset; *circuit breakers are
resettable protective devices*

residual [rɪ'zɪdjuəl] *adjective* referring
to the residue of something; *after de-icing
operations, external surfaces should be
examined for signs of residual ice or snow*

residue ['rezɪdjuː] *noun* the remainder of
something after the removal (often by use)
of the main part; what is left; *the leaking
oil left a sticky residue on the ground*

resin ['rezɪn] *noun* materials (such as
polyesters, epoxies, and silicones) that are
used with fillers and other components to
form plastics; *to make a composite it is
necessary to combine the reinforcing
glass fibres with some form of special glue
or resin*

resist [rɪ'zɪst] *verb* to fight off the effects
of something; *a tube resists bending in any
direction but beams are designed usually
to resist bending in one or two directions
only; in order for an aeroplane to fly, lift
and thrust must resist and overcome the
forces of gravity and drag*

resistance [rɪ'zɪstəns] *noun* **(a)** force
that opposes; aerodynamic resistance =
drag; impact resistance = the ability of a
material to withstand an impact **(b)**
(electricity) the opposition of a body or
substance to current passing through it; *the
shunt coil is made of fine wire which gives
a high resistance and small current flow*

resistant [rɪ'zɪstənt] *adjective* referring
to something which is unaffected by a
force, process or substance; *some alloys
are less resistant to corrosion than others;
crash resistant and heat resistant
materials*

resistive [rɪ'zɪstɪv] *adjective* **(a)**
(electricity) referring to resistance;
*windscreen heating and electrical
de-icing systems are resistive load circuits*
(b) *(force)* referring to resistance; *the
resistive force of drag*

resistor [rɪ'zɪstə] *noun* *(electricity)*
device used to control current in an electric
circuit by providing a (known) resistance;
*components such as resistors, rectifiers
and internal switches are all embedded in*

micro-size sections of semi-conductor material

respect [rɪ'spɛkt] *noun* in some respect = in some way; *the flat chart inevitably misrepresents the earth's surface in some respect*; with respect to = concerning or with reference to; *frost point is the temperature to which air must be cooled at constant pressure in order to reach a state of saturation with respect to ice*

respective [rɪ'spɛktɪv] *adjective* referring to two or more persons or things regarded individually; *the passengers returned to their respective seats*; *the temperature and pressure of the fuel supply are electrically transmitted to their respective indicators i.e. temperature to the temperature gauge and pressure to the pressure gauge*

respond [rɪ'spɒnd] *verb* (a) to reply or to answer (b) to react; to act in return; *the aircraft responds to the controls* = the aircraft attitude changes as a result of the pilot's movements of the flying controls

response [rɪ'spɒns] *noun* (a) answer or reply; *transponder response*; *despite repeated ATC (air traffic control) transmissions, there was no response from the pilot* (b) reaction; in response to = as a reaction to; *the primary function of the outflow valves is to regulate the discharge of cabin air in response to the pressure signals received from the controller*

responsibility [rɪspɒnsɪ'bɪlɪti] *noun* the condition of being responsible; *it is the responsibility of the captain to order an evacuation*

responsible [rɪ'spɒnsəbl] *adjective* (a) being a source or cause; *frontal systems are responsible for much of the weather and clouds which occur in temperate latitudes* (b) directing or being in charge, and open to blame is something goes wrong; *cabin crew are responsible for the well-being of passengers*; responsible to someone = answerable for one's actions to someone highly placed

restore [rɪ'stɔː] *verb* to return to original or normal condition; *loss of engine power should be fully restored when the control is returned to the cold air position*

restrict [rɪ'strɪkt] *verb* (a) to limit; to make free movement difficult; *the narrow aisles of the aircraft restrict the rapid movement of people* (b) to limit; during the bomb-scare, entry to the airport was restricted to authorized people only = only authorized people could enter the airport

restriction [rɪ'strɪkʃn] *noun* (a) narrowing; partial blockage; *any restriction in a pipeline will increase liquid velocity and produce turbulence* (b) limitation; *there are restrictions on the taking of photographs in the vicinity of the airport*

restrictor [rɪ'strɪktə] *noun* restrictor valve = valve designed to permit limited flow in one direction and full flow in the other direction; *the extent to which the oil pressure will fall depends on the size of the restrictor*

result [rɪ'zʌlt] **1** *noun* consequence, outcome; *engine oil and cylinder temperature will also increase as a result of higher combustion temperatures* **2** *verb* to result from = to happen as a consequence; *the structural weakness resulted from a minor collision while taxiing two years previously*; to result in = to produce as an effect; *failure to secure seat belts could result in serious injury*

resultant [rɪ'zʌltənt] **1** *adjective* resulting or consequential; *the temperature of the land rises, causing the layer of air in contact with it to warm up and expand with a resultant decrease in density* **2** *noun* one vector that is the equivalent of a set of vectors; *when two or more velocities act simultaneously on a body, the aircraft movement is called the resultant velocity due to the two or more component velocities*

retain [rɪ'teɪn] *verb* to keep or to hold; *retentivity is the ability of a material has to retain magnetism*; *when fuel-dumping,* *sufficient fuel must be retained for landing*

retard [rɪ'tɑːd] *verb* **(a)** to cause to occur later; to delay; *on most modern engines the spark is retarded to top dead centre, to ensure easier starting and prevent kick-back* **(b)** to move (throttles) backwards; *when reducing power, always retard the throttles before reducing RPM (revolutions per minute) with the propeller levers*

retentivity [rɪten'tɪvɪti] *noun (of metal)* ability to remain magnetized after the magnetizing force has gone; *steel has high retentivity, but soft iron has low retentivity*

retract [rɪ'trækt] *verb (undercarriage)* to move back (into); to raise; *mechanically operated sequence valves ensure that the landing gear does not extend until the doors are open and that the landing gear is retracted before the doors close*

retractable [rɪ'træktəbl] *adjective* which can be pulled back or raised; **retractable undercarriage** = undercarriage which can be raised into the fuselage or wings after use; *early aircraft had non-retractable undercarriages*

retraction [rɪ'trækʃn] *noun* the act or instance of pulling back or raising; **retraction of the undercarriage** = raising of the undercarriage into the fuselage after use; *retraction of the undercarriage may be electrical or hydraulic*

return [rɪ'tɜːn] **1** *noun* coming back or going back to a place; *we're waiting for the return of the aircraft*; **radar return** = radar echo **2** *adjective* **return flight** = flight back to the point of departure; **return valve** = valve which allows flow of fluid in both directions **3** *verb* **(a)** (to cause) to come back or to go back to an earlier position or place; *fly from A to B and return; the auto-control will return the ailerons to neutral as the aircraft returns to level flight*

reveal [rɪ'viːl] *verb* to bring to view; *radiographic inspection of the aircraft structure is able to reveal fatigue cracks without the need to dismantle the aircraft*

reversal [rɪ'vɜːsl] *noun* change to the opposite position, direction, or order; *stationary eddies can be hazardous, not only because of the down currents but also because an aircraft encountering the reversal of direction might have its airspeed momentarily reduced below stalling speed*; **thrust reversal** = setting of throttle levers to provide thrust in the opposite direction to decelerate the aircraft after landing

reverse [rɪ'vɜːs] **1** *noun* opposite; *one would expect a unit of humid air to be heavier than a similar unit of dry air but, in fact, the reverse is true* **2** *adjective* which goes backwards or in the opposite direction; **reverse flow** = flow of a fluid in the opposite direction to normal; **reverse panic** = a form of shock which makes passengers unable to comprehend the need for urgency; **reverse thrust** = thrust in the opposite direction to normal in order to decelerate the aircraft after landing **3** *verb* to go backwards or in the opposite direction; **to reverse a vehicle** = to make a vehicle go backwards

reverser [rɪ'vɜːsə] *noun* **thrust reverser** = device to change the direction of thrust so that it operates in the opposite direction to the normal direction; *in many turbo-jet thrust reversers, clamshell doors direct the exhaust gases forward*

reversible [rɪ'vɜːsəbl] *adjective* which can be made to go backwards or to change direction; *a reversible electric motor*; **reversible pitch propeller** = propeller which allows the aircraft to be propelled backwards when taxiing

reversion [rɪ'vɜːʃn] *noun* a return to an earlier condition or state; *in smaller aircraft, reversion to manual control is possible if complete loss of hydraulic power occurs*

revert [rɪ'vɜːt] *verb* to return to an earlier condition or state; *the elevator system has the ability to revert to manual control after a hydraulic failure*

revolution [revə'luːʃn] *noun* rotation or turn about an axis; **a revolution of the earth** = 360° turn of the earth about its axis; **a revolution of the crankshaft** = a 360° turn of the crankshaft; **revolutions per minute (r.p.m. or rpm)** = speed of an engine or the number of rotations of the crankshaft per minute; *rpm is the number of revolutions per minute that the engine crankshaft is making; the actuator control is sensitive to engine rpm*

revolve [rɪ'vɒlv] *verb* to turn about an axis; to rotate; *the earth revolves around the sun*

revolving [rɪ'vɒlvɪŋ] *adjective* **tropical revolving storm** = intense depressions which develop over tropical oceans; *tropical revolving storms originate within 5-15° of the equator; tropical revolving storms generally occur from June to October*

rhumb [rʌm] *noun* one of the points of a compass; **rhumb line** = a regularly curved line on the surface of the earth which cuts all meridians at the same angle; **rhumb line direction** = the average of all the great circle directions between the two points; *because the great circle direction between two points on the surface of the earth is not constant, it is often more convenient to consider the rhumb line direction*

rib [rɪb] *noun* one of many cross pieces of the airframe that provide an aircraft wing with shape and strength; *additional strength is required for the rib sections which are placed in the area of the undercarriage mountings, flaps and power plant attachment point*

rich [rɪtʃ] *adjective* (a fuel/air mixture) in which the ratio of fuel to air is greater than usual; *moving the mixture control lever forward to the rich position increases the amount of fuel mixing with the air*

ridge [rɪdʒ] *noun* (a) long narrow hill with a crest; *the mountain ridge stretches for miles* (b) long zone of relatively high atmospheric pressure; *a ridge of high pressure; on average, the wind backs with the passage of a ridge*

rigging position ['rɪgɪŋ pə'zɪʃn] *noun* attitude of the aircraft in which the lateral axis and (usually) the longitudinal axis are horizontal; *the aircraft was put into the rigging position*

rigid ['rɪdʒɪd] *adjective* unbending, inflexible; **rigid pipes** = pipes which do not bend easily; **a rigid structure** = a firm unbendable structure; *the areas between the ribs are utilized to house fuel tanks which can be either rigid or flexible* (NOTE: the opposite is **flexible**)

rigidity [rɪ'dʒɪdɪti] *noun* inflexibility, stiffness; *extra strength and rigidity must be provided in the tail section for aircraft with a tail wheel unit* (NOTE: the opposite is **flexibility**)

rim [rɪm] *noun* circular outer edge (of wheel, etc.); *creep marks are painted on the tyre and the wheel rim; the rim of the air intake is prone to icing*

rime [raɪm] *noun see* ICE

ring [rɪŋ] *noun* circle; *around the impeller is a ring of stationary vanes called a diffuser ring*

rise [raɪz] **1** *noun* **(a)** increase; *a rise in temperature* **(b) to give rise to** = to cause; *hills and mountains may give rise to particularly severe turbulence* **2** *verb* **(a)** to move upwards; *air rises* **(b)** to increase; *the temperature is rising; see Note at* RAISE

risk [rɪsk] **1** *noun* possibility of suffering harm or injury; danger; *when starting an engine, it is bad practice to pump the throttle lever as there is a risk of fire in the carburettor air intake* **2** *verb* to take a dangerous chance; **to risk the lives of passengers** = to put the lives of passengers in danger by taking a particular course of action

rivet ['rɪvɪt] **1** *noun* type of metal bolt or pin with a head on one end, inserted through one of the aligned holes in the parts to be joined and then compressed on the plain end to form a second head; *tensile or compressive loading makes the joined materials tend to slide and break the rivet* or bolt **2** *verb* to join with rivets; *the skin is riveted to both stringers and frames*

rod [rod] *noun* thin straight piece of metal; *aluminium rods and bars can readily be employed in the high-speed manufacture of parts*; **connecting rod** = engine part which connects the piston to the crankshaft

role [rəʊl] *noun* function; *movement of air plays a major role in the development of weather patterns*; **the role of the aircraft** = the type of operation the aircraft is required to perform

roll [rəʊl] **1** *noun* **(a)** rotation about the longitudinal axis of the aircraft, created by movement of the ailerons; *roll is produced by moving the stick to the left or right*; *see also* BANK **(b)** flight manoeuvre with 360° rotation about the longitudinal axis of the aircraft; *loops and rolls are aerobatic manoeuvres* **(c)** **roll cloud** = cloud created in the rotor zone on the downwind side of mountain ranges **2** *verb* to rotate the aircraft around its longitudinal axis; *move the control column to the left to roll the aircraft to the left*; **to roll into a turn** = to roll or bank the aircraft so that it turns left or right; *by rotating the yoke the ailerons are moved and the aircraft rolls into a turn*

COMMENT: the difference between roll and bank is that roll is movement whereas bank suggests a fixed attitude of the aircraft. Consequently, a turn might be expressed in angles of bank: 'turn at a bank angle of 30°', and the movement to obtain the bank might be expressed as roll: 'roll the aircraft to the left'

roller ['rəʊlə] *noun* cylindrical metal device which rotates; *the most common bearings used in gas turbine engine are the ball or roller type*

root [ruːt] *noun* **(a)** **wing root** = part of the wing where it meets the fuselage **(b)** **the root of the problem** = the cause of the problem **(c)** *(mathematics)* **square root** = divider of a quantity that, when multiplied by itself gives the quantity; *three is the*

square root of nine; *the square root of 64 is 8*

rose [rəʊz] *noun* **compass rose** = the compass card or its marking (of 32 points) on a map; *an arc of the compass scale, or rose, covering 30° on either side of the instantaneous track, is at the upper part of the display*

rotary ['rəʊtəri] *adjective* which rotates; **rotary inverter** = DC (direct current) motor driving an AC (alternating current) generator, the output of which must be regulated to give constant voltage and frequency; **rotary motion** = rotating movement; **rotary wing aircraft** = aircraft with a rotor which provides lift (such as a helicopter)

rotate [rəʊ'teɪt] *verb* to turn around on an axis or centre; *in the event of flame extinction in flight, the engine will continue to rotate, due to the airflow through it caused by the forward speed of the aircraft*; *the aircraft should be rotated to the recommended nose-up attitude for touch down*; *counter-rotating propellers rotate in opposite directions*; *see also* COMMENT *at* ROTATION

rotation [rəʊ'teɪʃn] *noun* **(a)** moving the control yoke or stick aft to raise the nose of an aircraft during the take-off run to facilitate the aircraft becoming airborne; *rotation should begin at about 60 knots* **(b)** turning around an axis or centre; *the rotation of the earth*; *crankshaft rotation*; *the speed of rotation determines the frequency of the generator output*

COMMENT: the aircraft rotates around three axes: **pitch** = rotation around the lateral axis; **roll** = rotation around the longitudinal axis; **yaw** = rotation around the vertical axis

rotational [rəʊ'teɪʃnl] *adjective* which rotates; rotary; *rotational movement of the camshaft*; *the rotational movement of the propeller blades creates lift at right angles to the blade*

rotor ['rəʊtə] *noun* device which turns about an axis or centre; *the rotor blade of a* *compressor*; **helicopter rotor** = two or more rotating blades (known as the main rotor) which provide lift (and thrust) for a helicopter; the smaller tail rotor, rotating in a vertical plane, prevents the helicopter from rotating in the opposite direction to the main rotor

rough [rʌf] *adjective* (a) not smooth; having an irregular surface; **rough air** = turbulent air; **rough running** = referring to a piston engine which is not operating correctly; **rough terrain** = uneven ground (NOTE: the opposite is **smooth**) (b) not fully detailed; **a rough estimate** = an approximate calculation, good enough for a given purpose; **a rough drawing** = a quick drawing usually used to illustrate or explain

roughness ['rʌfnəs] *noun* unevenness (of a surface); *the strength of turbulence near the earth's surface depends largely on the surface temperature, the surface wind, and the roughness of the surface*

round [raʊnd] **1** *adjective* circular; *a round life raft* **2** *adverb* in a circular movement; *the pointer swings round*

route [ruːt *US* 'raʊt] *noun* course of travel; *the purpose of charts is to plan and fly a safe route to a destination*; **escape route** = passengers' way out of an aircraft after an emergency landing; **route flight plan** = detailed information concerning an intended flight, provided to an ATC (Air Traffic Control) facility in written or oral form; **en route** = on the way; during the flight; *en route weather conditions*; *see also* EN ROUTE

routine [ruː'tiːn] **1** *noun* standard procedure; *meteorological information for scheduled flights will be passed to the operations department as a matter of routine* **2** *adjective* standard and regular; **routine servicing** = servicing carried out in the normal way at regular, scheduled intervals

row [raʊ] *noun* (a) series of objects in a line; *each row of rotating rotor blades is followed by a row of stationary stator*

blades (b) series of seats in an aircraft; *there are no empty seats in Row 8*

r.p.m. *or* **rpm** = REVOLUTIONS PER MINUTE

R/T = RADIOTELEPHONY (NOTE: **R/T** is frequently used in spoken language whereas **RTF** is the ICAO abbreviation)

RTF *(ICAO)* = RADIOTELEPHONY

rudder ['rʌdə] *noun* control surface on the fin which rotates the aircraft about its vertical axis to produce yaw; *the A320 retains a backup mechanical linkage for elevator trim and rudder to allow control in the unlikely event of complete electrical failure*; *see also* PEDAL

COMMENT: the rudder does not turn the aircraft. It is used, together with aileron deflection, to initiate turns, to balance forces in turns and to counteract yawing motions created by the propeller during flight. The rudder pedals are mounted on the floor of the cockpit

rule [ruːl] *noun* (a) standard and authoritative instruction or guide; *according to the rules, your ticket must be paid for two weeks in advance*; *a useful rule for the application of variation is - variation east magnetic least, variation west magnetic best*; **as a rule** = usually; *as a general rule, radio signals travel in straight lines*; **rule of thumb** = easily remembered, useful guide to a more complex principle (b) instrument for determining length; ruler; **slide rule** = device consisting of sliding scales for mathematical calculations

run [rʌn] **1** *noun* route; distance; **landing run** = distance on the runway from the touchdown point to the stopping point or taxiing speed; **take-off run** = distance from the start of take-off to the point where the wheels leave the ground; *acceleration forces can be felt as the aircraft begins its take-off run* **2** *verb* (a) to extend; *magnetic lines of force run from the north magnetic pole to the south magnetic pole* (b) to operate an engine; *an engine should be run at low r.p.m. (revolutions per minute)* *after flight to allow engine components to cool to a more uniform temperature*

run up ['rʌn 'ʌp] *noun* **engine run-up** = testing of piston engine at high power, in a light aircraft, just before take-off; *make certain that the parking brake is on before doing engine run-up checks*

runway ['rʌnweɪ] *noun* strip of level, usually paved ground on which aircraft take off and land; *Heathrow airport has four terminals and two main runways; to achieve a safe landing, an aircraft has to be controlled so that its wheels make contact with the runway smoothly; the aircraft lined up perfectly on the runway extended centre line*

runway visual range (RVR) ['rʌnweɪ 'vɪʒjuəl 'reɪnʒ] *noun* distance along a runway at which selected lights can be seen - adjusted to simulate approach visibility; *runway visual range is obtained by an observer standing at the side of the runway in the vicinity of the threshold counting the number of markers or lights visible along the side of the runway*

rupture ['rʌptʃə] **1** *noun* process of breaking open or bursting; *pressure in the fuel tanks must be controlled to prevent rupture or collapse* **2** *verb* to break open or burst; *the impact ruptured the fuel tank*

RVR = RUNWAY VISUAL RANGE

R/W = RUNWAY

RWY = RUNWAY

Ss

S = SOUTH

safe [seɪf] *adjective* free from danger; *approach to land must be made at a safe speed*; **safe landing** = landing which does not endanger people or damage the aircraft; *see also* FAIL SAFE

safeguard ['seɪfgɑːd] **1** *noun* something done as a precaution; *a propeller is feathered after engine failure, or as a safeguard when low oil pressure or excessive temperature have indicated the development of a possible defect* **2** *verb* to take action to make sure that something is protected from harm; *a pressure maintaining valve is generally used to safeguard operation of important services, such as flying controls and wheel brakes*

safety ['seɪfti] *noun* freedom from danger, injury or risk; *turbulence can have serious effects on aircraft safety and performance and makes air travel uncomfortable*; **acceptable level of safety** = good enough standard of safety; **safety awareness** = being familiar with and prepared for any situation in which safety is important; **safety conscious** = being aware at all times of the importance of safety and the means by which it is achieved and maintained; **safety pilot** = pilot present in the cockpit to ensure the safety of the flight (as when a student is practising instrument flying); **safety regulations** = rules or laws which must be followed to make a place safe; *equipment and furnishings of modern jet transports must comply with safety regulations*

SALR = SATURATED ADIABATIC LAPSE RATE

salvage ['sælvɪdʒ] *verb* to save items of property which may be in danger of being lost; *in the event of a crash landing in a remote area on land, an attempt should be made to salvage all items of survival equipment from the wreckage including beacons, rafts and raft equipment*

sample ['sɑːmpl] *noun* small amount which is representative of the whole; *if a sample of fuel taken from a tank was found to be hazy or cloudy in appearance, this would indicate the presence of water in suspension; if fuel contamination by water is suspected, a sample of fuel should be drained from the tank for inspection*

sandwich ['sændwɪʃ] *noun* construction of three layers, the material of the one in the middle being different from the two on each side; *standard connectors consist of a metal coupling with a rubber sandwich joint*

SAR = SPECIAL AERODROME REPORT; *(ICAO)* SEARCH AND RESCUE

satellite ['sætəlaɪt] *noun* object launched to orbit the earth, usually receiving and transmitting signals, pictures and data; *satellite communications improve the effective distribution of world area forecasts*; **satellite navigation (system)** = system of navigation which uses orbiting satellites to determine the position of an aircraft or point, in relation to the earth's surface; *see also* GEOSTATIONARY

satisfactory [sætɪs'fæktri] *adjective* adequate, good enough; *for satisfactory operation, an engine requires an adequate supply of oil*

QUOTE during the engine run-up, check that the use of carburettor heat gives a satisfactory drop in rpm or manifold pressure
Civil Aviation Authority, General Aviation Safety Sense Leaflet

satisfy ['sætɪsfaɪ] *verb* **(a)** to meet a particular prescribed standard; *Shell Avgas 100LL satisfies British specification* **(b)** to meet the needs or requirements of something; *to satisfy the requirements of aviation there are three types of meteorological offices for aviation, each with a specific role to fulfil*

saturate ['sætʃəreɪt] *verb* to cause a substance to combine with the greatest possible amount of another substance; *when a sample of air contains the maximum amount of water vapour for its particular temperature, it is said to be saturated; see also* LAPSE RATE

saturation [sætʃə'reɪʃn] *noun* being filled with the maximum amount of something which can be absorbed (such as a sample of air which contains the maximum amount of water vapour for its temperature); *the various types of fog are classified by the manner in which saturation is reached*; saturation point = level at which no more of a substance can be absorbed; **the moisture in the air reached saturation point and fell as rain** = the air could absorb no more water

save [seɪv] *verb* to prevent unnecessary use of; *electro-magnetic switches are generally used to control high current devices by means of a small current thus saving heavy duty cable and therefore weight*

scale [skeɪl] *noun* **(a)** marks at fixed intervals used as a reference standard in measurement; *this ruler has scales in inches and centimetres* **(b)** graded system of classification; **Beaufort scale** = scale (form 0 - 12) used to refer to the strength of wind; *wind speeds can be estimated by using the Beaufort scale of wind force* **(c)** proportion used in determining distance on charts; *many aeronautical charts use a scale of 1:500,000*

scan [skæn] *verb* **(a)** to look at quickly and systematically; *the pilot is trained to scan the instrument panel* **(b)** to move a radar beam in a systematic pattern in search of a target; *some radars scan in azimuth and glideslope*

scatter ['skætə] *noun* deflection of radiation; *high frequencies are freer of ionospheric scatter and are relatively free of noise*

schedule ['ʃedjuːl US 'skedjuːl] **1** *noun* **(a)** list of times of departures and arrivals; timetable; *an airline schedule* **(b)** printed or written list of items in the form of a table; *inspection schedule or maintenance schedule* **2** *verb* **(a)** to plan for a certain time or date; *the meeting is scheduled for 3 o'clock* **(b)** to enter on a schedule; *calculate and schedule each item on the proper form*

scheduled ['ʃedjuːld US 'skedjuːld] *adjective* **scheduled flights** = flights - as opposed to charter flights - listed in the airline timetable; **scheduled landing** = an arrival at a timetabled destination

schematic [skiː'mætɪk] *adjective (of diagram)* which shows the function of a device or system without trying to create a realistic image; *figure 3 shows a schematic diagram of the autopilot*

screen [skriːn] *noun* surface of a TV or computer monitor on which the image is seen; *the airborne weather radar (AWR) allows the range of cloud to be estimated from range markers displayed on the screen*; **radar screen** = CRT (cathode ray tube) screen on which radar information is displayed

screw [skruː] *noun* type of threaded connector used to fix things together by rotating it; **bleed screw** = small screw in highest point of a hydraulic system to allow for the removal of air or vapour; **screw jack or screwjack** = lifting device working with rotary input

sea [si:] *noun* **(a)** body of salt water between land masses; *Swissair flight 111 crashed into the sea*; **sea level** = the average level of the surface of the sea, used for measuring barometric pressure; **mean sea level** = average level of the sea taking tidal variations into account; *altitude is the vertical distance between an aircraft - or a point or a level - and mean sea level* **(b)** particular area of the body of salt water; *the North Sea*; *the South China Sea*; *see also* OCEAN

seaboard ['si:bo:d] *noun US* coast; seashore; *the eastern seaboard of the USA*

seal [si:l] **1** *noun* **(a)** device that joins two parts and prevents leakage; *an oil seal reduces the clearance between the rotating and static members*; **dynamic seal** = seal which is part of a moving component (eg in a hydraulic system); *dynamic seals require lubrication to remain effective*; **static seal** = seal which is part of a non-moving component; *static seals, gaskets and packing are used in many locations* **(b)** way in which a liquid or a gas may be prevented from escaping; *static seals, gaskets and packing effect a seal by being squeezed between two surfaces* **2** *verb* to join two parts in such a way as to prevent leakage; *in pressurized aircraft, bulkheads are provided at the front and rear ends of the fuselage to seal off the crew compartment and the passenger cabin*

sealant ['si:lənt] *noun* substance painted or sprayed onto a surface to prevent the escape of a liquid or gas; *the integral fuel tank may be completely coated on the inside with a layer of sealant*

sealing compound ['si:lɪŋ 'kɒmpaʊnd] *noun* = SEALANT

search [sɜːtʃ] **1** *noun* act of looking for, in order to find; *the aircraft reduced altitude and carried out a visual search for survivors* **2** *verb* to look for in order to find something; *the investigators searched the scene of the crash for the flight data recorder*

season ['si:zn] *noun* one of the four natural divisions of the year - spring, summer, autumn (or US: fall), and winter; *the amount of solar radiation received by the earth depends on the season*

seasonal ['si:znl] *adjective* **(a)** referring to the four natural divisions of the year; characteristic of a particular time of the year; *seasonal temperatures*; *seasonal winds*; **seasonal variation** = change occurring according to the season **(b)** which only lasts for a season

seat [si:t] *noun* **(a)** place for sitting; *pilot's seat*; *(for passengers)* **aisle seat** = seat which is by an aisle; **window seat** = seat next to a window **(b)** **valve seat** = metal rim on which the valve rests when it is closed; *the cylinder forms the combustion chamber, provides a working bore for the piston, and also has apertures for spark plugs, valve guides and valve seats*

seated ['si:tɪd] *adjective* sitting; on your seat; *passengers should remain seated*; *see also* SIT

secondary ['sekəndri] *adjective* **(a)** of the second rank in importance, etc., not primary; **secondary radar** = radar system in which the active target replies to the interrogation unit **(b)** *(electricity)* an induced current that is generated by a primary source

secondary surveillance radar (SSR) ['sekəndri sə'veɪləns 'reɪdɑː] *noun* radar which uses ground equipment called interrogators and airborne equipment called transponders to identify aircraft, determine altitude and range, etc.; *secondary surveillance radar (SSR) is normally used to supplement data from primary systems*

section ['sekʃn] *noun* **(a)** component or part of a structure; *tail section and nose section of the aircraft*; *the non-smoking section of the aircraft* **(b)** part of a text; *the book is divided into four sections - the first* *four chapters form the first section* (c) diagram of a solid object as it would appear if cut, so that the internal structure is displayed; cross-section; *see also* CROSS-SECTION

sectional ['sekʃnl] *adjective* (a) referring to a section; composed of sections (b) cross-sectional; *see also* CROSS-SECTIONAL

sector ['sektə] *noun* (a) part of the flight between an aircraft moving under its own power until it next stops after landing in its allocated parking position; *on certain sectors, because of fuel costs at the destination, it can be economical to carry excess fuel* (b) portion of a circle inside two radii and the included arc

secure [sɪ'kjʊə] 1 *adjective* fastened or locked; safe; *overhead baggage lockers must be secure* 2 *verb* to attach firmly, to fasten; to make safe; *if the onset of turbulence is sudden, crew must immediately secure themselves in the nearest available seats*

security [sɪ'kjʊərəti] *noun* safety, protection (against unauthorized person); **airline security area** = area in which measures are taken by an airline to ensure the safety of people and property; **airport security officer** = person employed by an airport authority to check people (passengers) and baggage for illegal substances or devices (eg drugs, guns, etc.)

seldom ['seldəm] *adverb* not often, rarely; *aircraft are seldom hit by lightning; the wet sump system of lubrication is seldom used on modern aircraft*

select [sə'lekt] *verb* to choose (such as a particular instrument or system setting); *a reverse thrust lever in the crew compartment is used to select reverse thrust; the cabin pressure controller is used to select cabin altitude*

selection [sə'ləkʃn] *noun* (a) choice (such as of a particular instrument or system setting); *by manual selection of the heating switch, the formed ice can be*

dispersed (b) collection of carefully chosen things; *a selection of photographs*

selector [sə'lektə] *noun* manually operated device like a switch, which offers a choice of settings; *turn the selector control; the purpose of this selector is to direct fluid to the appropriate side of an actuator*

self-contained ['selfkən'teɪnd] *adjective* independent; which constitute a whole; *the auxiliary power unit is a self-contained unit*

semi- ['semi] *prefix meaning* half

semicircle ['semisɜ:kl] *noun* half a circle; *most mathematical protractors are made of plastic in the shape of a semicircle*

semicircular [semi'sɜ:kjʊlə] *adjective* in the shape of half a circle; *most mathematical protractors are semicircular in shape*

semiconductor [semikən'dʌktə] *noun* solid crystalline substance with electrical conductivity greater than insulators but less than good conductors; *semiconductor material is used to make many electronic devices*

senior ['si:niə] *adjective* more important in rank; older; *senior cabin supervisor*

sense [sens] 1 *noun* (a) way; *after turning the aircraft, the auto-control will operate in the opposite sense and return the ailerons to neutral as the aircraft returns to level flight* (b) any of the physiological means by which we experience our surroundings: sight, hearing, smell, taste and touch; *when flying in cloud, a pilot must rely on his instruments and not on his senses* (c) wisdom or natural intelligence; *he has a lot of (common) sense* (d) meaning (of a word); *the word 'bearing' is used in a lot of different senses* 2 *verb* to detect automatically; *the fire warning system is designed to sense two levels of temperature - overheat and fire; see also* SENSOR

sensitive ['sensətɪv] *adjective* able to register very small differences or changes in conditions; *oscillating outputs from the alternators could cause sensitive equipment to malfunction or trip off; the actuator is sensitive to engine rpm*

sensitivity [sensə'tɪvəti] *noun* quality or state of being able to register very small differences or changes in conditions; *monitors detect disturbances which are below the sensitivity level of the gyros*

sensor ['sensə] *noun* device which receives and responds to a signal or stimulus; *pressure sensor; temperature sensor; the inlet pressure is sensed by a single pitot-type sensor probe which is situated just in front of the compressor*

separate 1 ['seprət] *adjective* existing as an independent thing; *propellers consist of a number of separate blades mounted in a hub* **2** ['sepəreɪt] *verb* to set or keep apart; *dry chemical extinguishants separate the oxygen element from the fire thus retarding combustion*

separation [sepə'reɪʃn] *noun* **(a)** condition of being spaced apart; *airways and advisory routes provide a high degree of safety by maintaining separation between aircraft* **(b)** removal from a mixture or combination; *the oil and air mixture flows over the de-aerator tray in the oil tank, where partial separation takes place*

separator ['sepəreɪtə] *noun* device which removes something from a mixture or combination; *the water separator will extract a percentage of free moisture from the air*

sequence ['siːkwəns] *noun* series of things or events which follow one another; order; *the ignition system provides a rapid series of sparks timed to fire in each cylinder in the correct sequence; sequence valves* = fluid flow controller which performs a number of actions in a particular order; *sequence valves are often fitted in a landing gear circuit to ensure correct operation of the landing gear doors and actuators*

series ['sɪəriːz] *noun* a number of things or events which come one after the other in a particular order; *a series of photographs; a series of switches;* **series circuit** = electric circuit connected so that current passes through each component of the circuit in turn without branching

serious ['sɪəriəs] *adjective* important; not slight, not insignificant; giving cause for great concern or worry; **serious damage** = very bad damage; **serious injury** = very bad injury

serve [sɜːv] *verb* **(a)** to act or to function as; *in some aircraft, pressure gauges also serve as a maintenance check on leakage* **(b)** to be used for a purpose; *different colour-coded warning lights serve to alert the observer that something is wrong with the system*

QUOTE a recent incident in Argentina serves to highlight some of the many safety problems in Latin America
INTER PILOT

service ['sɜːvɪs] **1** *noun* **(a)** facility; *a pressure reducing valve is often used to reduce main system pressure to a value suitable for operation of a service such as the wheel brakes* **(b)** work done for others as a profession; *cabin crew provide a commercial service to passengers; automatic terminal information service (ATIS)* **(c)** maintenance or repairs carried out; **service area** = area where maintenance and repairs are carried out; **service bay** = space in the structure of an aeroplane where equipment can be located for maintenance or repairs; *in most modern aircraft a number of the major components are grouped together in a hydraulic service bay which is easily accessible for routine servicing operations* **2** *verb* to do maintenance or repairs; *jet engines are simpler to dismantle and service than piston engines*

serviceability [sɜːvɪsə'bɪləti] *noun* ability to function as required; *when carrying out engine checks, it is usual to turn off the magnetos in turn to check their serviceability*

serviceable ['sɜːvɪsəbl] *adjective* able to function as required; *the pilot must make sure that the radio equipment is serviceable prior to take-off*

servicing ['sɜːvɪsɪŋ] *noun* carrying out maintenance and repairs; *accessibility of components and equipment during servicing enables work to be done more quickly*

servo ['sɜːvəʊ] = SERVOMECHANISM

servo-assisted ['sɜːvəʊə'sɪstɪd] *adjective* partially operated by a servomechanism; *servo-assisted brakes*; *servo-assisted steering*

servo-control unit ['sɜːvəʊkən'trəʊl 'juːnɪt] *noun* the unit - a combined selector valve and actuator - which moves a control surface; *a servo-control unit is part of the system which relieves the effects of aerodynamic forces on the flight controls*

servomechanism ['sɜːvəʊ'mekənɪzm] *noun* device to convert input forces into much larger output forces; *two phase motors are normally used for very small or miniature motors in servomechanisms*

set [set] 1 *noun* group of things which belong together; *a set of instruments*; *a set of figures* 2 *adjective* fixed or established; *a set procedure* 3 *verb* (a) to adjust to a particular point or figure; *the aircraft receiver is set to the required frequency* (b) to put in a particular position; *set the throttle fully closed* (c) to harden; *the resin sets*; *cold setting materials* = (resinous, etc.) materials which do not need heat to harden (NOTE: **setting - set**)

setting ['setɪŋ] *noun* (a) particular figure or position which a device is adjusted to; **altimeter setting** = adjustment of the sub-scale of the altimeter to read QFE, QNH, etc.; **standard pressure setting** = 1013.25 millibars (mb); **throttle setting** = particular position of the throttle which gives a required r.p.m. (revolutions per minute) or power (b) action of adjusting a device to a particular position, etc.; *the*

setting of the altimeter is done prior to take-off

settle ['setl] *verb* to move into a final position; *when wheels are first fitted to an aircraft, the tyres tend to move slightly as they settle down on the rims*

several ['sevrəl] *adjective* a number of but not many; more than a few; *there are several types of instrument landing systems (ILS) in use*; **several minutes** = a number of minutes

severe [sə'vɪə] *adjective* extreme or intense; **severe icing** = bad icing; **severe turbulence** = violent turbulence (NOTE: generally speaking, weather conditions can be described as light, moderate or severe, depending on the amount or intensity of the condition)

severity [sə'verəti] *noun* amount, intensity or seriousness of a condition; *when the wind is strong the vertical currents become quite vigorous with the resultant increase in the severity of turbulence*

shade [ʃeɪd] *noun* (a) intensity or richness of colour; *shades of colour of the landscape become lighter in misty conditions* (b) cover or shelter from the sun; *surface air temperature is the temperature recorded in the shade at a height just above ground level*

shadow ['ʃædəʊ] *noun* area which is not affected by full radiation because of partial or full blocking of rays by something between the area and the source of the radiation; *solar radiation does not exist at night when the rotation of the earth creates a shadow zone from the sun*; *line-of-sight transmission path means that obstacles and terrain can create shadow zones*

shaft [ʃɑːft] *noun* long, generally cylindrical bar, especially one that rotates and transmits power; *engine drive shaft*; *propeller shaft*

shaker ['ʃeɪkə] *noun* device which shakes or vibrates violently; *large aircraft* *use a stick shaker to supplement the natural stall warning of buffet*

shallow ['ʃæləʊ] *adjective* not deep; **shallow depression** = area of slightly lower atmospheric pressure; **shallow angle** = small angle

shape [ʃeɪp] *noun* form; *the shape of an aircraft is determined by the requirement to provide an aerodynamic lift force great enough to support the weight of the aircraft and payload whilst in flight*

sharp [ʃɑːp] *adjective* (a) thin and capable of cutting or piercing; *if a piece of thermosetting plastic is hit hard enough, it breaks into pieces with straight sharp edges* (b) *(signal)* clear; *the sharp setting means the bandwidth is reduced to 1kHz (kilohertz) to minimize noise or interference* (c) *(image)* clear and distinct; *cumulus clouds have sharp outlines* (d) sudden and acute; **a sharp increase** = a sudden large increase

sharpening ['ʃɑːpnɪŋ] *noun* **beam sharpening** = making a radio or light beam narrower; *any system employing beam sharpening is vulnerable to side lobe generation at the transmitter*

shatter ['ʃætə] *verb* to break into a number of pieces when hit; *clear ice is hard to shatter and break off*

shear [ʃɪə] *verb* to break by lateral movement; **shear stress** = stress that resists the force tending to cause one layer of a material to slide over an adjacent layer as in riveted and bolted joints; **wind shear** = change in wind direction and speed between slightly different altitudes; *wind shear, if strong enough can produce clear air turbulence*

shed [ʃed] *verb (weight)* to get rid of; *non-essential loads may need to be shed in order to reduce weight*

sheet [ʃiːt] *noun* (a) large, thin, flat piece of material; *aluminium sheet* (b) relatively large piece of paper; **instruction sheet** = piece of paper on which special instructions are written or printed; **load**

sheet = detailed list of cargo on a flight; load manifest

shield [ʃiːld] **1** *noun* protective covering; *heat shield* **2** *verb* to protect by covering; *the beacon should be sited on the highest ground to prevent the transmitted signal from being shielded*

shift [ʃɪft] **1** *noun* (a) movement from one place to another; *a shift in position* (b) change; *when a radio transmission is made from a moving platform, there will be a shift in frequency between the transmitted and intercepted radio signals*; **datum shift trim system** = trim system which varies the incidence of an all-moving tailplane without moving the cockpit controls; *in some aircraft, the datum shift is operated automatically* **2** *verb* to change the position of something; *to shift a load*

shock [ʃɒk] *noun* (a) sudden violent impact; *on all undercarriages some form of accepting the shock of landing must be included*; **shock absorber** = device to minimize the shock to the main structure of the aircraft when it lands; **shock wave** = compression wave caused by supersonic motion; *as sonic speed is approached, the efficiency of the intake begins to fall, because of the formation of shock waves at the intake lip* (b) disturbance of mental functions caused by a terrible experience or injury; *crew should be aware of reverse panic, a form of shock which makes passengers unable to comprehend the need for urgency*

shore [ʃɔː] *noun* land at the edge of the sea or a lake, etc.; *at a height of 3,000 feet it was possible to see the shore*; *see also* OFFSHORE; ONSHORE

shorten ['ʃɔːtn] *verb* to make short or shorter in length or duration; *mishandling of aero-engines during operation can cause considerable damage and wear which can shorten the life of the engine*; *the length of the mercury column shortens when cooled* (NOTE: the opposite is lengthen)

short-haul ['ʃɔːt'hɔːl] *adjective* short-haul **flight** = flight over a short distance (up to 1,000km); *on short-haul flights, a policy of non smoking has been adopted by some airlines*

shot [ʃɒt] *noun* discharge; *extinguishing of a fire in an auxiliary power unit (APU) compartment is normally done by a single-shot fire extinguisher*

shower ['ʃaʊə] *noun* short period of rain or snow; *showers are forecast for the evening; snow showers are expected in the area*

shunt [ʃʌnt] *noun* low-resistance connection between two points in an electric circuit that forms an alternative path for a portion of the current, also called a bypass; *the shunt-wound generator, used in conjunction with a voltage regulator, is the most common type of DC (direct current) generator system for aircraft*

shutter ['ʃʌtə] *noun* hinged door which control flow of air; *oil cooler shutters; radiator shutters*

sight [saɪt] **1** *noun* something seen; view; *the fog cleared and the mountain came into sight;* sight **glass** = simple fluid-level gauge **(b)** with **the airfield in sight** = transmission to ATC (air traffic control) to confirm that the pilot can see the landing airfield **(c)** line of **sight** (LOS) = a clear path between sending and receiving antennas **2** *verb* to see a long way away; *sea marker dyes can only be used once and should only be used when a search aircraft is sighted*

SIGMET = SIGNIFICANT METEOROLOGICAL INFORMATION

sign [saɪn] **1** *noun* **(a)** small quantity or amount of a something which may suggest the existence of a much larger quantity; *any sign of smoke or fire outside a wing exit makes it unusable* **(b)** display with letters and/or numbers, sometimes lit up; *the 'fasten seat belt' sign; 'no-smoking' sign* **(c)** *(mathematics)* symbol (such as: -, +, x, ÷) which represents an operation **2**

verb to put one's signature on a document, a letter, etc.; *remember to sign the letter*

signal ['sɪgnl] *noun* **(a)** device, action or sound which passes information; **signals area** = area on an aerodrome used for displaying ground signals; **signals mast** = vertical pole on an airfield from which signal flags are flown; **hand signals** *or* **marshalling signals** = signal used by the marshaller; *see also* MARSHAL; MARSHALLER **(b)** radio wave transmitted or received; *as a general rule, radio signals travel in straight lines*

signature ['sɪgnətʃə] *noun* name of a person written (in a special way) to show that a document has been authorized or to show who is the author of a letter, etc.; *look at the signature to see who wrote the letter*

significance [sɪg'nɪfɪkəns] *noun* importance; *except near a coastline where the sea breeze may augment the upslope motion, anabatic winds are of little significance*

significant [sɪg'nɪfɪkənt] *adjective* important (therefore noticeable); *a significant change in temperature; the vertical currents and eddies formed by the flow of air over hills and mountains have a significant effect on aircraft encountering them*

significant meteorological information (SIGMET) [sɪg'nɪfɪkənt miːtʃərə'lɒdʒɪkl ɪnfə'meɪʃn] *noun* weather advisory concerning weather conditions important to the safety of all aircraft, such as severe or extreme turbulence

significant weather chart [sɪg'nɪfɪkənt 'weðə 'tʃɑːt] *noun* weather chart with important weather information marked on it

signify ['sɪgnɪfaɪ] *verb* to indicate, to suggest, to mean; *buffet signifies the approach of a stall*

silence ['saɪləns] **1** *noun* absence of sound; **total silence** = the complete absence of sound **2** *verb* to make quiet; to stop (something) making a noise; *when an* engine fire warning is received on the flight deck, the first action should be to silence the warning bell

silencer ['saɪlənsə] *noun (on engine exhaust, air duct, gun, etc.)* device to reduce the noise; *in order to reduce the level of noise from the blower, silencers are incorporated in the main supply ducting*

similar ['sɪmlə] *adjective* nearly the same (as); *turbo-shaft engines are similar to turboprop engines*

similarity [sɪmɪ'lærətɪ] *noun* being similar; having features nearly the same (as); *there are points of difference and similarity between the two aircraft*

simple ['sɪmpl] *adjective* **(a)** basic, not complex; *a simple fuel system consists of a gravity feed tank, a filter, a shut-off valve and pipes* **(b)** easy; *a simple mathematical problem* = a mathematical problem which is not difficult

simplicity [sɪm'plɪsətɪ] *noun* having a basic, uncomplicated design or concept; *because of its lightness, cheapness and simplicity, a fixed pitch propeller is often fitted to single-engine aircraft*

simplify ['sɪmplɪfaɪ] *verb* to make easy, to make less complex or complicated; *repair procedures are being further simplified by increasing use of cold setting resins*

simulate ['sɪmjuleɪt] *verb* to imitate the conditions or behaviour of something; *the computer program simulates the action of an aircraft*; simulated instrument flight = instrument flight carried out in a simulator on the ground or in a specially prepared aircraft with screens on the windows

simulation [sɪmju'leɪʃn] *noun* imitation of a real situation, created often for training purposes; *a simulation of an engine fire*; *the computer animation showed a simulation of the events which followed the explosion on board the aircraft*

simulator ['sɪmjuleɪtə] *noun* flight simulator = device or computer program

which allows a user to pilot an aircraft, showing a realistic control panel and moving scenes, used as training programme

simultaneous [sɪməl'teɪnɪəs] *adjective* happening at the same time; *most aircraft are now fitted with remote magnetic indicator displays which can be selected to show two simultaneous bearings from different radio navaids*

single ['sɪŋgl] *adjective* one only; a single-engined *or* single-engine aircraft = an aircraft with one engine only

sink [sɪŋk] **1** *noun* downdraft of air; rate of sink = rate of descent of glider, etc.; *in order to achieve a safe landing, a glider has to be controlled so that it makes contact with the runway smoothly at a very low rate of sink* **2** *verb* to move downwards as in a fluid; *if water enters the fuel tank, it will sink to the bottom of the tank where it can be drained off*

sit [sɪt] *verb* to be resting with your behind on a seat (such as a chair); *the pilot sits in the cockpit* (NOTE: **sitting - sat**)

site [saɪt] **1** *noun* selected area of land; *landing site* **2** *verb* to position or to put in a particular place; *where it is impossible or inadvisable to site the localizer antenna on the runway centreline, it may be positioned to one side*

sitting ['sɪtɪŋ] *adjective* sitting position = position of a person who is on a seat; *the correct technique of using the escape slides is to assume a sitting position*

situate ['sɪtʃueɪt] *verb* to put in a particular place, to locate; *the inlet pressure is sensed by a single pitot-type probe which is situated just in front of the compressor*

situation [sɪtʃu'eɪʃn] *noun* **(a)** location, place where something is; *the situation of the flight controls is important* **(b)** conditions; circumstances; *the synoptic chart is a graphical representation of the general weather situation over a given area at a given time*

sixty ['sɪksti] number 60; *see also* ONE-IN-SIXTY RULE

size [saɪz] *noun* extent of a thing; how big something is; *whether or not an object can be seen by aircrew at a given distance will depend on factors such as size, shape and colour of the object*

skid [skɪd] 1 *noun* (a) slide on slippery ground; *anti-skid braking systems units are designed to prevent the brakes locking the wheels during landing, thus reducing the possibility of wheel skid* (b) movement of the aircraft away from the centre of a turn; *to correct a skid, use the ailerons to increase the bank, or apply rudder pressure on the same side as the deflected ball in the turn coordinator; deflection of the ball in the turn coordinator indicates a slip or a skid; see also* ANTI-SKID 2 *verb* to slide on slippery ground; *if you brake too hard on a wet surface, you might skid*; **to skid to a halt** = to slide *or* skid until you stop (NOTE: **skidding - skidded**)

skill [skɪl] *noun* expertise, excellent ability in something; *skill in accurate flying can only be achieved by constant practice*

skin [skɪn] *noun* outer layer of a body; outer layer of an aircraft; *the aircraft skin is riveted to stringers and frames*

skip [skɪp] *noun* **skip distance** = shortest distance at which a sky wave can be received; *the higher the layer in which a direct wave signal is totally refracted and returns as a sky wave, the greater the skip distance*

sky [skaɪ] *noun* atmosphere and outer space as seen from the earth; *the higher the sun is in the sky, the more intense is the radiation per unit area*; **sky wave** = that part of a radiated wave which is returned to earth by refraction from the ionosphere

slack [slæk] *adjective* (a) not tight; a **slack cable** = a loose cable (b) not busy; *early afternoon is a slack period of the day* (c) *(of isobars)* widely spaced; *throughout the tropics and sub-tropics, where pressure gradients are normally slack, the*

sea breeze is a regular feature; land and sea breezes occur in coastal areas when there is a slack pressure gradient

slant [slɑːnt] 1 *noun* slope or inclination; *Distance Measuring Equipment (DME) is a radio aid which measures aircraft slant range to a ground beacon* 2 *verb* to slope; *the wing slants upwards from the root to the tip*

slat [slæt] *noun* movable device on the leading edge of a wing which, when extended, creates a gap that allows air to pass smoothly over the top of the wing thus reducing the possibility of a stall; *the Socata Rallye is one of the few light aircraft with leading edge slats*

sleet [sliːt] 1 *noun* (a) *GB* melting snow or a mixture of rain and snow falling together (b) *US* frozen rain in the form of clear drops of ice or glaze ice covering surface objects (NOTE: care should be taken to avoid any ambiguity) 2 *verb (of precipitation)* to fall in the form of sleet; *it is sleeting*

slide [slaɪd] 1 *noun* (a) device which allows continuous movement over a smooth surface; **escape slide** = device which allows passengers to safely exit an aircraft in an emergency; **slide raft** = escape slide which can also be used as a raft in the event of a ditching (b) **slide rule** = graduated device with sliding parts for performing complex mathematical operations; **circular slide rule** = calculating device on which all manner of conversions and complex calculations can be made to assist in flight planning 2 *verb* to move continuously over a smooth surface; *shear stress is the stress that resists the force tending to cause one layer of a material to slide over an adjacent layer* (NOTE: **sliding - slid**)

slight [slaɪt] *adjective* small; minor; a **slight increase** = a small increase; a **slight drop in temperature** = a small decrease in temperature

slip [slɪp] 1 *noun* (a) condition of uncoordinated flight when the aircraft moves towards the inside of a turn;

deflection of the ball in the turn coordinator indicates a slip or a skid; slip is indicated by deflection of the ball in the turn and slip indicator **(b)** **blade slip** = loss of propulsive power from a propeller caused by the difference between geometric and effective pitch **(c)** **slip ring** *or* slipring = conducting ring rotating with rotor of electrical machine; *slip rings and brushes are fitted on the rear of the propeller hub and engine front casing* **2** *verb* to move sideways towards the inside of a turning manoeuvre as a result of excessive bank (NOTE: **slipping - slipped**)

> COMMENT: to correct a slip, the pilot should decrease the bank, or increase rudder pressure on the same side as the deflected ball in the turn coordinator. Slips are often used in aircraft with no flaps to increase the rate of descent without increasing the airspeed

slippery ['slıpri] *adjective* which is difficult to grip firmly because of wetness, smoothness, etc.; *a slippery surface such as a wet or snow-covered runway*

slipring ['slıprıŋ] *see* SLIP

slope [sləup] **1** *noun* **(a)** slanting surface or slanting piece of ground; an incline; *a slope of the runway may increase or decrease the take-off and landing runs* **(b)** state in which one end of an aircraft is higher than the other; **glide slope** *or* glideslope = part of the ILS (Instrument Landing System) which provides a radio beam at an angle of approximately 3° to the point of touchdown from the outer marker thus giving the pilot information about the height of the aircraft on final approach; **Visual Approach Slope Indicator** (VASI) = arrangement of red and white lights each side of the runway touchdown point to give the pilot information about his height on final approach **2** *verb* to be inclined, to be at an angle; *when the runway slopes upwards, away from the aircraft, the approach may appear to be higher than it actually is*

slot [slɒt] *noun* **(a)** groove or channel into which something can be fitted; *the float*

engages with a slot cut in the tube, so that, as the fuel level changes, the float moves up and down **(b)** particular time at which an aircraft is scheduled to depart; *flight GF 506 missed its slot and will have to wait 45 minutes for another*

sm = STATUTE MILE

smog [smɒg] *noun* mixture of smoke and fog; *smog is now rare because of pollution control*

smoke [sməuk] **1** *noun* white, grey or black product formed of small particles given off by something which is burning; *the weather associated with visibility reductions by particles suspended in the atmosphere is classified either as fog, mist, haze or smoke*; **smoke alarm** = warning system that will ring or light up if there is smoke somewhere; *washrooms are fitted with smoke alarms* **2** *verb* **(a)** to give off smoke; *someone noticed that one of the engines was smoking* **(b)** to breathe in smoke from a cigarette, cigar, etc.; *passengers are not allowed to smoke in the toilets*

smoking ['sməukıŋ] *noun* action of breathing in smoke from a cigarette, cigar, etc.; **the airline has a no-smoking policy** = the airline does not allow passengers to smoke during a flight ≠

smooth [smu:ð] *adjective* **(a)** even and without lumps or dents; *a smooth surface* **(b)** not rough or turbulent; *high ground will disturb the smooth, horizontal flow of air*; **a smooth running engine** = an engine which is operating well (NOTE: **the opposite is rough**)

snow [snəu] *noun* atmospheric water vapour frozen into ice crystals and falling to earth as white flakes; *snow cover tends to persist on north-facing slopes of mountainous regions after it has melted on south-facing slopes*; **snow plough** = vehicle built to push the snow from roads, tarmac, etc.

snowfall ['snəufɔ:l] *noun* quantity of snow which comes down at any one time; *a heavy snowfall*

snowflake ['snəʊfleɪk] *noun* small piece of snow formed from a number of ice crystals; *the size of a snowflake depends on the temperature*

snowstorm ['snəʊstɔːm] *noun* heavy fall of snow accompanied by wind; *the airport is closed because of the snowstorm*

soft [sɒft] *adjective* not hard; *thermoplastic materials become soft when heated*

soften ['sɒfn] *verb* to make soft; *thermoplastic materials are softened by many aircraft fluids*

solar ['səʊlə] *adjective* referring to the sun; **solar radiation** = total electromagnetic radiation given off by the sun; **solar system** = the sun and the planets, etc., whose motion the sun controls

solar-powered ['səʊlə'paʊəd] *adjective* powered by energy derived from the suns rays

QUOTE a 210-240-foot wingspan solar-powered aircraft for flight at 100,000 feet, is being designed in California
Pilot

sole [səʊl] *adjective* only; *the sole survivor of the air crash*

solenoid ['sɒlənɔɪd] *noun* cylindrical coil of wire acting as a magnet when carrying electric current; *fuel is metered from the aircraft fuel system by a solenoid-operated control valve*

solid ['sɒlɪd] 1 *adjective* (a) referring to something which is not liquid or gaseous; *visibility is reduced by the presence of solid particles, such as dust or sand, in the atmosphere* (b) **solid line** = unbroken line 2 *noun* substance which is not a liquid or a gas; *ice is a solid, water is a liquid and vapour is a gas*

solid-state ['sɒlɪd'steɪt] *adjective* referring to semiconductor devices; **solid-state device** = electronic device that operates by using the effects of electrical or magnetic signals in a solid semiconductor material; **solid-state technology** =

technology using the electronic properties of solids to replace those of valves

solo ['səʊləʊ] *adverb* done by one person alone; *to go solo or to fly solo; he flew solo across the Atlantic*

solution [sə'luːʃn] *noun* (a) act or means of solving a problem or difficulty; answer; *the navigation computer or slide rule is suitable for the solution of many different types of mathematical problem* (b) change of a solid or gas into a liquid by dissolving in water or some other liquid; *spillage from a lead acid battery may be neutralized by washing with a dilute solution of sodium bicarbonate*

solve [sɒlv] *verb* to find the answer to, or a way of removing, a difficulty or problem; *the triangle of velocities is used to solve navigation problems*

somewhat ['sʌmwɒt] *adverb* to some extent, a bit; *the usefulness of pure aluminium as a structural material is somewhat limited*

sonic ['sɒnɪk] *adjective* (i) referring to sound; (ii) within the human hearing range; **sonic boom** = noise, due to shock waves, produced when an aircraft travels through the air faster than the speed of sound; **sonic speed** = the speed of sound

sophisticated [sə'fɪstɪkeɪtɪd] *adjective* highly developed and complex; *the electronic flight instrument system, commonly known as EFIS, is a highly sophisticated type of flight director system; the A340 is a sophisticated aeroplane*

sortie ['sɔːti] *noun* an operational flight by one aircraft; *the test programme has accumulated 1,146 sorties*

sound [saʊnd] 1 *adjective* strong; *a stressed skin structure is used on modern aircraft which gives a sound structure with relatively low weight* 2 *noun* noise; something that can be heard and is caused by vibration of the surrounding air; *FM (frequency modulation) gives a wide range of sounds or a very high data rate* 3 *verb* (a) to make a noise; *if the trim* *position is incorrect, a warning horn will sound when number three thrust lever is advanced for take off; see also* SONIC **(b)** to seem; *it sounds as if the pilot is having trouble*

source [sɔːs] *noun* supply; *under emergency conditions, the battery may be the only source of electrical power; jet aircraft have a ready source of compressed air from the compressor sections of their engines*

south [sauθ] **1** *noun* compass point on the mariner's compass 180° clockwise from due north and directly opposite north; *fly towards the south;* **south facing mountain side** = face of a mountain which looks towards the south **2** *adjective* **(a)** referring to areas or regions lying in the south; referring to the compass point 180° from north; *the south side of the river* **(b)** a **south wind** = a wind blowing from or coming from the south (NOTE: a wind is named after the direction it comes from) **(c)** southern part of a region or country; *South America; South Dakota* **3** *adverb* towards the south; *the aircraft is flying south*

southbound ['sauθbaund] *adjective* travelling towards the south; *a southbound flight*

south-east ['sauθ'iːst] **1** *noun* direction between south and east; *a region in the south-east of Canada* **2** *adjective* **(a)** situated in the south-east; *the south-east coast of England* **(b)** *(wind)* which blows from or which comes from the south-east; south-easterly (wind) **3** *adverb* towards the south-east; *we were heading south-east*

south-easterly ['sauθ'iːstəli] *adjective* **(a)** *(wind)* which blows from or which comes from the south-east; *a south-easterly wind* **(b)** *(direction)* which moves towards the south-east; *we were following a south-easterly direction*

south-eastern ['sauθ'iːstən] *adjective* referring to the south-east; situated in the south-east; *the south-eastern coast of Spain*

southerly ['sʌðəli] **1** *adjective* **(a)** situated towards the south; *the most southerly point of a country* **(b)** (wind) coming from the south; *a southerly wind was blowing* **(c)** *(direction)* moving to or towards the south; *we were flying in a southerly direction* **2** *noun* wind which blows from the south

southern ['sʌðən] *adjective* situated in the south; *the southern hemisphere; the southern Atlantic*

South Pole ['sauθ 'pəul] *noun* point which is furthest south on the earth; *to fly over the South Pole*

southward ['sauθwəd] **1** *adjective* going towards the south; *to go in a southward direction;* **a southward flight** = a flight heading towards the south **2** *adverb* US = SOUTHWARDS

southwards ['sauθwədz] *adverb* towards the south; *the aircraft was flying southwards*

south-west ['sauθ'west] **1** *noun* direction between south and west; *a region in the south-west of France* **2** *adjective* **(a)** situated in the south-west; *the south-west tip of England* **(b)** *(wind)* which blows from or which comes from the south-west; south-westerly (wind) **3** *adverb* towards the south-west; *we were heading south-west*

south-westerly [sauθ'westəli] *adjective* **(a)** *(wind)* which blows or which comes from the south-west; *a south-westerly wind* **(b)** *(direction)* which moves towards the south-west; *we were following a south-westerly direction*

south-western [sauθ'westən] *adjective* referring to the south-west; situated in the south-west; *the south-western corner of England includes Cornwall and Devon*

space [speis] **1** *noun* **(a)** empty area; *a major problem with fuel storage is finding space within the airframe* **(b)** physical universe outside the earth's atmosphere; *VHF (very high frequency) waves tend to* *pass through the layers of the ionosphere into space*

span [spæn] *noun* distance between two points; **wing span** *or* **wingspan** = distance between the tip of one wing to the tip of the other wing

spar [spɑ:] *noun* main longitudinal beam of an aircraft wing; *designing a wing skin, a rib or a spar as a single big item rather than assembling it from many smaller components minimizes the number of structural parts*

spark [spɑ:k] **1** *noun* a light produced by a sudden electrical discharge; *an electric spark from an igniter plug starts combustion and the flame is then continuous* **2** *verb* to suddenly start a process or action; *crew must quickly establish control to ensure panic does not spark a premature evacuation*

sparking plug = SPARK PLUG

spark plug ['spɑ:k 'plʌg] *noun* device screwed into each cylinder head (of spark ignition engines), which initiates fuel combustion by an electric spark; *see also* AIR GAP (NOTE: also called **sparking plug**)

spat [spæt] *noun* streamline covering for wheel fitted on a light aircraft to reduce drag - also called 'wheel fairing'; **detachable wheel spats** = streamlined coverings for the wheels of light aircraft which can be taken off to allow inspection and repairs of tyres

speaker *noun see* LOUDSPEAKER

special ['speʃəl] *adjective* particular, specific; not ordinary; *to make a composite, it is necessary to combine the reinforcing glass fibres with some form of special glue*; **special aerodrome report (SAR)** = report used if there are significant weather changes since the last meteorological aerodrome report

specific [spə'sɪfɪk] *adjective* **(a)** clearly defined and definite; *flight levels are specific pressure altitudes; the airframe has to be built to very specific requirements* **(b)** *(old name for 'relative*

density') **specific gravity** = density of a substance compared with that of water which is 1.00

specification [spesɪfɪ'keɪʃn] *noun* detailed description (of the composition of something, of what is needed, of what is involved, etc.); *fluids are coloured for recognition purposes and fluids of different specifications must never be mixed*

specify ['spesɪfaɪ] *verb* to name in detail; *the minimum values for decision heights are specified by the national licensing authorities for various types of aircraft and for various airports; pressure must be maintained within specified limits during all phases of flight*

specimen ['spesəmɪn] *noun* part taken as an example of the whole; *by testing specimen structures and components to destruction a safe life can be assessed for all such structures and components*

speed [spi:d] *noun* rate of motion over a distance in time; **cruising speed** = speed of the aircraft during the main part of the flight between top of climb after take-off and top of descent, usually giving fuel economy and long engine life; **ground speed** = speed of the aircraft in relation to the ground over which it is flying; **idling speed** = rpm (revolutions per minute) of the engine when it is idling; *after start-up, the engine accelerates up to idling speed; before the engine is stopped, it should normally be allowed to run for a short period at idling speed to ensure cooling*; **positive idling speed** = idling speed selected with the throttle to ensure that the engine runs correctly without spark plug fouling; *an adjustable stop on the throttle control ensures a positive idling speed*; **never-exceed speed (Vne)** = speed which must not be exceeded; **stalling speed** = speed at which the angle of attack is such that lift over the wing surface breaks down; **wind speed** = speed of the wind which, if combined with a direction, is called velocity; *see also* AIRSPEED

COMMENT: traditionally, an aircraft

can stall at any airspeed, providing the angle of attack is great enough. Stalling speed is often used to refer to the speed below which the aircraft cannot remain airborne

sphere [sfɪə] *noun* object in the shape of a ball; *the earth is not a perfect sphere*; *a circle drawn on the surface of a sphere, whose plane passes through the centre of the sphere is called a great circle*

spherical ['sferɪkl] *adjective* shaped like a sphere; *the earth is almost spherical in shape*; *drain cocks are generally simple, manually operated spherical valves*

spill [spɪl] **1** *noun* running out of a liquid from a container especially when it is unintentional; *an oil spill; a fuel spill; see also* VALVE **2** *verb* to cause liquid to run out of a container, usually unintentionally; *if fuel is spilt, it creates a fire hazard* (NOTE: **spilling - spilled** *or* **spilt**)

spillage ['spɪlɪdʒ] *noun* spilling of a liquid; *any fuel spillage must be cleaned up immediately* (NOTE: the word **spillage** is used in a more general sense than the word **spill**)

spin [spɪn] **1** *noun* **(a)** fast rotation; *the spin axis of the earth* **(b)** continued spiral descent of an aircraft with the angle of attack of one wing is greater than the stalling angle **2** *verb* **(a)** to rotate rapidly; *the earth is spinning on its axis* **(b)** to put an aircraft into a continued spiral descent with the angle of attack of the mainplane greater than the stalling angle; *it is prohibited to spin general purpose light aircraft which are not equipped with a suitable harness*

COMMENT: the Moroccan aerobatic team 'La Marche Verte' perform a formation manoeuvre with three aircraft spinning through multiple rotations while inverted

spindle ['spɪndl] *noun* pin or bar which rotates or on which something rotates; *of a cup anemometer has three cups, mounted on a spindle, that are driven by the wind causing the spindle to rotate*

spine [spaɪn] *noun* the longitudinal central part (of the engine); *annular inner and outer air casings form a tunnel around the spine of the engine*

spiral ['spaɪrəl] *adjective* winding continuously as if along a cylinder; *a spiral dive* = dangerous uncontrolled turning descent of an aircraft in which rate of descent and speed increase

spline [splaɪn] *noun* groove in a shaft for meshing, or engaging with another component; *for satisfactory operation, an engine requires an adequate supply of oil at all bearings, gears and driving splines*

split [splɪt] **1** *noun* **(a)** division **(b)** break along a line, especially in wood, plastic or rubber; *a split in a tyre* **2** *verb* **(a)** to divide; *retractable undercarriages can be split into three groups* **(b)** to break along a line; *one of the tyres split on impact* (NOTE: **splitting - split**) **3** *adjective* which has been divided; which has been broken along a line; **split bus system** = electrical system in which there are two separate power generation systems; *the parallel system and the split bus system are both used to distribute electrical power*

spoiler ['spɔɪlə] *noun* hinged surface on upper wing which, when opened, decreases lift and increases drag; *if a problem occurs in the spoiler system a master caution light illuminates*

COMMENT: spoilers are sometimes called 'speed brakes'. They are used during the descent prior to landing and immediately after landing to decrease lift and increase braking effect

spontaneous [spɒn'teɪnɪəs] *adjective* happening without external cause; *spontaneous ignition may occur if oxygen is allowed to come into contact with oil or grease*

spool [spuːl] **1** *noun* one complete axial-compressor rotor; *the single spool compressor consists of one rotor assembly and stators* **2** *verb* to **spool down** = to allow (turbofan) engine r.p.m. (revolutions per minute) to decrease; **to spool up** = to increase (turbofan) engine r.p.m.

spot [spɒt] *noun* **(a)** special place; small place; *charts should be kept in a convenient spot in the cockpit* **(b)** small roundish mark or piece; *a spot of oil on a shirt*; **spot height** = height, such of a mountain peak, marked on a chart

spotlight ['spɒtlaɪt] *noun* powerful, often moveable light which illuminates a small area; *a spotlight is mounted on the roof*

spray [spreɪ] **1** *noun* **(a)** body of liquid in fine drops; *the generator is cooled by oil spray delivered by the constant speed drive section* **(b)** container that sends out liquid in fine drops **2** *verb* to apply or to send out liquid in the form of fine drops; *some engines have the coolant sprayed directly into the compressor inlet, but for axial flow compressor engines, it is more suitable to spray the coolant into the combustion chamber inlet*

spread [spred] **1** *noun* extending the area of something; *measures are taken to prevent the spread of fire* **2** *verb* to extend the area of something; *strong jets of water should not be used on a liquid fire as this may cause the fire to spread*; *the system sprays a quantity of fluid onto the windscreen, which is then spread by the wipers* (NOTE: **spreading - spread**)

spring [sprɪŋ] *noun* **(a)** metal device which, when under tension, tries to resume its previous position; *the pitch lock piston is held in the forward position by a spring* **(b)** season between winter and summer

squall [skwɔːl] *noun* sudden increase in wind speed above the average speed lasting for several minutes; *surface squalls are due to the spreading out of strong down draughts at the surface*; *even with a light mean wind speed, squalls of 50 kt (knots) or more can occur with sudden changes in direction*

square [skweə] **1** *noun* **(a)** shape with 4 equal sides and 4 right angles **(b)** signals square = area on an aerodrome from which

ground signals are displayed **2** *adjective* **(a)** shaped like a square; *a square panel* **(b)** square foot = measurement of area in feet (length x width); **square metre** = measurement of area in metres (length x width); *the room is 5m x 9m so the area is 45 square metres ($45m^2$)*; **square root** = divider of a quantity that, when multiplied by itself gives the quantity; *3 is the square root of 9*

squawk [skwɔːk] **1** *noun* identification code; *see also Comment at* TRANSPONDER **2** *verb* to activate specific modes, codes or functions on a transponder; *garbling occurs when two signals are received simultaneously and can be resolved either technically or by making one of the aircraft squawk*

squeeze [skwiːz] *verb* to press hard from opposite directions; *static seals, gaskets and packing are used in many locations, and these effect a seal by being squeezed between two surfaces*

SSR = SECONDARY SURVEILLANCE RADAR

stabilitator [stə'bɪlɪteɪtə] *see comment at* STABILIZER

stability [stə'bɪləti] *noun* **(a)** being stable or steady; **aircraft stability** = (generally) tendency of an aircraft to return to its original attitude after being deflected; *the stability of the Cessna 150 makes it an ideal training aircraft* **(b)** state of the atmosphere in which air will resist vertical displacement; *when air moves away from its source region, the stability of the lower atmosphere changes*

COMMENT: stability can be classified as three types. **Positive stability** is the tendency of a body to return to its original state after being displaced. Light training aircraft have positive stability. **Neutral stability** is the tendency of a body to remain in the new position after displacement. **Negative stability** is the tendency of a body to continue moving away from its original position after displacement

stabilize ['steɪbəlaɪz] *verb* to become steady and unchanging; *after the engine has been started, engine speed is increased to 1,000 r.p.m. (revolutions per minute) until cylinder head and oil temperatures have stabilized at normal operating temperatures* (NOTE: also written **stabilise**)

stabilizer ['steɪbəlaɪzə] *noun* device to improve the tendency of an aircraft to return to its original attitude after being deflected; **horizontal stabilizer** = device to provide longitudinal stability about the lateral axis of the aircraft; **vertical stabilizer** = fin (NOTE: also written **stabiliser**)

COMMENT: some aircraft have an all-moving tailplane called a 'stabilator' (a combination of the words stabilizer and elevator)

stable ['steɪbl] *adjective* (a) steady (b) referring to an atmosphere in which there is little or no vertical movement; *layer cloud occurs in a stable atmosphere*

stack [stæk] *verb* to put one on top of the other; *by stacking rows of horizontal dipoles one above the other, a well-defined electronic glide path can be transmitted*

stage [steɪdʒ] *noun* (a) one of several points of a process; *there are three stages in the life cycle of a thunderstorm: process of formation, development and decay*; *calculate headings to steer for each stage of the flight*; **cruise stage of the flight** = phase of flight between top of climb after take-off and start of descent to land; **at a later stage** = at a later time (b) division or section; *in the axial flow compressor, many stages of moving and stationary blades are needed, each row of rotors and a row of stators forming a stage*

stall [stɔːl] **1** *noun* (a) loss of lift caused by the breakdown of airflow over the wing when the angle of attack passes a critical point; *in certain configurations it is possible for the buffet speed to be less than the required 7% margin ahead of the stall*; **stall warning system** = system to warn the pilot that the aircraft is about to stall (b) point at which opposing force overcomes that of the driving part; *compressor stall can be caused by ice formation in the air intake*; *see also* RECOVER; RECOVERY **2** *verb* to lose lift by the breakdown of airflow over the wing when the angle of attack passes a critical point; *many light aircraft stall when the angle of attack exceeds 15°*

COMMENT: a stall has nothing to do with the engine stopping. An aircraft can stall at any airspeed and in any attitude

standard ['stændəd] **1** *noun* quality or measure to which others should conform; *water is the standard for determining relative density*; **a high standard of skill** = a high level of skill **2** *adjective* normal; (officially) accepted; **standard atmosphere** = model atmosphere defined by pressure, temperature, density, etc., used in instrument calibration; **standard parallels** = in a conical projection, the parallels of latitude where the cone cuts the surface; **standard procedure** = normal procedure; **standard pressure setting (SPS)** = 1013.25 mb (millibars); **standard time** = universally adopted time for all countries based on zone time

standard rate turn ['stændəd 'reɪt 'tɜːn] *noun* turn made at a precise number of compass degrees per second

COMMENT: Rate 1 turn = 180° in 1 minute, Rate 2 turn = 360° in 1 minute, Rate 3 turn = 540° in 1 minute, Rate 4 turn = 720° in 1 minute. Standard rate turns are made using particular angles of bank for specific airspeeds and are used while flying under Instrument Flight Rules (IFR). The pilot can make accurate turns to given headings by banking at the standard rate and timing the turn

standby ['stændbaɪ] *adjective* (a) which is used as backup; secondary; *some aircraft use a ram air turbine that can be very useful as a standby power source in the event of failure of a complete main AC (alternating current) generating system*

(b) standby ticket = cheaper air ticket bought just before departure time; *there are no standby tickets to Montreal*

standing ['stændɪŋ] *adjective* **standing wave** = wave that remains stationary in a moving fluid; *clouds are visible indications of the presence of standing waves, but waves often occur without cloud formation*

starboard ['stɑːbəd] *noun & adjective (when inside the aircraft)* right-hand side of an aircraft when facing forwards; *the angle between heading and track of an aircraft is called drift and is expressed in degrees to the port or starboard side of aircraft heading* (NOTE: the opposite is **port**)

starter ['stɑːtə] *noun* device to start an engine; **starter motor** = in a piston engine, a small electrically operated device to turn the engine until ignition starts

start-up ['stɑːtˈʌp] *noun* procedure to start the engine(s); *after start-up, the engine accelerates up to idling speed*

state [steɪt] **1** *noun* existing condition of something; *a state of equilibrium*; *ice in a liquid state is called water*; *water in a gaseous state is known as vapour*; *a logic gate is a two-state device i.e. on/off*; **in a poor state** = in a bad condition **2** *verb* to say or to mention; to give information clearly; *it states in the information that you must not open the can near a flame*; *please, state your name and address*

statement ['steɪtmənt] *noun* something formally expressed in words; *after the crash, the president and chief executive of the company made a brief statement to the waiting news reporters*

static ['stætɪk] *adjective* not acting or not changing; passive or not moving; **static display** = display of parked aircraft on the ground; **static electricity** = electricity not flowing as a current; *when the aircraft travels through the air, friction causes a charge of static electricity to be built up on the airframe*; **static pressure** = pressure of a fluid acting on and moving with a body; **static ground running** = running of the

engine while the aircraft is stationery on the ground; *see also* SEAL

static port ['stætɪk 'pɔːt] *noun* small hole in the side of the aircraft which senses static pressure used in the operation of the altimeter, vertical speed indicator and airspeed indicator; *ensure that the static port is clear*

station ['steɪʃn] *noun* **(a)** particular assigned location; *the interphone system allows the flight deck to communicate with cabin crew stations* **(b)** location of radio transmitter; *a VOR (very high frequency omni-directional radio range) station*

stationary ['steɪʃnri] *adjective* not moving; *the aircraft was stationary on the ground with engine running*

stator ['steɪtə] *noun* fixed part of a rotary machine; *the LP (low pressure) compressor has large rotor blades and stator blades and is designed to handle a far larger airflow than the other two compressors*; *a temperature probe is embedded into the stator of the generator and a meter is provided, so that generator stator temperature can be monitored*

status ['steɪtəs] *noun* condition; *the centre-zero ammeter tells the pilot the status of the aircraft battery*

statute mile (sm) ['stætʃuːt 'maɪl] *noun* non-SI unit of length of 1.609 km (kilometres); *it is 20 statute miles to the airport*

steady ['stedi] *adjective* constant and unchanging; *the manual test will give a steady red light*; **a steady wind** = a wind of constant speed and direction

steam fog ['stiːm 'fɒg] *noun* fog formed when cold air moves over relatively warm water; *visibility was impaired because of steam fog*

steel [stiːl] *noun* metal alloy of iron, carbon and other compounds; **stainless steel** = steel with chromium and nickel; *tubing in parts of the system containing* *fluid at high pressure are usually made from stainless steel*

steep [stiːp] *adjective* **(a)** sloping sharply; **a steep angle of approach** = the angle formed by the aircraft approach flight path and the horizontal is greater than usual **(b)** *(of isobars)* closely spaced **(c)** referring to marked changes in pressure or temperature in a relatively short horizontal distance; *cooling of the air in contact with the ground at night can cause a very steep inversion of temperature at the surface*; *pressure gradients in anti-cyclonic curvature tend not to be steep*

steer [stɪə] *verb (car, aircraft, etc.)* to direct by using a wheel or control stick; *the aircraft is steered on the ground by using the rudder pedals*; **heading to steer** = gyro-compass point in which to direct the aircraft

steering ['stɪərɪŋ] *noun* **(a)** *(of car, aircraft, etc.)* guiding or directing; *steering is controlled by rudder pedals* **(b)** system for guiding or directing a car, aircraft, etc.; *most modern light aircraft have nose-wheel steering but older tail-draggers are steered on the ground by using differential braking*

step [step] *noun* **(a)** stage; *the first step in map reading is to orientate the chart* **(b)** one stair; *mind the step!*

steward ['stjuːəd] *noun* male member of airline staff who look after passengers during the flight; *see also* CABIN CREW; FLIGHT ATTENDANT; STEWARDESS (NOTE: different airlines use different terminology for their staff)

stewardess [stjuːə'des] *noun* female member of airline staff who look after passengers during the flight; *see also* CABIN CREW; FLIGHT ATTENDANT; STEWARD (NOTE: different airlines use different terminology for their staff)

stick [stɪk] **1** *noun (control column)* main hand control used by the pilot to control the aircraft roll and pitch; *using fly-by-wire technology, the stalling angle cannot be exceeded regardless of stick input* **2** *verb*

to become fixed, as if with glue; *ice crystals and snowflakes do not stick to airframes, and so icing is a problem only when super-cooled water droplets are present*

stiff [stɪf] *adjective* **(a)** rigid or inflexible; *Kevlar 49 is stiffer than glass, but only about half as stiff as carbon fibres* **(b)** not easily bent or turned; **control surfaces may become stiff as a result of icing** = control surfaces may become difficult to move **(c)** **a stiff wind** = a fairly strong wind

stiffen ['stɪfn] *verb* **(a)** to make rigid or inflexible; to make stiff; *beams can be additionally stiffened in a downward direction by vertical and diagonal members* **(b)** *(of wind)* to become stronger

stop [stɒp] *noun* **(a)** end of a movement; **to come to a stop** = to stop moving **(b)** component which limits the travel of a moving part; *an adjustable stop on the throttle control ensures a positive idling speed*

storage ['stɔːrɪdʒ] *noun* storing of something; *a reservoir provides both storage space for the system fluid, and sufficient air space to allow for any variations in the volume of the fluid in the system which may be caused by thermal expansion*

store [stɔː] *noun* **(a)** supply; *the maintenance section keeps a store of spare components* **(b)** **stores** = goods; *freight carrying aircraft have supporting members of greater strength to allow for the carriage of heavy stores* **(c)** *US* shop **2** *verb* to put away for future use; *a capacitor is a device with the ability to temporarily store an electric charge*

storm [stɔːm] *noun* violent weather disturbance with high winds and rain or snow; *storms produced by daytime heating are most frequently encountered in the afternoon and early evening*; **tropical storm** = violent wind system which forms over tropical oceans; *tropical storms often dissipate when they pass*

from sea to land; *see also*
THUNDERSTORM

stow [stəʊ] *verb* to place something in its correct position in the aircraft; *make sure the fire-extinguisher is stowed*

stowage ['stəʊɪdʒ] *noun* space for stowing things; *a multi-wheel combination has the advantage of smaller and lighter undercarriage structures, and wing stowage problems can be overcome by suitable mechanisms*

stowaway ['stəʊəweɪ] *noun* person who travels secretly by hiding in an aircraft, or a ship, not paying the fare; *the crew must be alert at all times to the possibility of hijacking, bombs and stowaways*

strain [streɪn] *noun* deformation caused by stress; *at low value of stress the plot of stress and strain is basically a straight line*

strap [stræp] **1** *noun* belt; a long narrow strip of fabric with a buckle; **safety straps** = device to keep a person in position in a seat **2** *verb* **to strap in** = to fasten seat or safety belt

stratocumulus [streɪtəʊ'kjuːmjʊləs] *noun* layer of small cumulus clouds lower than altocumulus, that is, below 3,000 m; *light rain may fall occasionally from stratocumulus*

stratosphere ['strætəʊsfɪə] *noun* the layer of the atmosphere which extends from the tropopause to about 50 km (kilometres) above mean sea level; *a cumulonimbus cloud may extend vertically, into the stratosphere*

stratus ['streɪtəs] *noun* low altitude layer cloud; *drizzle falls from shallow layer cloud such as stratus*

stream [striːm] *noun* steady current of a fluid; *thermocouple probes are positioned in the gas stream, so as to obtain a good average temperature reading*; **jet stream** = (i) band of high-altitude, strong winds; (ii) flow of gases from a jet engine; *the occurrence of the equatorial jet stream is due to a temperature gradient with colder air to the south*

strength [streŋθ] *noun* **(a)** being strong; the ability of a material to support a load; *aircraft wheels require great strength and are constructed in two halves which are bolted together after the tyre is fitted*; *magnesium does not possess sufficient strength in its pure state for structural uses , but when mixed with zinc, aluminium, and manganese it produces an alloy having the highest strength-to-weight ratio of any of the commonly used metals*; **high-strength materials** = materials which are very strong **(b)** degree of clarity and volume of a signal; *a radio wave loses strength as range increases* **(c)** degree of dilution of a liquid; *incorrect mixture strength may cause detonation* **(d)** intensity of radiation; *the strength of the sun's radiation varies with latitude* **(e)** force (such as of wind); *high ground will disturb the smooth horizontal flow of air, with the degree of disturbance depending upon the strength of the wind and the roughness of the terrain*

strengthen ['streŋθn] *verb* to make strong or stronger; *some alloys are hardened and strengthened by heat treatment*; **the wind is strengthening** = the wind is increasing in speed

strengthening ['streŋθnɪŋ] *noun* **(a)** making stronger; *aircraft which require large apertures in the fuselage for freight doors, etc., need increased strengthening around these areas* **(b)** becoming stronger; *strengthening of the wind*

stress [stres] *noun* **(a)** the internal force, or load per unit area of a body that resists distortion or change of shape of the body; *turbine blades in the average jet engine vibrate at frequencies of 1 million per minute, and in each cycle experience stress* **(b)** **psychological stress** = a mentally or emotionally upsetting condition which effects one's health **2** *verb* to emphasize; *it must be stressed that the description is a model and departures from it often occur*

stretch [stretʃ] **1** *noun* continuous unbroken length; *a stretch of coast* **2** *verb* to extend or enlarge beyond the proper limitsverb *tensile stress or tension is the resistance of a material to being stretched*

stretching ['stretʃɪŋ] *noun* extending or enlarging beyond the proper limits; *tensile stress is the resistance to pulling apart, or stretching, produced when two forces in opposition act along the same straight line*

strict [strɪkt] *adjective* precise, exact; *fuels for aircraft must conform to strict requirements*; *all generator voltages, frequencies and their phase sequence must be within very strict limits to ensure proper system operation*

strike [straɪk] **1** *noun* hitting; **a lightning strike** = the hitting of something by a discharge of lightning **2** *verb* to hit; *a high frequency wave will strike the ground more often than a low frequency wave* (NOTE: **striking - struck**)

stringer ['strɪŋə] *noun* thin metal or wood strips which go from one end of the fuselage to the other; *stringers are made of a light alloy material*

strip [strɪp] **1** *noun* long narrow piece, usually of the same width from end to end; *a strip of paper*; **landing strip** = specially prepared area of land for an aircraft to land on **2** *verb (engine, etc.)* to dismantle; *after the collision, the engine was stripped down to its component parts*

stroke [strəʊk] *noun* any of a series of movements of a piston from one end of the limit of its movement to another; *the connecting rod links the piston to the crankshaft and transmits the force of the power stroke from the piston to the crankshaft*; **four-stroke combustion engine** = engine which operates in accordance with the four-stroke cycle; *induction, compression, power and exhaust are the four phases of the four-stroke combustion engine*; *see also* CYCLE

structural ['strʌktʃrəl] *adjective* referring to the structure (of an aircraft); *as laid down in the flight manual, the structural limitations must never be*

exceeded; **structural failure** = a breaking of part of the aircraft structure

structure ['strʌktʃə] *noun* (i) something constructed; (ii) framework; *aircraft structure serves the same purpose for an aircraft as the skeleton for a human body*

strut [strʌt] *noun* bar or rod used to strengthen a structure against forces from the side; *a strut is designed to withstand compressive loads*

stub [stʌb] *noun* short rectangular extension; *the plan-form of a military air traffic zone is in the shape of a circle with a stub*

sub- [sʌb] *prefix meaning* (i) of less importance in rank; (ii) below

sub-beam ['sʌb'biːm] *noun (of directional radar)* less important or minor beam; *a lobe is one of two, four or more sub-beams that form a directional radar beam*

subject ['sʌbdʒɪkt] *noun* topic; matter for discussion or study; *a knowledge and understanding of the subject of ice accretion is essential in order that the hazard can be minimized*

subject to 1 ['sʌbdʒɪkt 'tuː] *adjective* prone or liable to; *the airspeed indicator is subject to error*; *turbine engines are subject to icing during flight through super-cooled droplet cloud* **2** [səb'dʒekt 'tuː] *verb* to **subject to** = to make something or someone experience something (often unpleasant); *the aircraft was subjected to rigorous tests*

subjected [səb'dʒetɪd] *adjective* **subjected to** = made to experience something; affected by; *to maintain the pressure difference between two internal engine sections, which are subjected to air pressures of different value, a multi-air seal is used* (NOTE: there is an important difference between **subject to** and **subjected to**)

sublimate ['sʌblɪmeɪt] *verb* to transform directly from the solid to the gaseous state or from the gaseous to the solid state without becoming a liquid; *for hoar frost to form on an aircraft the airframe temperature must be below 0°C (Celsius), so that the surrounding air is cooled to below its dew point and water vapour in contact with the aircraft skin is directly sublimated into ice crystals*

sublimation [sʌblɪ'meɪʃn] *noun* transformation directly from the solid to the gaseous state or from the gaseous to the solid state without becoming a liquid; *in sub-zero conditions sublimation will occur when air is cooled below the frost point, producing a deposit of ice crystals*

sub-scale ['sʌb'skeɪl] *noun* secondary, not main, scale on an instrument; *the barometric pressure is set on the sub-scale and the altimeter main scale displays height or altitude*

subsequent ['sʌbsɪkwənt] *adjective* following in time or order; which comes later; *a structural prototype is put through cycles of stressing far more severe than can be expected during the aircraft's subsequent operational life*; a subsequent occasion = a following occasion

subside [səb'saɪd] *verb* (a) to sink to a lower level; *cool air subsides* (b) to become less active or strong; *the storm subsided* = the storm grew quiet

subsidence [səb'saɪdəns] *noun* sinking to a lower level; *descending air occurs because of subsidence in the high pressure belts of the sub-tropics and poles*

substance ['sʌbstəns] *noun* material of a particular sort; matter; *specific heat is the amount of heat required to raise the temperature of a substance by 1°C (Celsius) compared to the amount of heat required to raise the temperature of water by 1°C*

substantial [səb'stænʃl] *adjective* considerable; important; **substantial damage** = a lot of damage; **substantial increase** = a big increase

subtend [səb'tend] *verb* to be opposite to and delimit; *the angle subtended by an arc equal to one 360th part of the circumference of a circle is called 1° (degree)*

subtract [səb'trækt] *verb* to deduct or to take away (from); *6 subtracted from 10 equals 4 (10 - 6 = 4)*

subtraction [səb'trækʃn] *noun* operation of taking away or deducting; *the major arithmetic operations are addition, subtraction, multiplication and division*

subtropical ['sʌb'trɒpɪkl] *adjective* referring to the areas between the tropics and the temperate zone; *in winter, the subtropical high retreats and gives way to cyclonic pressure patterns*

sub-zero [sʌb'zɪərəʊ] *adjective (temperature, conditions)* below zero degrees; *in sub-zero conditions sublimation will occur when air is cooled below the frost point, producing a deposit of ice crystals*

success [sək'ses] *noun* achievement of something wanted; *the key to success in navigation is pre-flight planning*

successful [sək'sesfl] *adjective* satisfactory, as wanted; *his second attempt at landing was successful*

succession [sək'seʃn] *noun* process of following in a particular order; *a succession of minor incidents created a more serious situation*

successive [sək'sesɪv] *adjective* following one after the other without interruption; *all aircraft remained grounded for three successive days because of fog; a day is the period between successive transits of a meridian by the sun*

such [sʌtʃ] (a) *adjective* of this kind; *an example of such a chart is shown on page 3* (b) of a large enough extent or amount; *the height of the cabin floor to the ground on large jet transports is such that serious injuries can occur by exiting through the doors when steps or ramps are not available*

suction ['sʌkʃn] *noun* force that causes a fluid or solid to be drawn into a space because of the difference between the external and internal pressures: *in a fuel injection system, fuel is induced into the inlet port or combustion chamber by a pump rather than the suction caused by the venturi of a carburettor*

sudden ['sʌdn] *adjective* immediate and without warning; unexpected; *a sudden change or a sudden drop in temperature*

suffer (from) ['sʌfə 'frɒm] *verb* to be prone or to be liable to; to experience; *piston engines suffer from icing in moist air when the ambient air temperature is well above 0°C (Celsius)*

sufficient [sə'fiʃnt] *adjective* enough; *during his pre-flight checks, the pilot must ensure that he has sufficient fuel for the flight*

suffix ['sʌfiks] *noun* addition to the end of a word creating a new word; *apart from cirrus and stratus, which are complete names, all layer cloud names consist of a prefix according to height of base, and a suffix according to shape* (NOTE: in the word cloudless, **-less** is the suffix meaning: without)

suggest [sə'dʒest] *verb* (a) to indicate a possibility; *a strong cloud echo on radar suggests that hailstones are present* (b) to mean, to imply; *heap clouds, as the name suggests, often have great vertical extent*

suit [suːt] *verb* to meet the requirements of; *on some engines the ignition can be varied as the engine is running and is moved to suit the engine speed and load*

suitable ['suːtəbl] *adjective* appropriate or right for a particular purpose; *taking into account the limits imposed by aircraft performance, a suitable route must be chosen*

sulfur [sʌlfə] *noun US* = SULPHUR

sulphur ['sʌlfə] *noun* yellow non-metallic chemical element; *turbine fuels tend to corrode the components of the fuel and combustion systems mainly*

as a result of the sulphur and water content of the fuel (NOTE: chemical symbol **S**; atomic number **16**. Note also that words beginning **sulph-** are written **sulf-** in American English)

sum [sʌm] *noun* result of two or more numbers added together; *when the component velocities act in the same direction, the resultant velocity is equal to the sum of their speeds in that direction*

summarize ['sʌməraɪz] *verb* to present something in a shortened, concise form; *the effects of ice deposits on aircraft can be summarized as follows ....* (NOTE: also written **summarise**)

summary ['sʌmri] *noun* brief account of something more detailed; *at the end of each chapter there is a summary*

sump [sʌmp] *noun* oil reservoir of a piston engine situated at its base; *the oil level in the sump or tank is normally checked after the engine has been stopped for a certain time*

sun [sʌn] *noun* very bright star around which the earth travels and which gives light and heat; *the sun was just rising when we landed*; *the sun and the planets governed by the sun form the solar system*; *see also* SOLAR

sunrise ['sʌnraɪz] *noun* time when the upper limb of the sun - the first part you see at sunrise - is on the visible horizon

sunset ['sʌnset] *noun* time when the upper limb of the sun just disappears over the horizon

super- ['suːpə] *prefix meaning* more than normal

supercharge ['suːpətʃɑːdʒ] *verb* to increase the power of an engine by using a supercharger; *a supercharged engine delivers greater power than a non-supercharged engine of the same size*

supercharger ['suːpətʃɑːdʒə] *noun* blower or compressor, usually driven by the engine, for supplying air under high pressure to the cylinders of an internal combustion engine; *the function of the supercharger is to increase the power output and maintain sea-level conditions at altitude*

super-cooled *or* supercooled

['su:pə'ku:ld] *adjective* cooled below freezing point without solidification; *supercooled fog; nimbo-stratus cloud is composed of liquid water droplets some of which are supercooled*

superimpose [su:pəim'pəʊz] *verb* to

lay or to place something over the top of something else; *the computer utilizes a technique in which each successive atmospheric layer is analyzed and superimposed on the previous ones*

supersonic [su:pə'sɒnik] *adjective*

faster than the speed of sound; *for sustained supersonic flight, tank insulation is necessary to reduce the effect of kinetic heating*

supervisor ['su:pəvaizə] *noun* person

in charge; *senior cabin supervisor*

supplement 1 ['sʌplimənt] *noun*

(mathematics) angle or arc that, when added to a given angle or arc makes 180° or a semicircle **2** ['sʌpliment] *verb* to add to in order to make more complete; *the main power plant fire detection system should contain an audible warning device to supplement the visual indication*

supplementary [sʌpli'mentri]

adjective **(a)** extra or additional; *supplementary information* **(b)** supplementary angle = angle that, when added to a given angle makes 180° or a semicircle

supply [sə'plai] **1** *noun* amount of

something available for use; *an engine requires an adequate supply of oil* **2** *verb* to make available for use, to provide; *a battery is designed to supply limited amounts of electrical power* (NOTE: supplying - supplied)

support [sə'pɔ:t] **1** *noun* **(a)** device to

hold something in position; *direct-reading indicators consist of a float contained within a metal support tube* **(b)** practical

assistance; **support facilities** = equipment and buildings used by ground staff when working on aircraft at an airport; **support services** = services provided to an aircraft while it is at an airport **2** *verb* to bear the weight of; *the wings support the aircraft in flight*

suppress [sə'pres] *verb* **(a)** to prevent

the development or spreading of something; **the fire crew suppressed the fire** = the fire crew brought the fire under control **(b)** *(electricity)* to prevent electrical interference of a radio signal, etc.; *R/T noise interference can be suppressed*

suppressed antenna [sə'prest

æn'tenə] *noun* antenna which is mounted under the airframe skin; *static interference can be reduced by installing suppressed antennas*

suppression [sʌ'preʃn] *noun* **(a)**

prevention of the development or spreading of something; *a fire suppression system* **(b)** *(electricity)* prevention of electrical interference of a radio signal, etc.

suppressor [sə'presə] *noun* device,

such as a resistor or grid, used in an electrical or electronic system to reduce unwanted currents; *a suppressor improves the quality of the signal*

surface ['sɜ:fəs] *noun* **(a)** outer covering

of something; top part of something; *the surface of the wing* **(b)** earth's surface or ground; **surface air temperature** = temperature recorded in the shade at a height just above ground level; **surface front** = (weather) front at the surface of the earth; *the cirrus cloud can be 900 miles ahead of the surface front with a rain belt as wide as 200 miles*; **surface heating** = heating of the ground by the sun; **surface wind** = wind which blows across the land surface (and not the wind higher in the atmosphere) **(c)** **control surfaces** = movable devices on the wings and tail unit which can be operated from the cockpit by the pilot to change aircraft attitude

surge [sɜ:dʒ] **1** *noun* sudden increase in

(electrical power); **engine surge** =

instability in the power output of an engine **2** *verb* to move with force like a wave; *if combustion pressure increases above compressor outlet pressure, the airflow will reverse in direction and surge forward through the compressor*

surplus ['sɜːpləs] *adjective* excess; more than is needed; *fuel penalties can be incurred if fuel surplus to requirements is carried*

surround [sə'raʊnd] **1** *noun* something which encloses or borders; *the design of windows, hatches or door surrounds is very critical* **2** *verb* to encircle or to enclose; *the earth is surrounded by the atmosphere*

surveillance [sə'veɪləns] *noun* watching or monitoring; **surveillance radar** = primary radar scanning often through 360°

survey 1 ['sɜːveɪ] *noun* detailed examination; *an aerodrome meteorological office maintains a continuous survey of meteorological conditions over the aerodromes for which it is designated to prepare forecasts* **2** [sɜː'veɪ] *verb* to determine the boundaries, area, or elevations of land by means of measuring angles and distances; *take care when using wooded areas to fix position because tree felling may have changed their shape since the area was surveyed by the map makers*

survival [sə'vaɪvl] *noun* remaining alive after an accident; *the survival of passengers in the sea depends on rapid location and rescue; see also* PACK

QUOTE survival training is a vital element of all aircrew knowledge. Just because modern aircraft are more reliable than their predecessors, the need for such training does not diminish
Civil Aviation Training

survivor [sə'vaɪvə] *noun* person who continues to live after an accident; *whilst awaiting rescue on land or at sea,*

survivors should avoid exposure and conserve energy; the aircraft crashed into the sea and there were no survivors*

susceptible [sə'septəbl] *adjective* prone to, likely to be affected by; *a rough surface is more susceptible to fatigue cracking than a smooth one and for this reason highly stressed members are often polished*

suspect 1 ['sʌspekt] *adjective* referring to something believed to be causing problems; *the magnetic flaw detection technique is to induce a magnetic field in the suspect part and then to brush over it an ink containing a magnetic powder* **2** [sə'spekt] *verb* to believe to be the case; *if fuel contamination by water is suspected, a sample of fuel should be drained from the tank for inspection*

suspend [sə'spend] *verb* **(a)** to hang freely from a point; *when it is freely suspended, a magnet will turn until one pole is towards the earth's magnetic north pole* **(b)** to float freely in the air or in a liquid; *the weather associated with visibility reductions by particles suspended in the atmosphere is classified as fog, mist, haze or smoke*

suspension [sə'spenʃn] *noun* **(a)** the act of state of hanging freely from a point **(b)** dispersion of particles in a liquid or gas; *if a sample of fuel taken from a tank is hazy or cloudy in appearance, this indicates the presence of water in suspension*

sustain [sə'steɪn] *verb* **(a)** to continue, to maintain; *for sustained supersonic flight, some measure of tank insulation is necessary to reduce the effect of kinetic heating* **(b)** to receive; to experience; to suffer; *the aircraft sustained major damage in the crash; the pilot sustained minor injuries*

sweep [swiːp] *verb* to move across quickly and with force; *cold arctic air sweeps over North America in wintertime*

swell [swel] *noun* long wave on water that moves continuously without breaking;

when ditching an aircraft the selection of a landing direction which will result in the minimum relative speed between the aircraft and sea swell will reduce impact forces and minimize structural damage

swing [swɪŋ] *verb* **(a)** *(of aircraft)* to move laterally with some force; *there is often a tendency for a propeller driven aircraft to swing or yaw on take-off* **(b)** to swing a **compass** = to calibrate compass deviation by recording its value on a compass base while rotating the aircraft through 360° **(c) to swing a propeller** = to turn a propeller by hand to start the engine

swirl [swɜːl] *noun* movement with a twisting motion; eddy; *swirls of smoke came out of the engine*; **swirl chamber** = small chamber in the cylinder head to promote swirl; *the usual method of atomizing the fuel is to pass it through a swirl chamber, so converting its pressure energy to kinetic energy*

switch [swɪtʃ] **1** *noun* device to open or break an electric current; *there is an on/off switch on the front panel*; **centrifugal switch** = switch operated by centrifugal force **2** *verb* to connect or disconnect two lines by activating a switch; **to switch on** = to start to provide power to a system by using a switch; *switch on the light*; **to switch off** = to disconnect the power supply to a device or system; *switch off the navigation lights*

symbol [ˈsɪmbəl] *noun* printed or written sign used to represent something; *the work done by an electrical circuit or the power consumed is measured in watts and is given the symbol P*

symbolic [sɪmˈbɒlɪk] *adjective* referring to symbols; *a symbolic code is used for synoptic charts*

symmetrical [sɪˈmetrɪkl] *adjective* referring to something which has an exact likeness of form on opposite sides of a central dividing line; *the area covered by the forecast is divided into a series of grid*

or reference points at approximately 300 km (kilometres) symmetrical spacing

symptom [ˈsɪmptəm] *noun* sign or indication of something, possibly a problem; *buffet caused by turbulent airflow acting on the tailplane is one of the first symptoms of the approaching stall*

synchronization [sɪŋkrənaɪˈzeɪʃn] *noun* occurrence at the same time or rate; *prior to engagement, when the aircraft is being flown manually, the autopilot system will be following the aircraft flight attitude, thus ensuring that synchronization is achieved* (NOTE: also written **synchronisation**)

synchronize [ˈsɪŋkrənaɪz] *verb* to cause to occur or operate at the same time or rate; *the aircraft must be trimmed for the desired flight attitude before engaging the autopilot, which must be synchronized to maintain that attitude when it is engaged* (NOTE: also written **synchronised**)

synchronous [ˈsɪŋkrənəs] *adjective* referring to something operating at the same time or rate; *synchronous motors will run at constant speed and are small and light in weight*

synoptic [sɪˈnɒptɪk] *adjective* referring to something which gives a brief outline or general view of something more complex; *with the addition of fronts and isobars, the synoptic chart provides a representation of the weather over a large area, at a particular time*

synthetic [sɪnˈθetɪk] *adjective* not natural; artificial; *mineral based fluids are normally coloured red, and must be used with synthetic rubber seals and hoses*

system [ˈsɪstəm] *noun* group of interdependent parts forming and operating as a whole; *a braking system; an electrical system*; **frontal system** = an arrangement of weather fronts; **solar system** = the sun and the planets governed by the sun

Tt

'T' piece adapter ['tiː 'piːs ə'dæptə] *noun* device for connecting two inputs to one output or vice versa

tab [tæb] *noun* hinged rear part of flight control surface used for trimming; *trim tabs remove the pilot's control loads by aerodynamically holding the control surface in the required position*

table ['teɪbl] *noun* set of facts or figures displayed in columns and rows; *charts are issued at UK meteorological offices and show, for selected locations, a table of winds and temperatures at selected flight levels*

tabular ['tæbjʊlə] *adjective (of facts and figures)* **in tabular form** = arranged in a table; *the most widely acceptable presentation of fuel data is in tabular form but graphical presentations may also be used*

tachometer [tə'kɒmɪtə] *noun* instrument for the measurement of r.p.m. (revolutions per minute) of a rotating shaft; *the pilot checks the tachometer and notes the resulting drop in r.p.m. for each magneto*

TAF = TERMINAL AERODROME FORECAST; *(ICAO)* AERODROME FORECAST

tail [teɪl] *noun* rear part of the aircraft; *the tail section is the aft part of the fuselage to which is fitted the tail unit, comprising the tailplane, elevators, fins and rudders*; **tail-dragger** = tailwheel aircraft; aircraft with a small wheel at the tail instead of a nosewheel; **tail wind** = TAILWIND

tailplane ['teɪlpleɪn] *noun* horizontal stabilizer, a horizontal aerofoil at the rear of the aircraft; *on most high performance aircraft the incidence of the horizontal stabilizer (or tailplane) can be varied in flight*

tailwheel ['teɪlwiːl] *noun* **tailwheel aircraft** = aircraft with a small wheel at the tail instead of a nosewheel; **tailwheel conversion course** = course which familiarizes qualified pilots with the differences in handling characteristics between nosewheel and tailwheel aircraft; *see also* NOSEWHEEL

tailwind ['teɪlwɪnd] *noun* wind which is blowing in the same direction as the direction of movement or flight; *because of the tailwind, the flight took only six hours*; *see also* HEADWIND

take off ['teɪk 'ɒf] *verb (of aircraft)* to leave the ground; *when flying speed is reached the aeroplane takes off*

take-off *or* **takeoff** ['teɪkɒf] *noun* procedure when an aircraft leaves the ground; *the aircraft has to accelerate before take-off; there is a tendency for propeller driven aircraft to swing or yaw on take-off;* **take-off run** = distance from the start of take-off to the point where the wheels leave the ground; *acceleration forces can be felt as the aircraft begins its take-off run*

tangent ['tændʒənt] *noun* straight line, curve or surface which meets another curve or curved surface at a point, but which, if extended, does not cut through at that point; *the glide path is at a tangent to the runway*

tank [tæŋk] *noun* large container for storing fluid; *an aluminium alloy fuel tank is housed in each wing*

taper [ˈteɪpə] *verb* to reduce in thickness towards one end; *fuel flowing from the float chamber passes through a jet, in which is positioned a tapered needle valve*; **tapered wing** = wing which becomes narrower in width from root to tip

target [ˈtɑːgət] *noun* the indication shown on a radar screen resulting from a primary radar return or a radar beacon reply; *in a secondary radar system, the target is active*

tarmac [ˈtɑːmæk] *noun* runway and taxiways (of an airport); *they were working fast to clear the snow from the tarmac*

TAS = TRUE AIR SPEED

task [tɑːsk] *noun* function; duty; *present day transport aircraft are required to fly accurately, in all weather, for long distances or long periods of time and, in order to carry out this task efficiently, an autopilot is used*

taxi [ˈtæksi] *verb* to move an aircraft along the ground under its own power before take-off or after landing; *light aircraft can be steered while taxiing via a direct link from rudder pedals to the nosewheel* (NOTE: **taxies - taxiing - taxied;** American spelling is also **taxying**)

taxiing [ˈtæksiɪŋ] *noun* the movement of an aircraft along the ground under its own power before take-off or after landing; *the landing and taxiing of an aircraft; the taxiing of tail-wheel aircraft is more difficult than nosewheel aircraft* (NOTE: American spelling is also **taxying**)

taxiway [ˈtæksiweɪ] *noun* (usually) tarmac surface connecting the ramp or apron with the runway(s); *an airfield i.e. an area given over to runways, taxiways and aprons*

TCA = TERMINAL CONTROL AREA

technical [ˈteknɪkl] *adjective* **(a)** referring to mechanical subjects or applied sciences; *a technical education* **(b)** referring to the mechanical, electrical, hydraulic or pneumatic systems of an aircraft; *a technical problem with the aircraft prevented it from taking off on time*

technique [tekˈniːk] *noun* special method for doing something; *the preparation of charts is done by computer using numerical forecasting techniques*

technology [tekˈnɒlədʒi] *noun* study and use of the mechanical arts or applied sciences; **new technology** = new electronic equipment; *the use of fly-by-wire in airliners was delayed to allow thorough development and encourage universal acceptance of the new technology*

TEMP = TEMPERATURE

temperate [ˈtemprət] *adjective (of climate)* mild; not extreme; *cold air in temperate latitudes is usually unstable*

temperature [ˈtemprətʃə] *noun* measurement, in degrees, of the intensity of heat of a body; **ground temperature is the temperature recorded by a thermometer placed at ground level**; *the altitude and temperature of the tropopause are of concern to aircrew*

tempo [ˈtempəʊ] *noun* speed (of activity); *the flow of passengers to exits and tempo of evacuation will be influenced by the number of exits available*

TEMPO *(ICAO)* = TEMPORARY

temporary [ˈtemprəri] *adjective* lasting for a short time; not permanent; *the indicator 'tempo', followed by a 4-figure time group indicates a period of temporary fluctuations to the forecast meteorological conditions which may occur at any time during the period given* (NOTE: the opposite is **permanent**)

tend [tend] *verb* to be apt or inclined to do something; to do something more often than not; *depressions tend to move around* *large anticyclones following the circulation of wind*; **the weather tends to be wet in the UK in the winter** = the weather is often, but not always wet

tendency ['tendənsi] *noun* an inclination, situation or condition which occurs more often than not; **he has a tendency to be late** = he is often late; **he has a tendency to forget things** = he is forgetful; *there is a tendency for propeller driven aircraft to swing or yaw on take-off*; **barometric tendency** = quantitative value of change in barometer reading

tensile ['tensaɪl] *adjective* referring to stretching or pulling out; *reinforced plastic may have to support a tensile load, a compressive load or a bending load*; **tensile strength** = the strength of a structure to resist forces pulling it apart from opposite directions; **tensile stress** = the forces that try to pull a structure apart from opposite direction; *see also* LOAD

tension ['tenʃn] *noun* strained condition resulting from forces acting in opposition to each other; *a rod which is bent is shortened or in compression on the inside of the bend and is stretched or in tension on the outside of the bend*; **surface tension** = tension of the surface film of a liquid

term [tɜːm] *noun* **(a)** word or expression; *the term 'payload' includes passengers, baggage and freight* **(b)** limited period of time; **a 5 year term** = a period of 5 years; **in the long term** = when considering a long period of time; **short term forecast** = weather forecast for the coming hours only

terminal ['tɜːmɪnl] **1** *adjective* referring to a limit or to a final point; **terminal airfield** = airfield at which a flight finishes **2** *noun* **(a)** departure and/or arrival building at an airport; *the flight leaves from terminal three at Heathrow airport* **(b)** electrical connection point; *the negative terminal of the battery is marked '—'*

terminal aerodrome forecast (TAF) ['tɜːmɪnl 'eəriə 'fɔːkɑːst] *noun* weather forecast for the area around an aerodrome; *in terminal aerodrome forecasts, the height of the cloud base forecast is above airfield level unless otherwise stated*

> COMMENT: TAFs are scheduled four times daily for 24-hour periods beginning at 0000Z, 0600Z, 1200Z, and 1800Z

terminal control area (TCA) ['tɜːmɪnl kən'trəʊl 'eəriə] *noun* air traffic control area established at the meeting place of a number of routes near one or more major airports; *in certain areas where there is a local concentration of traffic, terminal control areas are set up*

terminate ['tɜːmɪneɪt] *verb* to end or to finish; to bring to a close; *the flight terminates in New York*; **the transmission terminated abruptly** = the transmission stopped suddenly and unexpectedly

terminology [tɜːmɪ'nɒlədʒi] *noun* set of words or expressions used for a particular subject; *it is necessary to learn some of the terminology associated with aircraft navigation*

terrain [tə'reɪn] *noun* land, especially in relation to its physical geography; *special attention should be paid to wind flow when flights are made over hills or mountainous terrain* (NOTE: almost always singular)

terrestrial [tə'restriəl] *adjective* referring to the earth; *clear skies allow terrestrial radiation to escape*

territory ['terɪtri] *noun* extent of the surface of the earth governed by a particular country, ruler, state, etc.; *all places in the same territory, or part of the same territory, maintain a standard of time as laid down by the government responsible for that territory*

tertiary ['tɜːʃəri] *adjective* referring to something which is third in order of rank, behind primary and secondary; *tertiary radar systems are synonymous with* long-range navigation aids; *tertiary structures, for example, fairings, wheel doors, and minor component brackets are essential parts of the airframe*

test [test] **1** *noun* **(a)** *(for engines, etc.)* series of operations to find out if something is working well; *the manual test for the engine fire warning system will give a steady red light on all the fire control handles* **(b)** examination to assess the knowledge of a person; *there is a navigation test for students at 0800 hours*; **general flying test (GFT)** = test of aircraft handling skills for student pilots **2** *verb* **(a)** to operate in order to find out if something functions correctly; *oxygen under pressure is used to test the oxygen masks and equipment for fit and leakage* **(b)** to examine in order to assess the knowledge of a person; *the students are tested in five subjects*

theory ['θɪəri] *noun* system of ideas or principles explaining something; *the theory of navigation must be studied before any practical plotting exercises are done*

thereafter [ðeər'ɑːftə] *adverb* after that; beyond that; *meteorological visibility is given in metres up to 5,000 metres and thereafter in km (kilometres)*

thereby [ðeə'baɪ] *adverb* by that means; in that way; *the evacuation was carried out at a slower rate, thereby minimizing the risk of injury to passengers*

therefore ['ðeəfɔː] *adverb* as a result; consequently; *at small throttle openings, the depression at the choke is very small and therefore no fuel flows from the main jet*

thermal ['θɜːməl] **1** *adjective* referring to heat; *intense surface heating causes thermal currents to develop and create convection* **2** *noun* rising current of relatively warm air in the lower atmosphere; *glider pilots circle in thermals in order to gain height*

thermo- ['θɜːməʊ] *prefix meaning* heat

thermocouple ['θɜːməʊkʌpl] *noun* device for measuring temperature; *variation in temperature of the cooling air will give some indication of engine trouble through a thermocouple system to a temperature gauge*

thermometer [θə'mɒmɪtə] *noun* instrument for measuring temperature; *ground temperature is the temperature recorded by a thermometer placed at ground level*

thermoplastic [θɜːməʊ'plæstɪk] *noun* type of plastic which can be softened by heating then shaped, then softened again by heating; *thermoplastic materials become soft when heated*

thermosetting plastic ['θɜːməʊsetɪŋ 'plæstɪk] *noun* type of plastic which is heated while being shaped but which cannot be softened by reheating; *if a piece of thermosetting plastic is hit hard enough, it breaks into pieces with straight sharp edges*

thick [θɪk] *adjective* **(a)** of great or particular extent between two surfaces; *this sheet of aluminium is not very thick*; *a 1cm thick steel bar* **(b)** with a large diameter; *thick wire* **(c)** dense; *thick fog; thick cloud* **(d)** of a consistency which does not flow easily; *thick oil* (NOTE: the opposite is **thin**)

thickness ['θɪknəs] *noun* **(a)** extent between two surfaces; *in monocoque construction, there is no internal stiffening because the thickness of the skin gives strength and stability* **(b)** extent of the diameter of a wire **(c)** *(of fog, cloud; of oil)* state or condition of being thick

thin [θɪn] *adjective* **(a)** of small extent between two surfaces; *a thin layer of paint* **(b)** with a small diameter; *thin wire* **(c)** not dense; *thin mist; altostratus cloud is thin enough for the sun to be dimly visible* **(d)** of a consistency which flows easily; *thin oil* (NOTE: the opposite is **thick**)

thinness ['θɪnnəs] *noun* **(a)** small extent between two surfaces; *the thinness of the material makes it unsuitable* **(b)** small extent of the diameter of a wire **(c)** *(of fog, cloud; of oil)* state or condition of being thin

thorough ['θʌrə] *adjective* complete; *all cabin crew must have a thorough knowledge of fire fighting equipment and procedures*; **a thorough inspection** = a very detailed, comprehensive inspection

threshold ['θreʃhəʊld] *noun* beginning of the part of the runway usable for landing; *runway visual range is obtained by an observer standing at the side of the runway in the vicinity of the threshold counting the number of markers or lights visible along the side of the runway*

> COMMENT: the threshold is marked with a single white line on visual runways or by eight parallel white lines arranged longitudinally in two groups of four each side of the runway centerline for runways with instrument approach/landing facilities

throttle ['θrɒtl] **1** *noun* (i) throttle lever; (ii) throttle valve; **throttle lever** = device operating the throttle valve; *when starting an engine, it is inadvisable to pump the throttle lever because of the risk of fire*; **throttle quadrant** = device for housing the throttle levers; **throttle valve** = device controlling the flow of fuel in an engine **2** *verb* **to throttle back** = to reduce engine power; *throttle back to increase the rate of descent*

> COMMENT: the verbs 'open' or 'advance' (= to increase engine power) and 'close' or 'throttle back' (= to decrease engine power) are frequently used by instructors to explain the required movement of the throttle lever in the cockpit

throughout [θruːˈaʊt] *adverb* from the beginning to the end of a time or place; *emergency lighting is provided throughout the cabin*; *heavy snow fell throughout the night*; **throughout the life of the aircraft** = during the entire life of the aircraft; **throughout the world** = all over the world; **throughout the year** = from January 1st to December 31st

thrust [θrʌst] **1** *noun* force produced by a propeller, jet or rocket; *a propeller is a means of converting engine power into a propulsive force known as thrust; in order for the aircraft to increase speed, thrust must overcome drag; in constant unaccelerated flight, thrust equals drag*; *see also* REVERSAL; REVERSER **2** *verb* to push suddenly with force; *a nozzle is an opening at the rear of a jet engine through which exhaust gases are thrust* (NOTE: **thrusting - thrust**)

thunder ['θʌndə] *noun* noise created by the violent expansion and contraction of air momentarily heated by a lightning discharge; *thunder immediately following the flash of lightning usually indicates that the storm is overhead*

thunderstorm ['θʌndəstɔːm] *noun* violent weather condition in which wind speeds increase, rain or hail falls and there is lightning activity; *thunderstorms occur in well-developed cumulonimbus clouds; the process of formation, development and decay of a thunderstorm*; *see also* ACTIVITY

thus [ðʌs] *adverb* **(a)** in this way; *this device fits with the other thus* **(b)** therefore; as a result; *the glideslope and localizer beam signals control the aircraft about the pitch and roll axes, thus maintaining alignment with the runway; anti-skid braking systems are designed to prevent the brakes locking the wheels during landing, thus reducing the possibility of wheel skid*

tie [taɪ] *noun* basic structural member which is designed to withstand mainly tensile loads; *diagonal ties can be used to relieve tension and increase the effectiveness of the top boom*

tight [taɪt] **1** *adjective* closely or firmly put together; fitting closely; **a tight fit** = situation when there is just about enough space to fit; **air-tight** = (joint, etc.) which does not leak air; **water-tight** = (joint, etc.) which does not leak water or other fluid **2** *adverb* closely or firmly (shut); *the door must be shut tight*

tilt [tɪlt] **1** *noun* sloping position; *land creates a drag effect on an electro-magnetic wave-front, reducing the velocity of the wave thereby causing a tilt* **2** *verb* to be at an angle to the vertical or horizontal; to slope; *the earth tilts on its axis*

timetable ['taɪmteɪbl] *noun* printed list which shows the times of departure from and arrival to various destinations; *all the scheduled flights are listed in the airline timetable*

timetabled ['taɪmteɪbld] *adjective* listed in a timetable; *a scheduled landing is an arrival at a timetabled destination*

tip [tɪp] *noun* end of a small or tapering thing; **propeller tip** = part of the blade of a propeller furthest from the central hub; **wing tip** = outermost part of the wing; *as an aircraft takes off, the forces on the wing tip and wing surfaces start reversing direction and instead of being only downward forces of weight, they become upward forces of lift*

tire ['taɪə] *noun US* = TYRE

titanium [tɪ'teɪniəm] *noun* light metal used to make strong alloys; *the fatigue resistance of titanium is greater than that of aluminium or steel* (NOTE: symbol **Ti**; atomic number **22**)

TKOF *(ICAO)* = TAKE OFF

T/O = TAKE OFF

toggle ['tɒgl] *noun* short piece of wood, etc., attached with a string (to a life jacket); *pull the toggles downwards to inflate the life jacket*

tolerance ['tɒlərəns] *noun* allowable variation in something which can be measured; *a tolerance of 2°; a tolerance of 1mm (millimetre);* **damage tolerance** = the ability of a material or structure to withstand or resist damage; *the structural efficiency of bonded and machined structure is not achieved at the expense of damage tolerance*

tone [təʊn] *noun* sound of one pitch; *the ground transmits a code in two short bursts each of which is modulated with two tones*

top [tɒp] *noun* highest point or part; *if cumulonimbus clouds cannot be avoided then flight through the top is less hazardous than through the centre or bottom of the cloud*; **top-dead-centre** = position of the piston at the extreme top of its stroke in a piston engine; *ignition should occur just before top-dead-centre*

topic ['tɒpɪk] *noun* subject of something heard, said, written or read; *the first section in the book deals with the topic of airmanship*

topographical [tɒpə'græfɪkl] *adjective* referring to topography; *an advantage of using airfield QNH is that altimeter readings can be compared directly with heights represented on topographical maps*

topography [tə'pɒgrəfi] *noun* **(a)** representation of detailed natural and man-made features of the earth's surface as represented on a map; *the chart shows the topography of the area* **(b)** relative elevations of the earth's surface; features of a geographical area; *the general circulation is complicated because the earth tilts and its surface is neither level, because of topography, nor uniform due to areas of land and sea*

tornado [tɔː'neɪdəʊ] *noun* violent storm of small extent, with rotating winds; *the winds of a tornado are of hurricane force*

torque [tɔːk] *noun (mechanics)* moment of forces causing rotation; *torque forces try to bend the propeller against the direction of rotation*; *high current flows through both the field and armature windings producing the high torque required for engine starting*

torquemeter ['tɔːkmitə] *noun* device for measuring forces (torque) causing rotation; *engine torque is used to indicate the power that is developed by a turboprop* *engine and the indicator is known as a torquemeter*

torsion ['tɔːʃn] *noun* twisting, especially of one end of a body while the other is fixed; *rivets are subjected to torsion and may break*

total ['təʊtl] *adjective* complete, whole; *of the total amount of radiation emitted by the sun, the earth receives only a very small part*; **total system failure** = complete system failure; **total seating capacity** = maximum number of passengers who can be accommodated on seats

touch down ['tʌtʃ 'daʊn] *verb* to make controlled contact with the landing surface after a flight; *if the atmospheric pressure at an airfield is 1,000 millibars (mb) and that pressure is set on the sub-scale of an aircraft altimeter, when the aircraft touches down at the airfield, the altimeter will read zero*

touchdown ['tʌtʃdaʊn] *noun* moment, after a flight, when the aircraft makes controlled contact with the landing surface; *if the atmospheric pressure at an airfield is 1,000 millibars (mb) and that pressure is set on the sub-scale of an aircraft altimeter, at touchdown at the airfield, the altimeter will read zero*

tow [təʊ] *verb* to pull an aircraft using a bar, rope etc. and a motor vehicle; *the glider was towed into the air by a Rollason Condor*

tower ['taʊə] *noun* tall airport or airfield air traffic control (ATC) building; *wait for permission from the tower before crossing an active runway*

track [træk] **1** *noun* projection on the earth's surface of the path of an aircraft, which can be expressed in degrees from north; *where an aircraft track and wind direction are the same, there will be a headwind component acting on the aircraft*; *the actual track does not necessarily follow the planned track and is given the name track made good* **2** *verb* to follow a line of the flight path of an

aircraft, as projected on the earth surface; *on final approach, track the imaginary extended centre line of the runway*

trade winds ['treɪd 'wɪndz] *noun* steady winds which blow on the side of the sub-tropical highs nearest to the equator; *trade winds maintain their direction over the oceanic areas, especially the Pacific, more than over land areas*

traffic ['træfɪk] *noun* quantity of aircraft in operation; *standard instrument routes are structured to provide the safest and most efficient flow of traffic from entry and exit points to the airfield*; *see also* PATTERN

trailing ['treɪlɪŋ] *adjective* referring to something which comes after something else; *the trailing brush is positioned behind the main brush on the rotor arm, thereby giving a retarded spark*; *the trailing edge of the wing is the section behind the rear spar and is of light construction, because the aerodynamic loads on this area are relatively low*

train [treɪn] **1** *noun* series of connected parts or wheels in machinery; *the turboprop turbine transmits increased power forward through a shaft and a gear train, to drive the propeller* **2** *verb* to teach a person a particular skill; *the student pilot is trained to scan an instrument panel, whilst at the same time listening to the aircraft radio and flying the aircraft*

trainee [treɪ'niː] *noun* person who is being taught; *a trainee pilot*

transducer [trænz'djuːsə] *noun* device which converts a non-electrical signal into an electrical one; *the manifold is connected into the pressure ratio transmitter, which consists of a transducer, to sense the pressure ratio, and an associated electrical circuit, providing signals to the servo indicator in the cockpit*

transfer ['trænzfə] **1** *noun* passing or moving to another place; *external cooling of the engine is necessary to prevent the transfer of heat to the aircraft structure* **2**

verb to pass or to move to another place; *it is sometimes necessary to transfer fuel from one tank to another tank* (NOTE: transferring - transferred)

transform [trænz'fɔːm] *verb* to change completely; *the purpose of an actuator is to transform fluid flow into motion i.e. it converts pressure energy into mechanical energy; friction results in some of the power available from a pump being transformed into heat*

transformer [trænz'fɔːmə] *noun (electricity)* device for changing the voltage or current amplitude of an AC (alternating current) signal; *current transformers differ from voltage transformers in that the primary circuit consists of a supply feeder cable rather than a coil connected across a supply*

transient ['trænziənt] *adjective* of short duration; passing, temporary; *transient loads can be absorbed by the busbar with a minimum of voltage fluctuations*

transit ['trænzɪt] *noun* (a) act of moving; **in transit** = moving; *a green light indicates the undercarriage is locked down, and a red light is displayed when the undercarriage is in transit*; **transit passenger** = traveller who is changing from one aircraft to another; **transit route** = route taken by one aircraft through controlled airspace (b) *(at an airport)* **transit lounge** = room where transit passengers wait for connecting flights

transition [træn'zɪʃn] *noun* passing from one place, state or condition to another; **transition altitude** = altitude in the vicinity of an airport, at or below which, the vertical position of the aircraft is controlled by reference to altitudes above mean sea level; *when a flight takes place above the transition altitude, the standard pressure setting of 1013.25 mb (millibars) is used*

transmission [trænz'mɪʃn] *noun* sending of a radio signal; *the combination of loop and sense antennae can determine the direction from which a transmission is made*

transmit [trænz'mɪt] *verb* (a) to pass, to convey; *as the camshaft rotates, the cam will transmit a lifting force through rods and pivots to open the valve; the charts are transmitted from one station to another by fax* (b) to send out a radio signal; *survival beacons transmit a signal which enables search aircraft to rapidly locate survivors in the water* (NOTE: transmitting - transmitted)

transmitter [trænz'mɪtə] *noun* device for sending out radio signals; *although continuous wave radars operate continuously, separate transmitter and receiver antennae must be used; signal strength is inversely proportional to the distance from the transmitter*

transparency [træn'spærənsi] *noun* condition of being transparent; *meteorological visibility gives information on the transparency of the atmosphere to a stationary ground observer*

transparent [træn'spærənt] *adjective* allowing light to pass through so that things can be seen; *aircraft windows and canopies are usually made from transparent acrylic plastic*

transponder [træn'spɒndə] *noun (aircraft)* device for receiving a radio signal and automatically transmitting a different signal so that an ATC (air traffic control) station can identify the aircraft; *the transponder in the aircraft comprises a transmitter and a receiver*

COMMENT: the pilot sets an identification code, or 'squawk', assigned by ATC, on the transponder in the aircraft

QUOTE flight trials began recently of a low-cost hand-held IFF transponder
Pilot

transport ['trænspɔːt] *noun* system of moving people, freight, baggage, etc., from one place to another; *on a large transport aircraft, the safety of hundreds of passengers is involved*

trap [træp] *verb* to catch and prevent from escaping; *if there is a failure of the pressurized air supply, the check valve will close and trap pressurized air in the cabin*; *smog is smoke or pollution trapped on the surface by an inversion of temperature with little or no wind*

tread ['tred] *noun* series of patterns moulded into the surface of a tyre to provide grip; *the risk of aquaplaning increases as the depth of tyre tread is reduced*

treat [triːt] *verb* (a) to behave or act towards something or somebody in a particular way; *in order to prolong the life of the engine, pilot's should treat the engine carefully* (b) to apply a process to something in order to get a particular result; **treated water** = water which has been made drinkable; **heat-treated alloys** = alloys which have undergone a process of hardening by using heat

treatment ['triːtmənt] *noun* subjection to the action of a chemical or physical process; *anti-corrosion treatment; heat treatment*

trembler ['tremblə] *noun* automatic vibrator for making and breaking an electrical circuit; *an ignition unit may be supplied with direct current and operated by a trembler mechanism*

trend [trend] *noun* (a) general direction or tendency; *continuous VOLMET, which is normally broadcast on a designated VHF (very high frequency) channel, contains current aerodrome reports and trends where available* (b) up-to-date or modern way of doing things; *warning systems can take the form of lights, captions, and aural signals, and the modern trend is to incorporate them into a central warning system*

triangle ['traɪæŋgl] *noun* (a) plane figure with three sides and three angles; *the triangle of velocities is a vector solution of what happens to an aircraft when wind causes drift; see also* WIND (b) **fire triangle** = the illustration of the chemistry of fire as the three sides of a triangle

representing fuel, oxygen and heat; *if fuel, oxygen or heat is removed from the fire triangle, combustion will cease*

trigger ['trɪgə] *verb* to cause to operate; to set off; *normally, both the captain's and first officer's airspeed indicator trigger an aural warning if the airspeed limits are exceeded*

trim [trɪm] **1** *noun* condition in which an aircraft is in static balance (usually) in pitch; *trim indicators have a green band, to show when the trim is correct for take-off*; **trim wheel** = wheel-shaped device, sometimes situated between the front seats of light aircraft, to trim the aircraft by hand (also called: trimmer); **(aircraft) out of trim** = when the aircraft is not in static balance in pitch, the result being that if the pilot removes his hands from the yoke or control stick, the aircraft will start to climb or descend; *see also* TAB (NOTE: some aircraft have rudder and aileron trim) **2** *verb* to adjust trimmers in order to get the required hands-off pitch attitude; *trim the aircraft for level flight*

trimmer ['trɪmə] *noun* trim wheel; *the trimmer is used to ease the loads imposed on the flying controls during flight*

trip [trɪp] *verb* to cause an electrical device to suddenly stop working; *oscillating outputs from alternators could cause sensitive equipment to malfunction or trip*

triple ['trɪpl] *adjective* consisting of three parts; *probes may be of single, double or triple element construction*

tropical ['trɒpɪkl] *adjective* referring to the area between the parallels of latitude 23° 26' north and south of the equator; *tropical air moving northwards is subjected to surface cooling and becomes increasingly stable in its lower layers*; *see also* STORM

tropics ['trɒpɪks] *noun* **the tropics** = the area between the parallels of latitude 23° 26' north and south of the equator; *throughout the tropics and sub-tropics, the sea breeze is a regular feature*

tropopause ['trɒpəpɔːz] *noun* meeting level of the troposphere and the stratosphere; *the altitude and temperature of the tropopause are of concern to aircrew because they affect aircraft performance*

troposphere ['trɒpəsfɪə] *noun* lowest region of the atmosphere; *the troposphere is at its deepest near the equator and shallowest near the poles*

trough [trɒf] *noun* long area of low barometric pressure; *severe icing and turbulence can be experienced when flying through a trough and the precipitation may be of hail, rain, snow or sleet*

true [truː] *adjective* referring to a calculation or reading which has been corrected for errors; **true airspeed** = airspeed corrected for instrument and position error in addition to altitude, temperature and compressibility errors; **true altitude** = actual height above sea level; **true bearing** = bearing with reference to true north (not magnetic north); **true degrees** *or* **degrees true** (°T) = degrees of direction measured from true north, not magnetic north; **true north** = direction towards north pole along meridian through the observer

tube [tjuːb] *noun* long, hollow cylindrical device for holding or carrying fluids; *a liquid-type fire detector consists of a tube and expansion chamber filled with liquid*

tubing ['tjuːbɪŋ] *noun* tubes (in general); *hydraulic tubing*

tubular ['tjuːbʊlə] *adjective* referring to something which is shaped like a tube; *diagonal members can be of angle section, box spar or tubular in shape*

tune [tjuːn] *verb* **(a)** to set a system at its optimum point by careful adjustment; *the engine has not been properly tuned* **(b)** *(radio receiver)* to adjust to the particular frequency of the required signal; *the RBI shows the bearing of the tuned radio beacon with reference to the aircraft's heading*

tuner ['tjuːnə] *noun (of radio receiver)* part which allows the operator to select the particular frequency of the required signal; *the tuner reduces interference*

turbine ['tɜːbaɪn] *noun* rotary motor or engine formed of a wheel driven by a flow of air or gas; *a turbocharger consists of a turbine wheel and impeller fitted on a common rotor shaft*

turbo- ['tɜːbəʊ] *prefix meaning* turbine

turbocharger ['tɜːbəʊtʃɑːʒə] *noun* supercharger driven by a turbine powered by exhaust gases; *the turbocharger significantly increases engine power; see also* SUPERCHARGER

turbofan ['tɜːbəʊfæn] *noun* jet engine in which most of the thrust is produced by air, accelerated by a large fan, which does not pass through the combustion chamber of the engine; *the Airbus A340 is powered by 4 CFM56 turbofans*

COMMENT: turbofan engines are much quieter than older turbojets and make a characteristic sound when in operation. The fan can be clearly seen in the front part of the engine. Modern airliners use turbofan engines produced by major manufacturers such as Rolls Royce, CFM or Pratt and Whitney

turbojet ['tɜːbəʊdʒet] *noun* jet engine which includes a turbine-driven compressor for the air taken into the engine; *the de Havilland Comet was the world's first turbojet commercial transport aircraft*

COMMENT: in recent years turbofan engines have taken over from turbojet engines. Frank Whittle (1907-1996) was an English engineer and RAF officer who invented the turbojet aircraft engine. Whittle developed a jet aircraft by 1941 and the first military jet aircraft, the Gloster Meteor, became operational in 1944

turbopropeller *or* **turboprop** [tɜːbəʊprə'pelə *or* 'tɜːbəʊprɒp] *noun* turbojet engine in which the turbine also drives a propeller; *the turboprop engine is often used in transport aircraft*

COMMENT: turboprop aircraft are efficient at lower speeds than turbojet aircraft and are often used for short haul operations

turbulence ['tɜːbjʊləns] *noun* irregular motion of the atmosphere; **clear air turbulence (CAT)** = turbulence encountered in air where no cloud is present; *clear air turbulence can be particularly dangerous because it is often unexpected due to lack of visual warning*; *see also* WAKE

turbulent ['tɜːbjʊlənt] *adjective* referring to the irregular motion of the atmosphere; *when flying in turbulent air conditions, an aircraft is subjected to upward and downward gust loads*

turn [tɜːn] **1** *noun* **(a)** angular change in track; *a 180° turn*; *the autopilot may be engaged during a climb or descent but not usually in a turn*; **turn coordinator** = instrument which shows the pilot if the aircraft is in coordinated flight, or if it is slipping or skidding; *see also* COORDINATED FLIGHT **(b)** section of a wire which is wound 360° around a centre; *the voltage in each winding is directly proportional to the number of turns in each winding* **(c) in turn** = (i) for its/their part; (ii) one after the other; *drag must be overcome with thrust, which requires engines, which in turn consume fuel*; *turn off the magnetos in turn to check their serviceability* **2** *verb* **(a)** to make an angular change in track; *turn right*; *turn to the west* **(b)** to rotate; *the crankshaft turns through 720° for every cycle of four strokes*; **turn the knob** = rotate the knob or control **(c) to turn (in)to** = to change state; *as it descends into warmer air, snow turns into rain* **(d)** to find a page, section, passage, etc., in a book; *turn to page 64*

turnaround ['tɜːnəraʊnd] *noun* US = TURNROUND

turn off ['tɜːn 'ɒf] *verb* **(a)** to switch an electrical device or system 'off'; *when*

carrying out engine checks, turn off the magnetos in turn to check their serviceability* **(b)** to stop the flow of something by using a valve; *turn off the fuel*

turn on ['tɜːn 'ɒn] *verb* **(a)** to switch an electrical device or system 'on'; *can you turn the light on or turn on the light* **(b)** to start the flow of something by using a valve; *turn on the fuel*

turnround ['tɜːnraʊnd] *noun* unloading, loading and preparing an aircraft for another flight and the time taken to do this (NOTE: the word **turnaround** is preferred in American English)

twin [twɪn] *adjective* **twin engine aircraft** *or* **twin-engined aircraft** = an aircraft with two identical engines

twist [twɪst] *verb* to turn against resistance; *centrifugal, bending and twisting forces act on a propeller during flight*; **blade twist** = (i) reduction in propeller blade angle from root to tip; (ii) the unwanted variation in propeller blade pitch from root to tip caused by aerodynamic loads

TWR = AERODROME CONTROL TOWER

type [taɪp] *noun* **(a)** sort or kind (in general); *temperature and oil pressure are critical to any type of system* **(b)** class of things having shared characteristics; *the type of undercarriage fitted to an aircraft is governed by the operating weight*; **type of aircraft or aircraft type** = all aircraft of the same basic design; **type rating** = authorization, usually entered on a licence, which allows the pilot to fly a particular aircraft type (NOTE: definition (b) is much stricter than definition (a))

typical ['tɪpɪkl] *adjective* **(a)** normal; standard; **a typical fuel system** = a standard type of fuel system **(b)** representative of a particular class of things; *the Piper Archer is a typical single-engine light aircraft*

tyre ['taɪə] *noun* rubber covering for a wheel; **tyre creep** = gradual rotation of the tyre in relation to the wheel, caused by landing; *aligned white marks on the wheel and tyre indicate that there is no tyre creep*; **tyre pressure** = air pressure in a tyre (NOTE: also written **tire** in American English)

COMMENT: tyre creep can lead to damage to the tyre valve and subsequent unwanted and possibly dangerous deflation of the tyre

Uu

ultimate ['ʌltɪmət] *adjective* (point) from which no further advance can be made; final; *to determine the ultimate load which a structure must be capable of withstanding, a multiplier, called the ultimate factor of safety is used*; *the ultimate responsibility for safety rests with the crew*

ultra- ['ʌltrə] *prefix meaning* beyond

ultrasonic [ʌltrə'sɒnɪk] frequencies in the range of 20,000 Hz which cannot be heard by the human ear; **ultrasonic inspection** = non-destructive inspection of materials using extremely high frequency vibrations

ultraviolet (UV)   [ʌltrə'vaɪələt] *adjective* **ultraviolet radiation** = invisible part of the light spectrum beyond violet

unaccompanied   [ʌnə'kʌmpnɪd] *adjective* **unaccompanied baggage** = baggage that travels on a different flight to the passenger who owns it; *see also* ACCOMPANIED

uncontrolled airspace [ʌnkən'trəʊld 'eəspeɪs] *noun* airspace in which air traffic control does not provide a service and in which an ATC clearance is not required to fly; *while first learning to handle an aircraft, student pilots fly in uncontrolled airspace*

undercarriage ['ʌndəkærɪdʒ] *noun* landing gear of an aircraft; *to reduce the effect of drag by fixed undercarriages a retractable type of undercarriage was*

introduced (NOTE: **undercarriage** is often called **landing gear** or simply **gear**)

undergo [ʌndə'gəʊ] *verb* to experience; to pass through a process; *when water changes from vapour to liquid, energy is released into the atmosphere which is thus warmed although the water itself does not undergo a change of temperature* (NOTE: **undergoing - underwent - has undergone**)

underlying [ʌndə'laɪɪŋ] *adjective* **(a)** being under; *thermal modifications occur when the temperature of the underlying surface differs from that of the source region* **(b)** forming the basis of a theory or principle; *the principle underlying the construction of a mercury barometer has not changed since 1643, when Torricelli first demonstrated that the atmosphere has weight*

undershoot [ʌndə'ʃuːt] *verb (of an aircraft)* to land before, or in front of the intended target; *because of the strong wind, the student pilot undershot the runway and landed before the runway threshold*

underside ['ʌndəsaɪd] *noun* surface underneath; *the underside of the wing should be carefully inspected for damage or leaks*

undertake [ʌndə'teɪk] *verb* to do; *in light aircraft, pilot/passenger communication can be satisfactorily undertaken verbally on a one to one basis* (NOTE: **undertaking - undertook - has undertaken**)

undulating [ˈʌndjuleɪtɪŋ] *adjective* rising and falling in gentle slopes; *flight over undulating terrain will result in changing indications of aircraft height on the indicator of the radio altimeter*

uniform [ˈjuːnɪfɔːm] *adjective* the same; not varying in quality, dimensions, etc.; *an engine should be run at low r.p.m. (revolutions per minute) after flight to allow engine components to cool to a uniform temperature*

unique [juːˈniːk] *adjective* the one and only of its sort; having no like or equal; *the pulse coded message contains a unique 4-number identification*

unit [ˈjuːnɪt] *noun* **(a)** quantity or amount used as a standard, accepted measurement; *the internationally agreed unit of pressure is the millibar; the higher the sun is in the sky, the more intense is the radiation per unit area* **(b)** person, group or device, complete in itself; *the operation of flying controls is by means of self-contained power flying control units (PFCUs);* **auxiliary power unit (APU)** = self-contained power unit on an aircraft, providing electrical and pneumatic power to aircraft systems during ground operations; **tail unit** = rear part of the aircraft, usually consisting of the fin and tailplane

universal [juːnɪˈvɜːsəl] *adjective* of all or everybody; affecting all or everybody; *the use of fly-by-wire systems in airliners was delayed to allow thorough development and encourage universal acceptance of the new technology; see also* COORDINATED UNIVERSAL TIME

unload [ʌnˈləʊd] *verb* to remove a load (from an aircraft); *it took three hours to unload the aircraft*

unloading point [ʌnˈləʊdɪŋ ˈpɔɪnt] *noun* place where an aircraft is unloaded; *after taxiing, a marshaller marshalls the aircraft to the disembarkation and unloading point*

unsaturated [ʌnˈsætʃəreɪtɪd] *adjective* **unsaturated air** = air that does not contain the maximum amount of water vapour for its temperature

unserviceable [ʌnˈsɜːvɪsəbl] *adjective* (aircraft, radio, etc.) which is not operative; *the aircraft cannot be flown because the radio is unserviceable* (NOTE: often abbreviated in spoken English as U (you) S (ess) [ˈjuː ˈes])

update [ʌpˈdeɪt] *verb* to bring up to date; to add the latest information to something; *forecasts are updated and reissued every four hours*

updraft [ˈʌpdrɑːft] *noun* US = UPDRAUGHT

updraught [ˈʌpdrɑːft] *noun* rising current of air; *in cumulonimbus clouds, there are updraughts of tremendous force* (NOTE: the opposite is **down draught**. Note also that it is written **updraft** in American English)

uplift [ˈʌplɪft] *noun* lifting of air by surface features; *thunderstorms are triggered off by convection and/or orographic uplift*

upper [ˈʌpə] *adjective* **(a)** at high altitude; *upper air; upper winds; in modern meteorological practice, upper air analysis and the construction of contour charts is carried out by computer* **(b)** top; **the upper surface of the wing** = surface of the wing facing upwards (as opposed to the underside) (NOTE: the opposite is **lower**)

upward [ˈʌpwəd] **1** *adjective* moving or directed up; *as the aircraft accelerates down the runway, the forces on the wing tips and wing surfaces start reversing direction and instead of being only downward forces of weight, they become upward forces of lift* **2** *adverb* US = UPWARDS (NOTE: the opposite is **downward**. Note also that in American English, **upward** is used as an adjective and as an adverb)

upwards [ˈʌpwədz] *adverb* towards the top; *heat is transferred from the earth's* *surface upwards by convection* (NOTE: the opposite is **downwards**)

upwind [ʌpˈwɪnd] *adverb* against the wind; *the glider was released from the aero-tow 3 miles upwind of the airfield* (NOTE: the opposite is **downwind**)

urgency [ˈɜːdʒənsi] *noun* importance; haste; need for prompt or fast action; *warnings, cautions and advisory messages are displayed only when necessary and are colour coded to communicate the urgency of the fault to the flight crew*

USA *or* **US** = UNITED STATES OF AMERICA

usable [ˈjuːzəbl] *adjective* capable of being used; *on receiving the evacuate order, cabin crew must assess if their exits are usable*

usage [ˈjuːsɪdʒ] *noun* using something; consumption; *fuel flight planning combines navigation data with fuel usage*

use 1 [juːs] *noun* being used; *it must be ensured that smoke masks are available for use by personnel within the aircraft*; **runway in use** = runway currently being used for take-offs and landings **2** [juːz] *verb* to employ; *gas turbine engines use low viscosity synthetic oil*

UTC = COORDINATED UNIVERSAL TIME

utilization [juːtɪlaɪˈzeɪʃn] *noun* making use of; *integral tanks are now favoured for aircraft owing to the high utilization of space and reduction in weight* (NOTE: also written **utilisation**)

utilize [ˈjuːtɪlaɪz] *verb* to make use of; *the most common type of barograph is one which utilizes an aneroid capsule mechanically connected to a pen* (NOTE: also written **utilise**)

UV = ULTRAVIOLET

Vv

vacuum ['vækjuəm]   *noun*   space completely empty of everything even air; *if the fuel tank vent pipe is blocked, a vacuum will form in the tank and fuel flow to the engine will be restricted*

valid ['vælɪd]   *adjective*   **(a)** having official force or effect; *all passengers should have valid passports* **(b)** which is acceptable because it is true; well-based, worth taking seriously; *significant weather charts use abbreviations and symbols to illustrate en route weather phenomena and are valid for a specified time*; **a valid assumption** = a well-based supposition

validity [və'lɪdɪti]   *noun*   being valid; *the period of validity of a visa*; *aerodrome forecasts included in VOLMET should have a validity period of 9 hours*

valley ['væli]   *noun*   area of low-lying land between mountains or hills; *an example of a valley wind is the Mistral*

value ['vælju:]   *noun*   **(a)** quantity shown as a number; *deviation is not a constant value but varies from one aircraft to another* **(b)** quality of being useful or desirable; **the value of doing something** = the usefulness or worth of doing something

valve [vælv]   *noun*   mechanical device for controlling the flow of a fluid; **exhaust valve** = valve in a piston engine which allows exhaust gases to leave the cylinder; **inlet valve** = valve in a piston engine which allows the fuel/air mixture to enter the cylinder; **non-return valve** = valve which allows a fluid to pass in one direction only; **poppet valve** = common valve type found in piston engine; *(as for pressure)* **relief valve** = valve which opens at maximum safe pressure and closes again upon return to normal operating conditions; **valve seat** = angled ring in cylinder head on which the poppet valve sits when closed; *see also* SEQUENCE

vane [veɪn]   *noun*   flat surface acted on by the wind or an airflow; *a centrifugal compressor consists of a disc on which is formed a number of radially spaced vanes*

vapor ['veɪpə]   *noun*   *US* = VAPOUR

vaporize ['veɪpəraɪz]   *verb*   to turn into vapour; *water vaporizes when heated* (NOTE: also written **vaporise**)

vapour ['veɪpə]   *noun*   gaseous form of a liquid; *over desert areas, the lack of water vapour in the atmosphere produces cold nights*; **vapour lock** = blockage of fuel flow from a tank caused by a bubble of vapour at a high point in the pipeline; **Reid vapour pressure test** = test to determine the pressure required above a liquid to hold the vapours in the liquid at a given temperature (NOTE: also written **vapor** in American English)

variable ['veəriəbl]   *adjective*   changing or changeable; *winds are more variable in the northern hemisphere than in the southern hemisphere*; **variable pitch propeller** = propeller with a mechanism to change the blade angle, to suit flight conditions

variation [veəri'eɪʃn]   *noun*   **(a)** change or amount of change; **seasonal variation** = change linked to a season; *map reading is often complicated by seasonal variations, such as the difference between the* *landscapes before and after snowfall* **(b)** angular difference between magnetic north and true north, which is measured in degrees and is named east or west according to whether the north seeking end of a freely suspended magnet lies to the east or to the west of the true meridian at that point; **variation east, magnetic least: variation west, magnetic best =** mnemonic to help someone remember whether to add or subtract variation

variety [vəˈraɪəti] *noun* a lot of different things; *display units provide a wide variety of information relevant to engine and other automated systems operation*

vary [ˈveəri] *verb* to change; to be different; *the tropopause over the UK can vary between 25,000 feet and 45,000 feet according to whether the country is covered by a polar or tropical air mass*

VASI = VISUAL APPROACH SLOPE INDICATOR

vast [vɑːst] *adjective* large, immense, huge; **a vast area** = a huge area; **the vast majority** = the great majority; *the vast majority of people*

vector [ˈvektə] *noun* **(a)** a quantity with magnitude and direction indicated by a line of a given length, representing magnitude and specific direction; *the triangle of velocities is a vector solution of what happens to an aircraft when wind causes drift* **(b)** heading given to a pilot to provide navigational guidance by radar; *wind velocity is indicated by a vector, identified by a single arrow, pointing in the direction the wind is blowing towards*

veer [vɪə] **1** *noun* shifting of the wind in a clockwise direction (in the northern hemisphere); *the passing of a weather trough is marked by a sharp veer in the direction of the wind* **2** *verb* **(a)** *(of wind)* to change in a clockwise direction (in the northern hemisphere); *winds veer and increase with height ahead of a warm front* (NOTE: the opposite is **to back**) **(b)** to change direction, especially as in an uncontrolled movement; *the aircraft veered off the runway into the grass*

velocity [vəˈlɒsəti] *noun* rate of change of position in a given direction which is composed of both speed and direction; *wind velocity*; *the anemograph gives a continuous recording of wind velocity which is displayed on a chart and reveals gusts, squalls and lulls*; *see also* TRIANGLE

vent [vent] *noun* hole serving as an inlet or outlet for a fluid, usually a gas such as air; *during the pre-flight inspection, check that the fuel tank vent pipe is not blocked*; *the vent/pressurization system must allow for the passage of air whenever a fuel tank is refuelled or defuelled or the aircraft climbs or descends*

ventilate [ˈventɪleɪt] *verb* to cause air to pass in and out freely; *the water separator is installed downstream of the cold air unit to extract a percentage of free moisture from the air, which subsequently ventilates and pressurizes the cabin*

ventilation [ventɪˈleɪʃn] *noun* passing of air in and out freely; *a constant supply of air for ventilation purposes is always available from the air conditioning system*

venturi (tube) [venˈtjʊəri ˈtjuːb] *noun* tube which narrows at the centre; choke tube; *when the temperature of the air passing through the carburettor is reduced below 0°C (Celsius), any moisture in the air forms into ice and builds up on the venturi and throttle valve*

verification [verɪfɪˈkeɪʃn] *noun* act or instance of establishing the truth or validity of something; *the document required verification*

versus [ˈvɜːsəs] *preposition* against, as compared with; *the diagram illustrates typical strength properties by plotting applied stress versus resulting strain*

vertical [ˈvɜːtɪkl] **1** *adjective* at right angles to the earth's surface or to another line or plane; *beams can be additionally stiffened in a downward direction by vertical and diagonal members*; *height is defined as the vertical distance of a level, point or object, considered to be a point,* *from a specified datum* **2** *noun* (imaginary) vertical line or plane; *the hot rod ice detector head consists of an aluminium alloy oblong base on which is mounted a steel tube detector mast, angled back to approximately 30° from the vertical*

vertical axis ['vɜ:tɪkl 'æksɪs] *noun* **(a)** imaginary line running through the fuselage at the centre of gravity from top to bottom, around which the aircraft rotates when it yaws; *the rudder is a control surface on the fin which rotates the aircraft about its vertical axis to produce yaw; see also* AXIS; YAW **(b)** vertical reference line (Y axis) of a graph; *the vertical axis shows engine power available; see also* AXIS

vertically ['vɜ:tɪkli] *adverb* in a vertical position; *the aircraft pitched up vertically*

vertical speed indicator (VSI)

['vɜ:tɪkl 'spi:d 'ɪndɪkeɪtə] *noun* flight instrument which indicates the rate of climb and descent

vessel ['vesl] *noun* boat or ship; *when flying over the sea you must not fly closer than 500 feet to a vessel*

VFR = VISUAL FLIGHT RULES

VHF = VERY HIGH FREQUENCY; *see* FREQUENCY

via ['vaɪə] *preposition* by way of; *the flight is from Cairo to Paris via Rome; after heating, the air passes into the cabin via a chamber through which cold air also flows*

vibrate [vaɪ'breɪt] *verb* to move rapidly and continuously backwards and forwards; *turbine blades in the average jet engine vibrate at frequencies of 1 million per minute*

vibration [vaɪ'breɪʃn] *noun* rapid and continuous movement; *according to the pilot, engine vibration was detected in engine number one*

vice versa ['vaɪsə 'vɜ:sə] *adjective* the other way around; *when engine demand is high, fuel pressure tends to be low and vice versa* = when the engine demand is low, fuel pressure tends to be high

vicinity [vɪ'sɪnəti] *noun* area nearby; *after an emergency evacuation, passengers should be directed to move away from the vicinity of the aircraft quickly*; **in the vicinity of the airport** = near the airport

view [vju:] *noun* **(a)** what you are able to see from a certain place; *cabin crew must have a clear view of the aisles from their stations* **(b)** picture of something presented in a particular way; *a cross-sectional view of an aerofoil* **(c)** personal opinion; *he expressed strong views on the subject of airport security* **(d)** **with a view to** = with the intention of; *he wrote the report with a view to improving in-flight services*; **in view of** = because of; *in view of the poor weather conditions, the flight will be delayed*

violate ['vaɪəleɪt] *verb* **(a)** to enter without permission; *the aircraft violated a danger area* **(b)** to break rules or regulations; *by not wearing a cap, the cadet is violating the dress code*

violent ['vaɪələnt] *adjective* with great force; *flying through atmospheric dust causes the airframe to build up a static electrical charge and the associated discharges can be violent*

virtually ['vɜ:tʃuəli] *adverb* almost; *resistance to alternating current remains virtually constant and is independent of frequency*

viscosity [vɪs'kɒsəti] *noun* a liquid's internal resistance to flowing; *excessive oil temperatures are dangerous, as the oil viscosity is reduced and inadequate bearing lubrication results*

visibility [vɪzə'bɪlɪti] *noun* (subject to atmospheric conditions) ability to see unlighted objects by day and lighted objects by night; *measurement of visibility by day is made by direct observation of objects at known distances and is therefore an estimated value*;

air-to-ground visibility = description of how easily seen an object on the ground is from the air; *glare caused by reflection of sunlight from the top of a layer of fog or haze can seriously reduce the air-to-ground visibility*; **poor visibility** = not being able to see things clearly because of fog, mist, smoke, etc.; **visibility-by-day values** = values which indicate how easily seen an object is in a horizontal line from an observer in daylight conditions

visible ['vɪzəbl] *adjective* referring to something that can be seen; *when the undercarriage is selected down it may be visible from the crew compartment, but it is not usually possible to tell if it is securely locked*; *if the sun is seen through cumulus cloud it will be clearly visible*

vision ['vɪʒn] *noun* (a) the power of seeing; ability to see; eyesight; *lightning at night may cause temporary loss of vision* (b) what you are able to see; ability to see; *in low wing aircraft, downward vision may be limited by the airframe*

visual ['vɪʒuəl] *adjective* referring to seeing; *the instrument landing system is to provide guidance in the horizontal and vertical planes to an aircraft on final approach into a position from which a safe visual landing can be made*; **visual inspection** = inspection using the eyes; **visual warning** = warning which can be seen as opposed to a audible warning which can be heard

visual approach slope indicator (VASI) = ['vɪʒuəl ə'prəʊtʃ 'sləʊp 'ɪndɪkeɪtə] *noun* arrangement of red and white lights each side of the runway touchdown point to give the pilot information about his height on final approach

visual flight rules (VFR) ['vɪʒuəl 'flaɪt 'ruːlz] *noun* rules set down by an authority for flight in visual conditions, such as flight visibility and distance from cloud, etc.; **special VFR (visual flight rules) flight** = a controlled VFR flight permitted by air traffic control to fly within

a control zone in meteorological conditions below visual meteorological conditions

COMMENT: particular requirements for VFR depend on the type of airspace, time of day, and height above terrain

visual meteorological conditions (VMC) ['vɪʒuəl miːtʃərə'lɒdʒɪkl kən'dɪʃnz] *noun* meteorological conditions concerning visibility, distance from cloud, etc., equal to, or better than specified minima; *the weather conditions enabling a pilot to fly VFR (Visual Flight Rules) is known as visual meteorological conditions (VMC)*

vital ['vaɪtl] *adjective* extremely important; *verbal commands from the crew are vital at all times but particularly so if smoke restricts cabin visibility; accurate measurements of atmospheric pressure and the rate of change of pressure are of vital interest to the meteorological forecaster*

viz ['vɪz] *adverb* namely; in other words; that is to say; *there are two types of inverter viz rotary and static*

VMC = VISUAL METEOROLOGICAL CONDITIONS

Vne = NEVER-EXCEED SPEED

volatile ['vɒlətaɪl] *adjective* (liquid) which easily changes into a gas or vapour; *to aid starting in cold weather, more volatile fuels can be used*

volatility [vɒlə'tɪləti] *noun* ease with which a liquid changes into a gas or vapour; *with kerosene-type fuels, the volatility is controlled by distillation and flash point, but with the wide-cut fuels it is controlled by distillation and the Reid Vapour Pressure test*

VOLMET ['vɒlmet] routine ground-to-air broadcast of meteorological information; *the meteorological Operational Telecommunications Network Europe (MOTNE) is provided for the exchange of meteorological information needed by meteorological* offices, *VOLMET broadcasting stations, air traffic service units, operators and other aeronautical users*

volt [vɔult] *noun* SI unit of electrical potential; *the system requires a power supply of either 115 volts AC (alternating current), 28 volts DC (direct current), or both*

voltage ['vɒltɪdʒ] *noun* electrical force measured in volts; *as an installed battery becomes fully charged by the aircraft generator, the battery voltage nears its nominal level and the charging current decreases*

volume ['vɒljuːm] *noun* (a) amount of space occupied by a solid, a liquid or a gas; *if the pressure of a given mass of gas is maintained constant, the volume of gas increases as its temperature is increased* (b) loudness of a transmission; **turn down the volume** = make the sound less loud by adjusting the volume control; **turn up the volume** = make the sound louder by adjusting the volume control; **volume control** = knob used to adjust the sound by making it louder or less loud

VOR = VERY HIGH FREQUENCY OMNI-DIRECTIONAL RADIO RANGE; *see* FREQUENCY

VSI = VERTICAL SPEED INDICATOR

vulnerable ['vʌlnrəbl] *adjective* liable to attack or damage; not protected; *some engines still retain the centrifugal type of compressor because it is simple, comparatively cheap to manufacture, robust in construction and less vulnerable to damage*

Ww

W = WEST

wake ['weɪk] *noun* **wake turbulence** = disturbance of the air remaining after the passage of an aircraft

wall [wɔːl] *noun (of something hollow)* side; *there is a film of oil between the piston and cylinder wall*

warn [wɔːn] *verb* to give notice of possible danger; *a light illuminates to warn the crew*

QUOTE ultrasonic technology which automatically warns pilots of ice build-up on aircraft may soon be approved for general use by carriers
Flight International 16-22 July 1997

warning ['wɔːnɪŋ] **1** *noun* notice of possible danger **2** *adjective* which gives notice of possible danger; *the main power plant fire detection system should contain an audible warning device to supplement the visual indication*; **warning light** = small light which lights up to show that something dangerous may happen

waste [weɪst] *noun* something which can no longer be used; *a smouldering fire in a toilet waste container or waste disposal bin could become very active due to pressure changes during descent*

washroom ['wɒʃrʊm] *noun* = TOILET

watt [wɒt] *noun* SI unit of measurement of electrical power; *the work done by an electrical circuit or the power consumed is measured in watts*

wave [weɪv] *noun* **(a)** motion by which heat, light, sound or electric current is spread; *the speed of propagation of radio waves is faster over sea than over land*; **carrier wave** = continuous transmission of a wave of constant amplitude and frequency **(b) ridge waves** = oscillations about the stable state of the undisturbed air flow with the range of hills providing the disturbance; **standing wave** = motion of air downwind of a steep hill or mountain in which the high and low points of the wave do not move **(c)** mass of water moving across the surface of a lake or the sea, rising higher than the surrounding water as it moves; *wind speeds increase with height, the speed of the wind at the crest of a wave being the greatest*

waveform ['weɪvfɔːm] *noun* shape of a repetitive wave; *a cycle is one complete sequence of the waveform, from any point, to the same value 360° later*

wavelength ['weɪvleŋθ] *noun* distance from the highest point of one wave to the highest point of the next; *short wavelength permits sharper beams for direction finding and more efficient reflections*

weak [wiːk] *adjective* **(a)** not strong; *a weak radio signal* **(b)** (over)diluted with water or air; **weak mixture** = fuel/air mixture in which there is more air than usual; *excessive cylinder head temperatures could be caused by prolonged use of a weak mixture, especially at high altitude* (NOTE: also called a **lean mixture**); **weak solution** = mixture of water and some other substance in which the amount of water is more than usual

weaken ['wiːkn] *verb* to make weak; *inflation of the de-icer boot weakens the bond between the ice and de-icer boot surfaces*

wear [weə] **1** *noun* damage or loss of quality by use; *mishandling of aero-engines during operation can cause considerable damage and wear which can shorten the life of the engine* **2** *verb* **(a)** to become damaged or to lose quality because of use; *the more the brakes are used, the more they wear* **(b)** *(clothes, etc.)* to have on the body; *the nature of modern jet transport does not require the pilot to wear an oxygen mask*

weather ['weðə] *noun* conditions of atmospheric temperature, pressure, wind, moisture, cloudiness, precipitation and visibility; *generally speaking, weather conditions can be described as light, moderate or severe depending on the intensity of the conditions*; **forecast weather** = predicted weather (not actual weather); **weather report** = official account of weather conditions; *see also* CONDITION; FORECAST

web [web] *noun* main vertical member of a beam; *the web connecting the upper and lower flanges of the beams must be rigid enough to withstand direct compressive loads without buckling*

weigh [weɪ] *verb* to measure how heavy something is; *a given quantity of lead weighs more than the same quantity of aluminium*

weight [weɪt] *noun* force with which a body is drawn towards the centre of the earth; measure of how heavy something is; *carry-on baggage is limited by regulations as to size and weight and items in excess of this should be stowed in the hold*; **operating weight** = total mass (weight) of aircraft ready for flight but excluding fuel and payload; *the type of undercarriage fitted to an aircraft is governed by the operating weight*

west [west] **1** *noun* **(a)** compass point on the mariner's compass 270° clockwise from due north and directly opposite east;

in Europe, snow occurs more frequently in the east than in the west **(b)** the direction of the setting sun **2** *adjective* **(a)** referring to areas or regions lying in the west **(b)** **west wind** = wind blowing from or coming from the west (NOTE: a wind is named after the direction it comes from) **(c)** western part of a country; *West Africa* **3** *adverb* towards the west; *the aircraft was flying west*

westbound ['westbaʊnd] *adjective* travelling towards the west; *a westbound flight*

westerly ['westəli] **1** *adjective* **(a)** situated towards the west **(b)** *(wind)* blowing or coming from the west; *a westerly wind is blowing* **(c)** *(direction)* moving to the west or towards the west; *he should fly in a westerly direction* **2** *noun* (predominant) wind which blows or comes from the west; *temperate westerlies occur on the side of the sub-tropical anti-cyclonic belts which is remote from the equator*

western ['westən] *adjective* situated in the west; *Western Europe*

westward ['westwəd] **1** *adjective* which is going towards the west; **a westward flight** = a flight heading towards the west **2** *adverb* US = WESTWARDS

westwards ['westwədz] *adverb* towards the west; *flying eastwards or westwards for long periods of time affects sleep patterns*

wheel [wiːl] *noun* circular, rotating, load-carrying part between the tyre and axle; the whole wheel and tyre assembly on which a vehicle rolls; **nose wheel** = undercarriage wheel at the front of the aircraft; **wheel bay** = space in the fuselage or wing structure in which the wheel is housed after retraction; *to avoid damage to the wheel bay, the nose wheel must be aligned in a fore and aft direction during retraction*; **wheel fairing** = streamline covering over a wheel of a light aircraft; *wheel fairings are also called spats*

whereas [weə'æz] *conjunction* but in contrast; but on the other hand; *in the piston engine, the cycle is intermittent, whereas in the gas turbine, each process is continuous; kerosine has a low vapour pressure and boils only at very high altitudes or high temperatures, whereas a wide-cut fuel will boil at a much lower altitude*

whereby [weə'baɪ] *adverb* by what, by which; according to which; *compression heating relies on the principle whereby the air temperature is increased by compression; in ram air supply systems, the cooling method is of the simplest type, whereby the cold air can be directly admitted to the cabin via adjustable louvres*

whereupon [weəʌ'pɒn] *adverb* at that point; after which; *pitch changes are achieved using the throttle lever, which is usually taken up and back through a gate in the quadrant whereupon fuel is added to increase power*

wherever [weə'evə] *adverb* wherever possible = in places where it is possible; **wherever possible, thunderstorms should be avoided by a wide margin** = thunderstorms should be avoided by a wide margin in situations or places where it is possible to avoid them

while ['waɪl] *conjunction* (a) during the time that; *the pilot is trained to scan an instrument panel, while at the same time listening to the aircraft radio and flying the aircraft* (b) although; in spite of the fact that; *while metal fatigue is not a modern phenomenon, it is only in recent years that much emphasis has been placed upon determining its causes* (NOTE: **whilst** is sometimes used in place of **while**)

whole [həʊl] *adjective* complete; *the whole aircraft should be inspected to ensure that it is free from deposits of ice, snow and frost*; **whole number** = undivided number, number which is not a fraction

wide [waɪd] *adjective* (a) referring to the distance (of something) measured from side to side (compared to length); *the localizer antenna array is normally about 80 feet wide and 12 feet high* (b) a **wide range of temperatures** = a large difference between the lowest and the highest temperature; **a wide variety of information** = a lot of different information; *see also* WIDTH

wide-cut fuel ['waɪdkʌt 'fjuːəl] *noun* general term for aviation turbine fuels made up of a wider variety of petroleum products than kerosine-type fuels; *kerosine has a low vapour pressure and boils only at very high altitudes or high temperatures, whereas a wide-cut fuel will boil at a much lower altitude*

widespread ['waɪdspred] *adjective* found or distributed across a large area; *the storm caused widespread damage*; **widespread precipitation** = rainfall or snowfall covering a large area

width [wɪdθ] *noun* distance (of something) measured from side to side (compared to length); *the polar front jet stream may have a width of up to 200 nm (nautical miles)*; *see also* WIDE

wind [wɪnd] *noun* horizontal movement of air in relation to the earth's surface; **surface wind** = wind which blows across the land surface (and not the wind higher in the atmosphere); **wind gradient** = rate of increase of wind strength with unit increase in height above ground level; **wind velocity** = wind speed and direction

wind [waɪnd] *verb* to move in a curving or twisting manner; *if a wire is wound as a coil, the field will be like that of a bar magnet* (NOTE: **winding - wound** [waʊnd])

winding ['waɪndɪŋ] *noun* series of 360° turns of wire; *the voltage in each winding is directly proportional to the number of turns in each winding*

windscreen ['wɪnskriːn] *noun* front window of an aircraft through which the pilot has forward vision; *the windscreen is a glass laminated construction with an* *electrical element, made of gold film, sandwiched between the layers; see also* WIPER

windshear *or* **wind shear** ['wɪnʃɪə] *noun* extreme local wind gradient which can be dangerous; *fly-by-wire technology can be very useful in windshear situations*

windshield ['wɪnʃiːld] *noun* US = WINDSCREEN

windspeed *or* **wind speed** ['wɪnspiːd] *noun* rate of wind movement, usually measured in knots; *wind direction is given in degrees true rounded to the nearest 10°, followed by the mean windspeed*

windward ['wɪndwəd] *adjective & adverb* (side) facing the direction from which the wind blows; **windward of a range of hills** = upwind of the range of hills; *if precipitation occurs, water will have been removed from the atmosphere thus causing the air on the lee side to be drier than that on the windward side* (NOTE: the opposite is **lee**)

wing [wɪŋ] *noun* main horizontal aerofoil or mainplane; *the wing supports the weight of the aircraft in flight*; **fixed wing aircraft** = aeroplane with rigidly attached wings (unlike helicopters); **rotary wing aircraft** = aircraft with a rotor which provides lift (such as a helicopter); **wing loading** = weight of an aircraft per unit wing area; **wing root** = part of the wing where it meets with the fuselage; **wing tip** = outermost part of the wing; *see also* LEADING EDGE; ROTOR

winglet ['wɪŋlət] *noun* upturned wing tip or small additional vertical aerofoil on wing tip; *the attachment of winglets improved the handling characteristics of the aeroplane*

wingspan ['wɪŋspæn] *noun* measurement from the tip of one wing to the tip of the other wing; *the wingspan of the aircraft is 7 metres*

wipe [waɪp] *verb* to clean or to dry by using a cloth; *in the event of hydraulic fluid spillage on paintwork, the affected area should be wiped clean immediately*

wiper ['waɪpə] *noun* device with a rubber blade which clears rain, snow, etc., from a windscreen; *in certain circumstances, such as heavy rainstorms, the windscreen wipers may not be able to cope and pilot's visibility is impaired*

wire ['waɪə] *noun* metal drawn out into the form of a thread or string; *while the shunt coil is made of fine wire which gives a high resistance and small current flow, the series coil is made of thick wire, which gives a low resistance and large current flow*

withdraw [wɪθ'drɔː] *verb* to pull back, to draw back; *instructions are given to the cabin crew to arm the escape devices immediately the boarding steps or airbridges are withdrawn*

within [wɪð'ɪn] *preposition* in; inside; not beyond; *great care must be taken to ensure that the aircraft operates within regulated or permissible weight limits*; **within two hours** = in about two hours or less (but not more)

withstand [wɪθ'stænd] *verb* to resist; to bear; *wings must be capable of not only withstanding the aircraft weight, but also the stresses and strains which are imposed during flight*

WMO = WORLD METEOROLOGICAL ORGANIZATION

work [wɜːk] *noun* (a) operation of a force to produce movement or some other physical change; *1 horsepower is defined as 33,000 foot-pounds of work accomplished in one minute (a foot-pound being the ability to lift a one pound weight a distance of one foot)* (b) something, such as maintenance, which has to be done; *work is being carried out on the auxiliary power unit (APU)* (c) something done to earn a living; *he enjoys his work* **2** *verb* (a) to operate, to function; *the computer doesn't work* = the computer doesn't operate as it should do because there is something wrong with it (b) to do something, such as maintenance; *engineers worked on the aircraft all night* (c) to do something to earn a living; *he works for a large airline* (d) to work out = to calculate, to solve a mathematical problem; *aircraft performance is a function of weight and therefore it is important that you can work out weight from volume and vice-versa*

Xx Yy Zz

yaw [jɔ:] **1** *noun* rotation of the aircraft around its vertical axis; *three-axis control of roll, pitch and yaw is effected by ailerons, elevators and rudder*; **adverse yaw** = yaw caused by aileron drag, in the opposite direction to the direction of the intended turn **2** *verb (of an aircraft)* to rotate around the vertical axis; *single engine, propeller driven aircraft tend to yaw on take-off*

yoke [jəʊk] *noun* **(a)** type of pilots control column in which the ailerons are controlled by rotating a device on top of the column to the left or right; *rotate the yoke to the left to roll the aircraft to the left* **(b)** supporting structure like the forked metal mounting for the nosewheel; *the yoke was damaged in the incident*

Z = ZULU (TIME)

zero ['zɪərəʊ] *noun* nought or the figure 0; *if the atmospheric pressure at an airfield is 1,000 millibars (mb) and this pressure is set on the sub-scale of an aircraft altimeter, then when that aircraft touches down at the airfield, the altimeter will read zero*

zonal ['zəʊnl] *adjective* referring to one of the five parts into which the earth's surface is divided by imaginary lines parallel to the equator; *the circulation of air around the earth is zonal in character*

zone [zəʊn] *noun* **(a)** area with particular features or purpose; **danger zone** = area where danger exists **(b)** administrative area of airspace; *control zone*; *aerodrome traffic zone (ATZ)* **(c)** **time zone** = one of the 24 parts of the earth in which the same standard time is used **(d)** one of five divisions into which the earth's surface is divided by imaginary lines parallel to the equator; *temperate zone*; *see also* CLIMATIC

Zulu time (Z) ['zu:lu: 'taɪm] *noun see comment at* GREENWICH MEAN TIME

For further details, please tick the titles of interest and return this form to:
Peter Collin Publishing, 1 Cambridge Road, Teddington, TW11 8DT, UK
fax: +44 181 943 1673   tel: +44 181 943 3386   email: info@pcp.co.uk
For full details, visit our web site: **www.pcp.co.uk**

Title	ISBN	Send Details
English Dictionaries		
English Dictionary for Students	1-901659-06-2	❏
Accounting	0-948549-27-0	❏
Aeronautical English	1-901659-10-0	❏
Agriculture, 2nd ed	0-948549-78-5	❏
American Business, 2nd ed	1-901659-22-4	❏
Automobile Engineering	0-948549-66-1	❏
Banking & Finance, 2nd ed	1-901659-30-5	❏
Business, 2nd ed	0-948549-51-3	❏
Computing, 3rd ed	1-901659-04-6	❏
Ecology & Environment, 3ed	0-948549-74-2	❏
Government & Politics, 2ed	0-948549-89-0	❏
Hotel, Tourism, Catering Management	0-948549-40-8	❏
Human Resources & Personnel, 2ed	0-948549-79-3	❏
Information Technology, 2nd ed	0-948549-88-2	❏
Law, 2nd ed	0-948549-33-5	❏
Library & Information Management	0-948549-68-8	❏
Marketing, 2nd ed	0-948549-73-4	❏
Medicine, 2nd ed	0-948549-36-X	❏
Military Terms	1-901659-24-0	❏
Printing & Publishing, 2nd ed	0-948549-99-8	❏
Science & Technology	0-948549-67-X	❏
Vocabulary Workbooks		
Banking & Finance	0-948549-96-3	❏
Business, 2nd ed	1-901659-27-5	❏
Computing, 2nd ed	1-901659-28-3	❏
Colloquial English	0-948549-97-1	❏
English for Students	1-901659-11-9	❏
Hotels, Tourism, Catering	0-948549-75-0	❏
Law, 2nd ed	1-901659-21-6	❏
Medicine	0-948549-59-9	❏
Professional/General		
Astronomy	0-948549-43-2	❏
Multimedia, 2nd ed	1-901659-01-1	❏
PC & the Internet, 2nd ed	1-901659-12-7	❏
Bradford Crossword Solver, 3rd ed	1-901659-03-8	❏
Bilingual Dictionaries		
French-English/English-French Dictionaries		❏
German-English/English-German Dictionaries		❏
Spanish-English/English-Spanish Dictionaries		❏

- -

Name: ....

Address: ....

....

....Postcode: ....Country: ....